eTELEMED 2020

The Twelfth International Conference on eHealth, Telemedicine, and Social Medicine

November 21 – 25, 2020

Valencia, Spain

eTELEMED 2020 Editors

Sandra Sendra, University of Granada, Spain

Yoshitoshi Murata, Iwate Prefectural University, Japan

Javier Civit-Masot, Cober SL. Gnomon Informatics SA, Spain

Arian Rajh, University of Zagreb, Croatia

Printed from e-media with permission by:

Curran Associates, Inc.
57 Morehouse Lane
Red Hook, NY 12571

CURRAN ASSOCIATES INC. proceedings .com

Some format issues inherent in the e-media version may also appear in this print version.

International Academy, Research, and Industry Association (IARIA)
412 Derby Way
Wilmington, DE 19810

Phone: (408) 893-6407
Fax: (408) 527-6351

petre@iaria.org

Additional copies of this publication are available from:

Curran Associates, Inc.
57 Morehouse Lane
Red Hook, NY 12571 USA
Phone: 845-758-0400
Fax: 845-758-2633
Email: curran@proceedings.com
Web: www.proceedings.com

Table of Contents

eTELEMED 2020

Forward

The Twelfth International Conference on eHealth, Telemedicine, and Social Medicine (eTELEMED 2020) considered advances in techniques, services, and applications dedicated to a global approach of eHealth.

Development of wireless homecare, of special types of communications with patient data, of videoconferencing and telepresence, and the progress in image processing and date protection increased the eHealth applications and services, and extended Internet-based patient coverage areas. Social and economic aspects as well as the integration of classical systems with the telemedicine systems are still challenging issues.

eTELEMED 2020 provided a forum where researchers were able to present recent research results and new research problems and directions related to them. The topics covered aspects from classical medicine and eHealth integration, systems and communication, devices, and applications.

We take this opportunity to thank all the members of the eTELEMED 2020 Technical Program Committee as well as the numerous reviewers. The creation of such a broad and high-quality conference program would not have been possible without their involvement. We also kindly thank all the authors who dedicated much of their time and efforts to contribute to the eTELEMED 2020. We truly believe that, thanks to all these efforts, the final conference program consists of top quality contributions.

This event could also not have been a reality without the support of many individuals, organizations, and sponsors. We are grateful to the members of the eTELEMED 2020 organizing committee for their help in handling the logistics and for their work to make this professional meeting a success.

We hope that eTELEMED 2020 was a successful international forum for the exchange of ideas and results between academia and industry and for the promotion of progress in eHealth and Telemedicine research.

eTELEMED 2020 Chairs

eTELEMED 2020 General Chair
Sandra Sendra, University of Granada, Spain

eTELEMED 2020 Steering Committee

Yoshitoshi Murata, Iwate Prefectural University, Japan

eTELEMED 2020 Advisory Committee

Lisette Van Gemert-Pijnen, University of Twente - Enschede, the Netherlands
Marike Hettinga, Windesheim University of Applied Sciences, the Netherlands
Åsa Smedberg, Stockholm University, Sweden
Conceição Granja, Norwegian Centre for e-health Research, University Hospital of North Norway, Norway

eTELEMED 2020
Committee

eTELEMED 2020 General Chair
Sandra Sendra, University of Granada, Spain

eTELEMED 2020 Steering Committee
Yoshitoshi Murata, Iwate Prefectural University, Japan

eTELEMED 2020 Advisory Committee
Lisette Van Gemert-Pijnen, University of Twente - Enschede, the Netherlands
Marike Hettinga, Windesheim University of Applied Sciences, the Netherlands
Åsa Smedberg, Stockholm University, Sweden
Conceição Granja, Norwegian Centre for e-health Research, University Hospital of North Norway, Norway

eTELEMED 2020 Technical Program Committee
Don Adjeroh, West Virginia University, USA
Giovanni Albani, Istituto Auxologico Italiano - IRCCS, Italy
Jan Alexandersson, DFKI GmbH,Germany
Prima Oky Dicky Ardiansyah, Iwate Prefectural University, Japan
Panagiotis D. Bamidis, School of Medicine - Aristotle University of Thessaloniki, Greece
Oresti Baños, University of Granada, Spain
Ivana Bartoletti, Gemserv, UK
Christian-Alexander Behrendt, GermanVasc Research Group/ University Medical Center Hamburg-Eppendorf, Germany
Riccardo Belluzzo, StethoMe®, Poland
Elisabetta Benevento, University of Rome Tor Vergata, Italy
Pedro Bibiloni, University of the Balearic Islands / Balearic Islands Health Research Institute (IdISBa), Spain
Vilmos Bilicki, University of Szeged, Hungary
Enrico Gianluca Caiani, Politecnico di Milano, Italy
Manuel Campos Martínez, University of Murcia, Spain
Bhargava Chinni, University of Rochester, USA
Javier Civit, Cober SL, Spain / Gnomon Informatics, Greece
Sandra Costanzo, University of Calabria, Italy
Pierpaolo Di Bitonto, Grifo multimedia S.r.l., Italy
Linying (Lin) Dong, Ryerson University, Canada
Audrey DunnGalvin, University College Cork, Ireland
Claudio Eccher, FBK Fondazione Bruno Kessler, Italy
Christo El Morr, York University, Canada
Sebastian Fudickar, Universität Oldenburg, Germany
Niels F. Garmann-Johnsen, University of Agder, Norway

Anthony Gelibert, Carbon Bee, France
Helena Germano, Polytechnic Institute of Setúbal -School of Health, Portugal
Wojciech Glinkowski, Polish Telemedicine Society / Center of Excellence "TeleOrto", Poland
Kuang Gong, Massachusetts General Hospital / Harvard Medical School, USA
Manuel González-Hidalgo, University of the Balearic Islands / Balearic Islands Health Research Institute (IdISBa), Spain
Conceição Granja, Norwegian Centre for e-health Research, University Hospital of North Norway, Norway
David Greenhalgh, University of Strathclyde, UK
Teresa Guarda, Universidad Estatal Peninsula Santa Elena - UPSE / Universidad de las Fuerzas Armadas – ESPE / ALGORITMI Research Centre | ESPE | UPSE, Ecuador
Katarina Gvozdanovic, Agency for Medicinal Products and Medical Devices, Zagreb, Croatia
Oliver Heinze, University Hospital Heidelberg, Germany
Marike Hettinga, Windesheim University of Applied Sciences, the Netherlands
Amin Hossein, Université libre de Bruxelles, Belgium
Ying-Feng Hsu, Osaka University, Japan
Yan Hu, Blekinge Institute of Technology, Sweden
Ming Huang, Mayo Clinic, USA
Femi Isiaq, Southampton Solent University, UK
Martijn Kiers, University of Applied Science FH JOANNEUM, Austria
Evdokimos Konstantinidis, Aristotle University of Thessaloniki, Greece / Nively, Nice, France
Frank Kramer, Faculty of Medicine/University of Augsburg, Germany
Vinay Kumar, Thapar University, Patiala,India
Tatjana Lončar-Turukalo, University of Novi Sad, Serbia
Gang Luo, University of Washington, USA
Rafael Maestre Ferriz, CETEM, Spain
Flora Malamateniou, University of Piraeus, Greece
Luis Marco-Ruiz, Norwegian Centre for E-health Research | University Hospital of North Norway, Tromsø, Norway /Peter L. Reichertz Institute for Medical Informatics of TU Braunschweig |Hannover Medical School, Germany
Robert Mischak, Graz University of Applied Sciences, Austria
António H. J. Moreira, 2Ai - Polytechnic Institute of Cávado and Ave, Barcelos, Portugal
Fernando Moreira, Universidade Portucalense, Portugal
Mário W. L. Moreira, Federal Institute of Education, Science, and Technology of Ceará, Brazil
Yoshitoshi Murata, Iwate Prefectural University, Japan
Nuria Ortigosa, Universitat Politecnica de Valencia, Spain
Anna Pastusiak, StethoMe® / Adam Mickiewicz University, Poznan, Poland
Hugo Peixoto, Algoritmi Research Center | University of Minho, Portugal
Vitor Pinheiro de Almeida, Pontifícia Universidade Católica do Rio de Janeiro (PUC-Rio), Brazil
Ivan Miguel Pires, Instituto de Telecomunicações | Universidade da Beira Interior, Covilhã, Portugal
Filipe Portela, University of Minho, Portugal
Sandhya Prabhakaran, Moffitt Cancer Center, Tampa, USA
Rüdiger Pryss, Ulm University - Institute of Databases and Information Systems, Germany

Copyright Information

Lower-Limb Falling Detection System Using Gated Recurrent Neural Networks

Francisco Luna-Perejón, Luis Muñoz-Saavedra, Javier Civit-Masot, Juan M. Montes-Sánchez,
Manuel Domínguez-Morales and Anton Civit
Robotics and Computer Technology Lab
Avda. Reina Mercedes s/n, E.T.S. Ingeniería Informática, Universidad de Sevilla, Sevilla, Spain
Email: {fralunper, luimunsaa, jcivit, jmontes, mdominguez, civit}@atc.us.es

Abstract—**Accidental falls are one of the most common causes of premature disability and mortality related to unnatural causes. This affects mainly the elderly population. With the current aging of the population, the rate of accidental falls increases. Computer systems for gait analysis and fast assistance in ubiquitous environments can be effective tools to prevent these accidents. In this article we present the advances in the creation of an intelligent device for detecting falls and risk situations based on accelerometer signals registered on the user's ankle. The proposed method makes use of Deep Learning techniques, specifically Gated Recurrent Neural Networks. The results show that the proposed model is a viable alternative to detect falls and fall risk, which can be implemented in low performance devices for greater autonomy, lower cost and comfortable portability. These results open the possibility of combining fall detection with a biomechanical analysis system to identify gait deficiencies and their relation with falls.**

Keywords-fall detection; fall prevention; accelerometer; Gated Recurrent Neural Networks.

I. INTRODUCTION

Elderly people commonly suffer falls, which can drastically reduce their quality of life, their capabilities for independent living and, in many cases, their sense of self efficiency. This is a very significant public health problem in a worldwide scale. World Health Organization (WHO) reports indicate that around 30% of seniors over 65 years old suffer at least one fall per year and the fall rate increases significantly with age [1]. There are many age related factors associated with this increased fall rate including cognitive, affective, sensory, musculoskeletal, neurological, and metabolic changes [2]. It is also well known that gender is also a key factor as women fall more often than men and sustain more injuries when they fall [3]. Many factors related to falls have also an important influence on the person gait. Thus, we can study a person gait in order to predict their future probability of falling [4] and, also, it would be interesting to study the capability of wearable devices to perform both gait analysis [5] and fall detection tasks [6][7].

An important factor with a significant impact in falls is that many elderly people lose confidence and adopt a more sedentary life, losing mobility and increasing the probability of falling because of their poor shape [8][9]. Direct consequences of falls include injuries to muscles, bone fractures and head trauma among others. Major injuries pose significant risk for post-fall morbidity and mortality and have strong economic impacts on family and public health providers [10].

Devices that monitor user activity and ideally alert when a fall has occurred are known as Fall Detection Systems (FDS). Their main objective is to distinguish between Activity of Daily Living (ADL) and fall events (alerting when this one happens) [11]. On one hand, context-aware systems use sensors installed in the environment: users don't need to wear any special device and they do not have significant computational or energy limitations. However, these solutions are limited to specific areas and are prone to privacy related ethical issues. On the other hand, wearable devices allow user monitoring without any dependence with environment-based sensors. These systems are usually based on simple sensors such as accelerometers or gyroscopes, that can be integrated in low-power devices to increase their battery life [12]. Wearable FDS require a periodic sensor monitoring process (several times per second) that may demand a significant power consumption; but, if fall detection is performed in the embedded device, the algorithm may reduce the detection accuracy and increase the response time.

Regarding fall detection algorithms two main families are widely used: threshold based and machine learning based algorithms. Delving into Machine Learning systems, Gated Recurrent Neural Networks (RNN), such as Long Short-Term Memory (LSTM) and Gated Recurrent Units (GRU) are Deep Learning networks specifically designed for sequence processing. Recent studies have shown the potential of RNNs for dynamic signal classifications [13] and specifically for accelerometer data [14][15].

Considering the other function that we want to embed into our wearable device, Gait Analysis Systems (GAS) have been available for a long time and are widely used in rehabilitation scenarios [16]. A good survey of the different technologies available for gait analysis can be found at [17]. As in the case of fall detector and GAS can either be environment based or wearable. Wearable GAS are popular in the fitness related communities. Moreover, our group developed and tested an embedded GAS placed in a wearable device [18].

Thus, analyzing the publications related to these devices it is clear that gait analysis can be performed using a sensor in the foot or the ankle. However, it is not clear that wearable FDS can perform adequately when they are attached to these locations.

The main goal would be to create a combined FDS and GAS device (Fall Detection and Gait Analysis System: FDGAS) so that we can use gait data to forecast possible falls. In this work, we will study the feasibility of a lower-limb FDS. As the feasibility of a lower-limb GAS has already been established, if we are able to prove that our intermediate goal is reachable then, we would be able to achieve the final goal of design and evaluating the FDGAS.

The article is organized as follows: Section II describes Gated Recurrent Neural Networks used for the Deep Learning classification algorithm; Section III describes the wearable device and the database used for the training process; Section IV presents the results obtained after the training and testing phases; and, finally, Section V includes the discussion regarding the effectiveness of the trained Deep Learning models and the final conclusions of the work.

II. GATED RECURRENT NEURAL NETWORKS

Recurrent Neural Networks are Deep Learning architectures specifically designed to extract features from sequential data, such as time varying signals. Gated Neural Networks are a newer version of RNN that solve traditional problems that affect the learning process of these networks. They add a vector as an information storage component called cell, which stores information acquired during the sequence analysis, relevant to process the rest of the sequence. Long Short-Term Memory (LSTM) and Gated Recurrent Units (GRU) are the most popular architectures of this type, which has demonstrated good performance in problems such as signal classification [13][14]. The main differences between them are the way they update and use the stored information in the cells during the sequence analysis (See Figure 1). An LSTM cell management system consists of three activation functions called gates to add new information to the cell (input gate), remove current information stored (forget gate) and pass information to the neural network that considers relevant to process the sequence in an specific stage (output gate). In case of a GRU cell, there are only two gates, which adds (update gate) and removes (reset gate) information from the cell respectively, allowing all stored information to be used by the neural network throughout the sequence analysis. Both kind of Recurrent Neural Networks have similar performance, although GRU is theoretically more efficient due to having two gates.

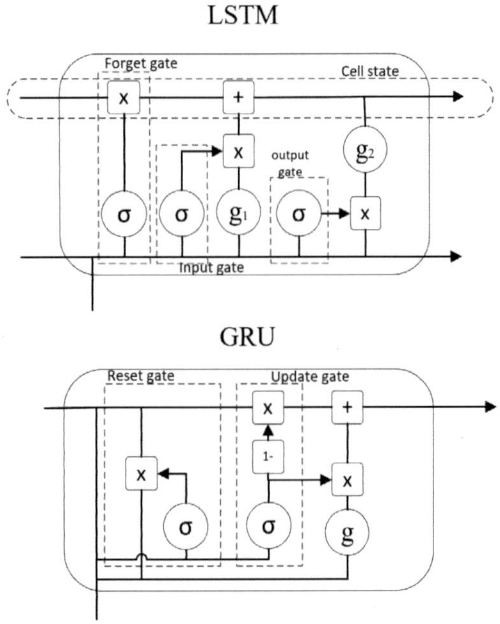

Figure. 1. LSTM and GRU cells.

III. Materials and Methods

A. Data register device

To collect data, we developed a data acquisition prototype using a ST-nucleo L432KC low-power microcontroller, that obtains tri-axial acceleration data from an ADXL345 accelerometer and transmits it via Bluetooth to check the correct sensing of the data in real time during the activity recording. The microcontroller frequency sampling and transmission was configured at 25Hz. The accelerometer was configured to work with a precision of $\pm 16g$. The x axis of the accelerometer was aligned with the horizontal line, y axis with the vertical, and z axis was in the march direction.

B. Dataset aqcuisition

A set of activities were selected for this work, including Activities of Daily Living, fall risk situation and falls. We proposed 7 ADL, two of which include a fall risk phase, and 4 different falls (Table I). The subjects who performed the activities were 7 volunteers, 6 males and 1 female. The volunteers had heights between 1.60 and 1.95 meters, weights between 70 and 110 kilograms and ages between 24 and 30. All the activities were registered with the device placed on the right ankle.

The ADL that the volunteers performed are described in Table I:

TABLE I. TYPE OF ACTIVITIES RECORDED.

Parameter	**Set of values for grid search**
ADL	
Activity 1	The subject sits down on a chair.
Activity 2	The subject gets up from a chair.
Activity 3	The subject walks calmly.
Activity 4	The subject goes down the stairs.
Activity 5	The subject goes up the stairs.
Activity 6	The user trips over with the right foot.
Activity 7	The user trips over with the left foot.
Falls	
Activity 8	The user falls backwards.
Activity 9	The subject falls to the left.
Activity 10	The subject falls to the right.
Activity 11	The subject falls forward.

C. Data labelling and block segmentation analysis

Each sample in the activity records was labelled according to what part of the activity it belongs to. Four classes were considered:

- **Fall**: if the sample is part of the moment of fall in an activity.
- **Risk**: if the sample is part of the moment of risk of falling, before a fall or recovering the balance in a stumble.
- **ADL**: if the sample is part of walking or another ADL contemplated in the activity set.
- **Background (BKG)**: if the sample does not belong to any of the previous classes. This class includes moments of stillness at the beginning or end of the record of each activity. It also contemplates the period of inactivity after a fall.

Two of the authors labeled the samples and verified the result. All the activities were also recorded in a video as support material for labeling.

The Gated Recurrent Neural Networks input should have a short fixed length for better performance. For this reason, each activity record should be split in segments having the same number of temporal samples. Henceforth, each segment will be referred to as a block, and the number of samples as width. Since fall and risk events last a short time, this fact is an advantage since it allows a detection in real time. Additionally, each block needs to be associated to a unique label. The criteria established to select a label was according to the percentage of samples of the most relevant class (fall, risk, ADL, BKG). BKG was the default class, that is, the label chosen when the block does not contain the minimum percentage of samples of each other class. In this study, different width and minimum percentage of fall, risk and ADL classes were used. Data augmentation was also used by establishing a window stride during the record split. Different stride values were also analyzed.

D. Model training and evaluation

Two basic models based on Recurrent Neural Networks were assessed. They are consisting of a batch normalization layer, a recurrent layer, LSTM and GRU respectively, and a dense layer with four nodes and softmax for class inference (See Figure 2). The results in [12] show that this architecture has a high performance with data acquired at the waist to infer three classes (fall, risk and ADL-BKG combined) and low computational complexity allowing a classification in real time. A weighted loss function was applied to offset the imbalance of the data.

Hold-out validation and grid-search were used for hyperparameter optimization. The hyperparameters consisted of the number of nodes of the recurrent layer (N), dropout rate (Dr), learning rate (lr) and batch size (bs). The set of values used for each hyperparameter is shown in Table II.

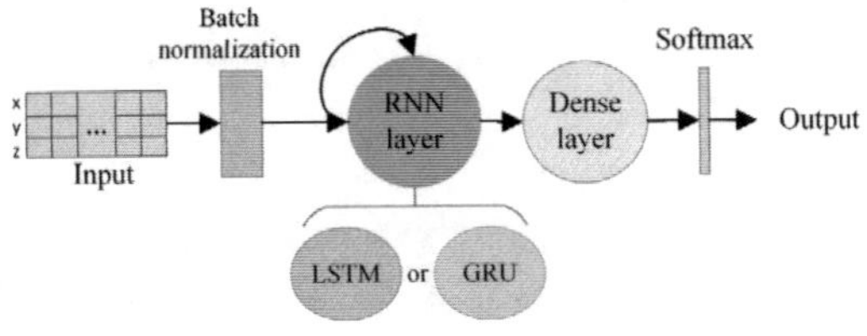

Figure. 2. Diagram of the Gated RNN architectures assessed.

TABLE II. GRID SEARCH VALUES FOR EXHAUSTIVE PARAMETERS OPTIMIZATION.

Parameter	Set of values for grid search
Learning rate	0.0001, 0.0005, 0.001
Batch size	32, 64
Number of nodes	16, 32, 48
Dropout	0.15, 0.25

To assess the effectiveness of the models, we use a set of metrics, that is, macro recall or sensitivity, precision and macro f1-score, and specificity. They are expressed as follows:

$$Precision_m = \sum_c \frac{TP_c}{TP_c + FP_c}, c \in classes \quad (1)$$

$$Recall_m(sensitivity) = \sum_c \frac{TP_c}{TP_c + FN_c}, c \in classes \quad (2)$$

$$Specificity = \sum_c \frac{TN_c}{TN_c + FP_c}, c \in classes \quad (3)$$

$$F1 - score_m = 2 * \frac{precision_m * recall_m}{precision_m + recall_m} \quad (4)$$

where m index refers to macro metric and $classes = \{BKG, ADL, risk, fall\}$. TP_c, FP_c and FN_c denote the number of true positives, false positives and false negatives of each class $c \in classes$, respectively. Finally, TN_c denotes the number of true negatives of each class $c \in classes$.

IV. RESULTS

The results obtained after optimization for the two models are presented in Table IV. The models were trained using the resulting dataset from 6 users and tested with the data from 1 user. The dataset distribution before data segmentation and augmentation is shown in Table III.

TABLE III. DATASET DISTRIBUTION FOR EACH SUBSET.

	Blocks				
Subset	**Total**	**ADL**	**BKG**	**Alert**	**Fall**
Training	5,621	1,179	3,553	348	541
Test	1,259	422	698	57	82

The confusion matrix can be observed in Figure 3. The metrics values obtained before grid search optimization are shown in table IV. The GRU model reached a better performance, but it does not get a significant difference, that matches with results in [19]. These similarity on result by comparing the two types of RNN layers are somewhat similar to [12]. Since the GRU layers present quite lower computational costs, it can be a better option for a low power energy device.

The performance obtained with these models and data is somewhat low compared to the current state-of-the-art. However, there are several limitations that influence the results. First, a classification problem with four classes has been carried out. A binary problem would probably have reached higher success rates but, with this problem we wanted to analyze the ability of the model to distinguish several classes. Second, the dataset created is quite small to train Deep Learning models, because it has many features to optimize. We have not contemplated the analysis of more complex models for this reason, since the number of features would be considerably higher. Finally, no preprocessing of the data values has been performed. Considering these factors, the results obtained show that these models have great potential to identify falls and risk situations with accelerometer data positioned on the ankle and without filtering.

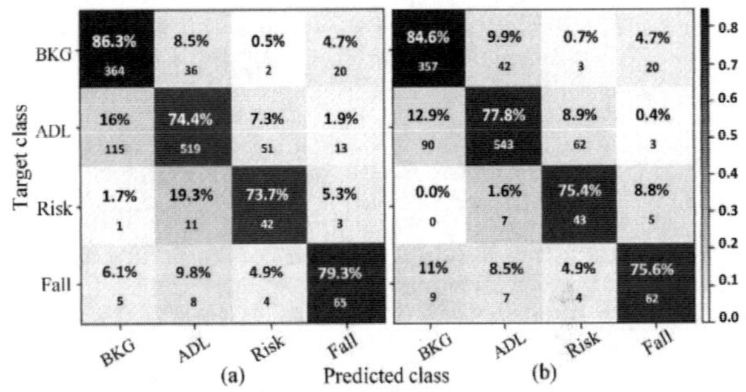

Figure. 3. Confusion matrices for the LSTM model in (a) and GRU model in (b).

TABLE IV. RESULTS OBTAINED AFTER GRID SEARCH OPTIMIZATION.

RNN type	$Precision_m$	$Recall_m$	$Specificity$	$F1score_m$
LSTM	0.681	0.784	0.920	0.729
GRU	0.691	0.790	0.925	0.737

The results obtained after these preliminary tests demonstrate the feasibility of using a fall detection system located in the lower limb. This fact is the main novelty of this work, since in the literature can only be found works that place the device on the waist or upper extremities. However, this study has limitations that must be detailed:

- The final number of volunteers used to obtain the database is limited: in future studies it is necessary to expand the number of people.
- Due to lack of time, a single data collection was carried out for each activity and person. It is important to expand the samples in the future to have, at least, 3 data collection of each activity from each person.
- The previous deficiencies mean that the results obtained in this preliminary study are not as good as those obtained in other fall detection systems (not located in the lower limbs). However, the improvement of the previous points in subsequent studies will improve those results.

V. CONCLUSION

The performance reached with simple LSTM and GRU models indicates the feasibility to extract features to identify ADL, falls and risk events (in falling detection systems). This work demonstrates that FDS can be placed in the lower limbs in order to combine the information obtained from it with Gait Analysis Systems (GAS). This is the main novelty of this work.

However, in order to improve the results obtained in this work, it is necessary to deepen into these systems with more complex architectures and larger datasets, as well as in the application of appropriate data preprocessing techniques.

In future works, this lower-limb FDS will be combined with the GAS to create a high-level system able to detect any abnormality during the daily activities of the patient.

ACKNOWLEDGMENT

The authors would like to thank the Robotics and Computer Technology Lab for the loan of materials and facilities to collect data and testing.

REFERENCES

[1] "Ageing; life course unit. who global report on falls prevention in older age," *World Health Organization*, 2008.

[2] N. Kronfol, "Biological, medical and behavioral risk factors on falls," *World Health Organisation http://www who int/ageing/project/falls_prevention_older_age/en/inded html*, 2012 [Last access on January 25, 2020].

[3] M. E. Tinetti, J. T. Doucette, and E. B. Claus, "The contribution of predisposing and situational risk factors to serious fall injuries," *Journal of the American Geriatrics Society*, vol. 43, no. 11, pp. 1207–1213, 1995.

[4] L. Wolfson, R. Whipple, P. Amerman, and J. N. Tobin, "Gait assessment in the elderly: a gait abnormality rating scale and its relation to falls," *Journal of gerontology*, vol. 45, no. 1, pp. M12–M19, 1990.

[5] J. Pineda-Gutiérrez, L. Miró-Amarante, M. Hernández-Velázquez, F. Sivianes-Castillo, and M. Domínguez-Morales, "Designing a wearable device for step analyzing," in *2019 IEEE 32nd International Symposium on Computer-Based Medical Systems (CBMS)*. IEEE, 2019, pp. 259–262.

[6] F. Luna-Perejon *et al.*, "An automated fall detection system using recurrent neural networks," in *Conference on Artificial Intelligence in Medicine in Europe*. Springer, 2019, pp. 36–41.

[7] F. Luna Perejón *et al.*, "Sampling frequency evaluation on recurrent neural networks architectures for iot real-time fall detection devices," in *IJCCI 2019: 11th International Joint Conference on Computational Intelligence (2019), p 536-541*. ScitePress Digital Library, 2019.

[8] L. Rubenstein, "Falls in older people: epidemiology, risk factors and strategies for prevention," *Age and ageing*, vol. 35, no. suppl_2, pp. ii37–ii41, 2006.

[9] M. Aschkenasy *et al.*, "Aschkenasy, miriam t and rothenhaus, todd c," *Emergency Medicine Clinics*, vol. 24, no. 2, pp. 413–432, 2006.

[10] J. A. Stevens, P. S. Corso, E. A. Finkelstein, and T. R. Miller, "The costs of fatal and non-fatal falls among older adults," *Injury prevention*, vol. 12, no. 5, pp. 290–295, 2006.

[11] R. Igual, C. Medrano, and I. Plaza, "Challenges, issues and trends in fall detection systems," *Biomedical engineering online*, vol. 12, no. 1, p. 66, 2013.

[12] F. Luna-Perejón, M. J. Domínguez-Morales, and A. Civit-Balcells, "Wearable fall detector using recurrent neural networks," *Sensors*, vol. 19, no. 22, p. 4885, 2019.

[13] C. Gao, D. Neil, E. Ceolini, S.-C. Liu, and T. Delbruck, "Deltarnn: A power-efficient recurrent neural network accelerator," in *Proceedings of the 2018 ACM/SIGDA International Symposium on Field-Programmable Gate Arrays*, 2018, pp. 21–30.

[14] S. Yu, "Residual learning and lstm networks for wearable human activity recognition problem," in *2018 37th Chinese Control Conference (CCC)*. IEEE, 2018, pp. 9440–9447.

[15] M. Musci, D. De Martini, N. Blago, T. Facchinetti, and M. Piastra, "Online fall detection using recurrent neural networks," *arXiv preprint arXiv:1804.04976*, 2018.

[16] R. Baker, "Gait analysis methods in rehabilitation," *Journal of neuroengineering and rehabilitation*, vol. 3, no. 1, p. 4, 2006.

[17] A. Muro-De-La-Herran, B. Garcia-Zapirain, and A. Mendez-Zorrilla, "Gait analysis methods: An overview of wearable and non-wearable systems, highlighting clinical applications," *Sensors*, vol. 14, no. 2, pp. 3362–3394, 2014.

[18] M. J. Domínguez-Morales, F. Luna-Perejón, L. Miró-Amarante, M. Hernández-Velázquez, and J. L. Sevillano-Ramos, "Smart footwear insole for recognition of foot pronation and supination using neural networks," *Applied Sciences*, vol. 9, no. 19, p. 3970, 2019.

[19] J. Chung, C. Gulcehre, K. Cho, and Y. Bengio, "Empirical evaluation of gated recurrent neural networks on sequence modeling," *arXiv preprint arXiv:1412.3555*, 2014.

Incremental Learning For Fundus Image Segmentation

Javier Civit-Masot, Luis Muñoz-Saavedra, F. Luna-Perejón, Juan M. Montes-Sánchez, M. Domínguez-Morales
Robotics and Computer Technology Lab
Avda. Reina Mercedes s/n, E.T.S. Ingeniería Informática, Universidad de Sevilla, Sevilla, Spain
Email: {jcivit, luimunsaa, fralunper, jmontes, mdominguez}@atc.us.es

***Abstract*—Automated Fundus image segmentation is traditionally done in the image acquisition instrument and, thus, in this case it only needs to be able to segment data from this acquisition source. Cloud providers support multi GPU and TPU virtual machines making attractive the idea of cloud-based segmentation an interesting possibility. To implement this idea we need to make correct predictions for fundus coming from different sources. In this paper we study the possibility of building a web base segmentation service using incremental training, i.e, we initially train the system using data from a single data set and, afterwards, perform retraining with data from other acquisition sources. We are able to show that this type of training is efficient and can provide good results suitable for web-based segmentation.**

***Keywords*—Deep learning; Incremental Learning; U-Net; Image Segmentation; Eye Fundus; Optic disc; Glaucoma Detection**

I. Introduction

Segmentation is the process of detection of limits within an image. Deep Learning methods for segmentation are limited by the number and the variety of the training images. For cloud services it is necessary to train with new data set samples from different acquisition sources periodically. The objective of this paper is to study the effects of incremental training where we initially train with single dataset and later refine the training process by adding data from additional datasets.

A. Cloud based Image Segmentation

There are two scenarios for image segmentation in a clinical set up. In the first case the segmentation tool is marketed together with the acquisition instrument. Thus, we can train using images provided by the instrument linked to the segmentation software. In a second case segmentation is implemented as a web-based service. In this case we need to segment images coming from different sources.

Some works on combined dataset training in [1] are available but they are not related to image segmentation. In [2], [3] different data sets are use to train and test with each data set independently. [4] Uses several datasets but training is performed only with a combined dataset.

In this paper we will first train the system using a single data set and perform predictions over that dataset and also on images acquired with other instruments. Later we will perform a small (5 epoch) retraining using images from the other dataset and study the influence in the results. We apply these techniques to the detection of optical disc in eye fundus images.

B. U-Net Networks

For this study we will a generalized U-Net. U-Net [5] has been used for many medical segmentation problems including glaucoma [2] and diabetic macular edema [6].U-Net can be considered as a type of deconvolutional network [7]. In these networks a set of convolutional layers outputs a deep small representation of the original image. This highly encoded representation is decoded to the original image size using a set of upsampling layers. This network is shown in Figure 1. Its structure is describe in the original paper but has been widely modified by many researchers (e.g. [2], [3]

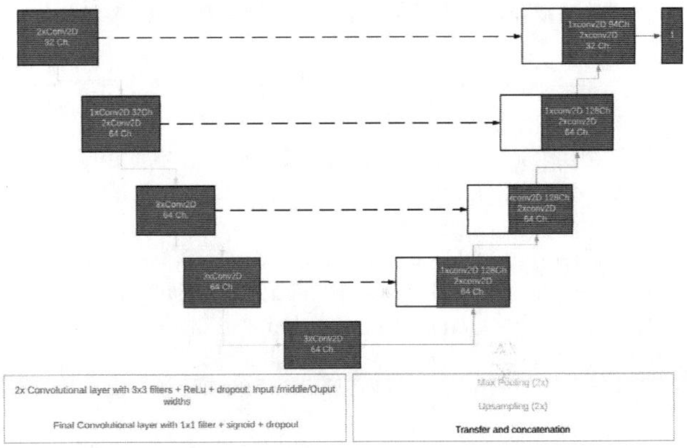

Fig. 1: U-Net Architecture

C. Fundus image glaucoma indicators

Glaucoma is a set of diseases that provoke damage to the optic nerve at the back of the eye causing loss of vision over time. [8]. Intraocular Hypertension (IH) is the most significant risk factor associated to glaucoma.

IH causes damage to the beginning of the the optic disc (OD) which is the beginning of the optic nerve. Optic disc can be visualized using many techniques including fundus color photography. The OD is made up by two subregions (Figure 2) a peripheral area (neuroretinal border) and a white central region (optic cup - OC).

As glaucoma develops, the OC increases occupying a large part of the OD. The ratio of the OC radium to that of the OD is known as CDR (Cup to Disc ratio) and is a glaucoma indicator [9]. The measurement of the OC and OD radii is needed to calculate the CDR. OD and OC human segmentation is difficult and leads to many errors. Machine based segmentation is, thus, attractive.Many machine learning (ML) alternatives are available for fundus image OD segmentation. [10], [2], [3], [11], [12].

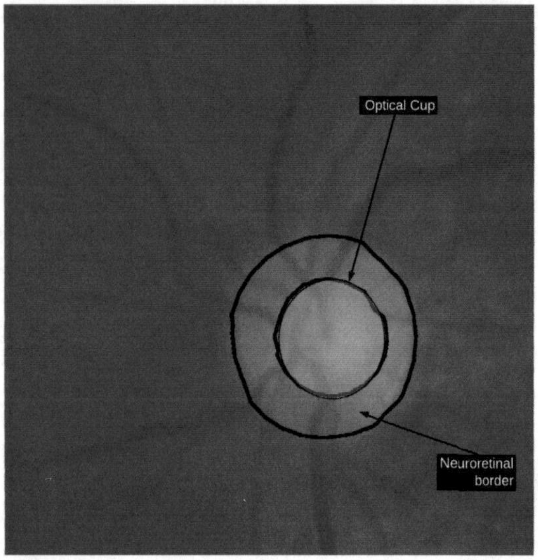

Fig. 2: Optic Disc and Cup

We do not introduce a new technique but studies the influence of retraining on the results. Thus we will only study the segmentation of the OD, although our technique is equally applicable to the OC case. The used approach based on [3] but with many major changes introduced in [4] to make it more adequate for a cloud implementation.

II. Materials and methods

In this paper we use a generalized 6 layer U-Net. It has only 40 channels in the first layer and the layer increment ratio, i.e. the ratio of the number of channels in a layer to that of the following one [13] is 1.1 instead of 2. Considerig that we resize the dataset images to 128x128 this reduces the number of trainable parameters to less than 1 million.

In this paper we test the feasibility of incremental training using the web resources and networks that we would deploy in a web service. This is ndifferent from other papers (e.g. [2], [3], [11], [12] where the data from a single source is used for training and testing.

We use Google Collaboratory python notebook environment for our implementation and apply a recursive flexible U-net model. We perform aggressive static and dynamic data augmentation modifying the approach proposed in [14].

For training and testing we use 96 image batches as this is suitable for GPU implementation. We use is 25 epochs, 256 training steps and 6 validation steps for each epoch. The employed optimizer is Adam with a 0.0007 learning rate. This values have proven adequate for training our U-Net and give excellent results with reasonable training times. We perform random sub-sampling based cross validation. Our loss function is the negative log of the Sørensen-Dice coefficient [15] (Dice).

We use the RIM-ONE v3 and DRISHTI data sets. RIM ONE [16] comes from the Spanish University of La Laguna and includes 159 images. DRISTI [17] comes from Aravind Eye Hospital, Madurai, India and includes 101 images. Both data sets have been tagged by expert ophtalmologists.

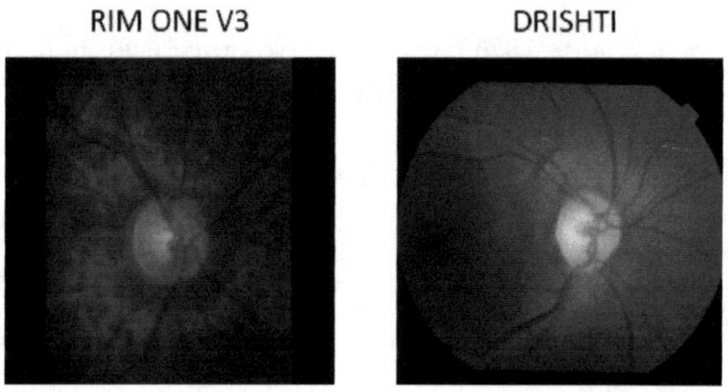

Fig. 3: Images from RIM ONE and DRISHTI datasets

Figure 3 shows that images in each dataset, which clearly have been captured using different devices and have, thus, different characteristics.Our processing approach is very similar to [4] but, in our case we don't crate a mixed dataset but train with a single dataset (DRISHTI) and then perform a small number of retrain steps using the RIM ONE dataset.

The ratio between OD and OC diameters (CDR), is one of the most popular glaucoma indicators. We define a clinically significant parameter (RRP -Radii Ratio parameter) based on the ratio between the radius of the machine segmented and that of the ground truth segmented discs. The RPP is defined as the percent of test images for which the radius error is below 10%.

We compare our work with the results from papers that use Deep Learning for optic disc segmentation and uses DRISHTI dataset. Zilly et. al. [12] use a light three layer CNN with sophisticatd pre and post-processing and apply it independently to both datsets. Sevastopolsky [3] uses a very light U-Net architecture and provides results for the RIM ONE data set. Al-Bander [2] uses a widely modified dense U-Net and provides results for both datasets. Shankaranarayana [11] uses a modified residual U-Net and provides results for the RIM ONE dataset.

III. Results

We want to find out how our system behaves when it is trained with a dataset and then lightly retrained with the other and compare our results with those obtained when a single dataset (i.e. RIM ONE or DRISHTI) is applied to train the system. We will compare our results to those by other authors who use a single data set for training and validation. The Dice coefficient is used to estimate the similarity between the correct and predicted disc. This figure of merit, also known as

F1 score, is widely used and allows us to compare our results with those from other work. Dice coefficient is defined as:

$$DC = \frac{2TP}{2TP + FP + FN} \quad (1)$$

In this equation TP indicates true positives, FP false positives, and FN false negatives.

In Table I results for Disc segmentation for our two study cases are shown. In the first one train using just the DRISHTI data set and validate using the part of that dataset not used for training and the other dataset. In the second scenario we so a short (5 epoch) retrain using the RIM ONE and data set. Thus our scenarios are the following:

- 75% of the DRISHTI dataset is used for training and after validation is carried out first with the rest of DRISHTI data set and then with the full RIM ONE data set.
- 75% of the RIM ONE data set is used to retrain the network and then we validate with the rest of the test part of both data sets.

TABLE I. OD segmentation Dice (Mean/Best/Worst) and RRP

	Dice-DRI	Dice-RIM
DRI-Trained	0.98	0.64
RIM-retrained	0.89	0.80

We can see in table I that when we train with the DRISHTI dataset results when testing with images from that dataset get good Dice coefficient values. Specifically we get a mean dice value of 0.98 (DRISHTI) but only 0.64 (RIM1) for OD segmentation. The situation is worst than it looks as in the worst case for RIM the segmentation thus not produce any pixel. When we retrain the network with the other data set results are 0.89 (DRISHTI) and 0.80 (RIM). For the worst segmentation case we get a Dice value of 0.69. Thus, we can see that with a very light retraining the network can quickly learn the specific characteristics of the second dataset.

TABLE II. OD segmentation RRP

	RRP-DRI	RRP-RIM
DRI-Trained	100%	23%
RIM-retrained	89%	80%

In Table II we show the the percentage of predictions that estimate the OD radius with an error smaller than 10% as a percentage. This data is clinically relevant as the CDR (ratio between the cup and disc radii) is a widely used glaucoma indicator. We can see that without retraining only 23% of the predicted disk segmentations have radium error bellow 10%. After the quick retrainning this value increases to 80%.

In table III we include results from other papers that have performed OD segmentation using Deep Learning methods and have trained with one of the datasets used in our study. These papers have, in every case, trained and tested with each data set independently.

Although we use networks with a small number of trainable parameters, when training with a single dataset we get results for that dataset that are similar to those obtained by other research papers. When training with the DRISHTI dataset we obtained a Dice value of 0.98 for OD segmentation. This value is slightly above 0.97 [12].

TABLE III. OD segmentation Dice comparison.

Author	DRI	RIM ONE
Zilly et al. [12]	0.97	-
Al-Bander [2]	0.95	0.90
Sevastopolsky [3]	-	0.94
Shankaranarayana et al. [11]	-	0.98
Drishti Trained	0.98	0.64
RIM Retrained	0.89	0.80

The most significant results in table III come after the retrainning that is not performed in the other studies. The results obtained when we do a quick retrain show that, in this casem we get good prediction results for all the test images. This demonstrates that it will not be feasible to create a service using training data captured with a single acquisition device.

IV. CONCLUSIONS AND FUTURE WORK

We have shown that by performing a fast retrain when adding data from a new dataset, and by preprocessing images and performing static and dynamic data augmentation, we can implement disc segmentation with an equivalent performance to that reported by researchers who use a single dataset both for evaluation and testing.

We also define a clinically significant parameter (Radii Ratio parameter- RRP) that can be useful to estimate the quality of the CDR estimations and thus, to give some confidence on the quality of the system for glaucoma prediction.

This work shows the importance of retraining when adding new image sources to the segmentation system. In a real clinical segmentation service scenario, we would have to start training the network with the initially available data and retrain it when new images from different instruments become available. The possibility of improving the network architecture by the inclusion of residual blocks [18] or the combination of a these blocks and a conventional U-Net [19] has been shown effective in several medical segmentation applications and could potentially improve the performance of our segmentation process. The robustness of these networks when analyzing images from many different instruments is still an open issue for the future.

ACKNOWLEDGMENT

Development in Cloud environment was supported by GoogleCloud platform research credit program.

REFERENCES

[1] Y. Choi, M. Choi, M. Kim, J.-W. Ha, S. Kim, and J. Choo, "Stargan: Unified generative adversarial networks for multi-domain image-to-image translation," in *Proceedings of the IEEE Conference on Computer Vision and Pattern Recognition*, 2018, pp. 8789–8797.

[2] B. Al-Bander, B. Williams, W. Al-Nuaimy, M. Al-Taee, H. Pratt, and Y. Zheng, "Dense fully convolutional segmentation of the optic disc and cup in colour fundus for glaucoma diagnosis," *Symmetry*, vol. 10, no. 4, p. 87, 2018.

[3] A. Sevastopolsky, "Optic disc and cup segmentation methods for glaucoma detection with modification of u-net convolutional neural network," *Pattern Recognition and Image Analysis*, vol. 27, no. 3, pp. 618–624, 2017.

[4] J. Civit-Masot, F. Luna-Perejón, S. Vicente-Díaz, J. M. Rodríguez Corral, and A. Civit, "Tpu cloud-based generalized u-net for eye fundus image segmentation," *IEEE Access*, vol. 7, pp. 142 379–142 387, 2019.

[5] O. Ronneberger, P. Fischer, and T. Brox, "U-net: Convolutional networks for biomedical image segmentation," in *International Conference on Medical image computing and computer-assisted intervention*. Springer, 2015, pp. 234–241.

[6] C. S. Lee, A. J. Tyring, N. P. Deruyter, Y. Wu, A. Rokem, and A. Y. Lee, "Deep-learning based, automated segmentation of macular edema in optical coherence tomography," *Biomedical optics express*, vol. 8, no. 7, pp. 3440–3448, 2017.

[7] H. Noh, S. Hong, and B. Han, "Learning deconvolution network for semantic segmentation," in *Proceedings of the IEEE international conference on computer vision*, 2015, pp. 1520–1528.

[8] H. A. Quigley and A. T. Broman, "The number of people with glaucoma worldwide in 2010 and 2020," *British journal of ophthalmology*, vol. 90, no. 3, pp. 262–267, 2006.

[9] H. A. Quigley, "Better methods in glaucoma diagnosis," *Archives of Ophthalmology*, vol. 103, no. 2, pp. 186–189, 1985.

[10] M. C. V. S. Mary, E. B. Rajsingh, J. K. K. Jacob, D. Anandhi, U. Amato, and S. E. Selvan, "An empirical study on optic disc segmentation using an active contour model," *Biomedical Signal Processing and Control*, vol. 18, pp. 19–29, 2015.

[11] S. M. Shankaranarayana, K. Ram, K. Mitra, and M. Sivaprakasam, "Joint optic disc and cup segmentation using fully convolutional and adversarial networks," in *OMIA 2017*, ser. Fetal, Infant and Ophthalmic Medical Image Analysis. Springer International Publishing, 2017, Conference Proceedings, pp. 168–176.

[12] J. Zilly, J. M. Buhmann, and D. Mahapatra, "Glaucoma detection using entropy sampling and ensemble learning for automatic optic cup and disc segmentation," *Computerized Medical Imaging and Graphics*, vol. 55, pp. 28–41, 2017.

[13] A. G. Howard and et al., "Mobilenets: Efficient convolutional neural networks for mobile vision applications," *arXiv preprint arXiv:1704.04861*, 2017.

[14] B. Zoph, E. D. Cubuk, G. Ghiasi, T.-Y. Lin, J. Shlens, and Q. V. Le, "Learning data augmentation strategies for object detection," *arXiv preprint arXiv:1906.11172*, 2019.

[15] L. R. Dice, "Measures of the amount of ecologic association between species," *Ecology*, vol. 26, no. 3, pp. 297–302, 1945.

[16] F. Fumero, S. Alayón, J. L. Sanchez, J. Sigut, and M. Gonzalez-Hernandez, "Rim-one: An open retinal image database for optic nerve evaluation," in *2011 24th international symposium on computer-based medical systems (CBMS)*. IEEE, 2011, Conference Proceedings, pp. 1–6.

[17] J. Sivaswamy, S. Krishnadas, G. D. Joshi, M. Jain, and A. U. S. Tabish, "Drishti-gs: Retinal image dataset for optic nerve head (onh) segmentation," in *2014 IEEE 11th international symposium on biomedical imaging (ISBI)*. IEEE, 2014, Conference Proceedings, pp. 53–56.

[18] P. Xiuqin, Q. Zhang, H. Zhang, and S. Li, "A fundus retinal vessels segmentation scheme based on the improved deep learning u-net model," *IEEE Access*, vol. 7, pp. 122 634–122 643, 2019.

[19] S. Kim, W. Bae, K. Masuda, C. Chung, and D. Hwang, "Fine-grain segmentation of the intervertebral discs from mr spine images using deep convolutional neural networks: Bsu-net," *Applied Sciences*, vol. 8, no. 9, p. 1656, 2018.

An Approach to Explainable AI for Digital Pathology

J. M. Montes-Sánchez , L. Muñoz-Saavedra , F. Luna-Perejón , J. Civit-Masot , S. Vicente-Diaz , A. Civit
Robotics and Computer Technology Lab. Reina Mercedes s/n, E.T.S. Ing. Informática, Universidad de Sevilla
Sevilla, Spain. Email: {jmontes, luimunsaa, fralunper, jcivit, satur, civit}@atc.us.es

Abstract—**Many medical diagnostics are based, at least, in part on medical imaging. The development of machine learning and, in particular Deep Learning (DL) based image processing in the last decade has led to the growth of diagnostic support aids based on these technologies. A problem regarding the adoption of this systems the lack of understandability of their diagnostic suggestions due to their Blackbox nature. Several approaches have been proposed to increase their explainability including evaluation of the internal layer contributions to outputs, network modifications to make these contributions more meaningful and model agnostic explanations. Medical systems are considered the paradigmatic case where understandability is of outmost importance. Digital Pathology (DP) is an especially difficult, but especially interesting case for image based diagnostic support aids. This is due, among other factors, to the fact that DP images are very large and multidimensional with the information not easily available at first sight. It is important to develop tools that let the pathologists apply their available knowledge easily while improving the diagnostic quality and their productivity. The design and evaluation of an interpretable digital pathology diagnosis aid would open the possibility for developing and deploying larger scale systems that would provide pathologists with reliable and trustworthy tools to help them in their daily diagnosis tasks.**

Keywords—**Digital Pathology; Explainable Artificial Inteligence; Deep learning;**

I. Introduction

Many medical diagnostics are currently based, at least, in part on imaging technologies. Currently the interpretation of these images is, in most cases done almost directly by medical professionals. The great development of machine learning (ML) based image processing applications in the last decade has significantly increased the research on diagnostic support aids based on these technologies.

Medical image processing will experiment a breakthrough when this type of ML based diagnostic assistance tools became widely available and accepted by the medical community. Clearly one of the problems regarding the adoption of this type of systems is related to the lack of understandability of their diagnostic suggestions due to their Blackbox nature. This aspect is especially relevant in the case of medical diagnostic aids. This fact has been recently highlighted by several recent articles [1]. In a recent presentation[2] Carlos Guestrin, Senior Director of AI and Machine Learning at Apple considers the transition from black box to inclusivity as one of the four challenges of ML systems for the next few years. In all these cases medical systems are considered the paradigmatic case where understandability is of out-most importance. Some authors even consider that GDPR [3] includes the requirement that companies should be able to give users an explanation for decisions that this type of systems produce. However, in the medical case no real automatic decision making is envisioned as the tools are just diagnostic assistance and are never responsible for any final decision.

Several approaches have been proposed to increase the understandability of CNN based image processing ML systems. The three main approaches are:

- Understanding the internal layer results and their contribution to the system global outputs [4].
- Modifying the system architecture to make the internal layer results more meaningful [5].
- Using a "model agnostic" component that provides complementary explanations [6].

Additionally, there is always the possibility of constructing networks that look for the individual characteristics that doctors use to make a diagnosis. This solution, although less elegant from a scientific point of view, could currently make sense when developing a product with a short time to market. Digital Pathology is an especially difficult, but also especially interesting case. This is due to several factors [7]:

- Digital pathology is not just a transformation of the classical microscopic analysis of histological slides to digital visualization, it is an innovation that is changing medical workflows greatly;
- Much information is hidden in high dimensional spaces, not easily accessible at first sight, thus we need AI systems to help the pathologists in accessing and interpreting this data.
- The new workflows should provide ways in which pathologists can easily use their existing knowledge.

Thus, the possibility of designing and evaluating a small scale interpret able digital pathology image diagnosis aid would open the possibility for developing and deploying, in the near future, larger scale systems that would provide pathologists with reliable and trustworthy tools to help them in their daily diagnosis. These systems would also have a significant potential in the education of pathology students.

II. Technology Benchmarking

Among the most widely used ML methods in medical image analysis are support vector machines, random forests, and deep learning (DL). Due to its commercial success DL is currently the most popular framework in ML. Most mapping tasks from input images to an output images can be accomplished via DL given a large enough data set of well labelled training and testing examples. In the medical domain, very good results have been achieved for cancer detection with an accuracy that is similar to that achieved by an average pathologist. Some recent examples related to breast cancer include the results of the Chamelyon 16 challenge for identifying metastatic breast cancer. Some of the participants were able to obtain a Receiver Operating Characteristic (ROC) area under the curve above 92% with and error rate that was below 0.52% when used as an assistance tool by a human pathologist [8], [9] . Other important works have used the DL approach to accurately quantify the tumor extent [10] or to review the real impact of diagnostic tool assistance [11]. To our best knowledge all proposed DL approaches for digital image pathology tools are basically black-box models. However, in the medical domain it is necessary to be able to open this models to a glass-box and to make the results transparent and explainable on demand. The main result of our proposed tool and its success criteria would be to provide a causal explanation that provides useful information to the pathologist, e.g. areas in blue are considered a type X tumor because characteristics A and B are present. This would greatly improve the trustworthiness of the tool and its acceptability by the medical professionals.

III. CONCLUSIONS AND FUTURE WORK

The potential of understandable (explainable, visible, glass-box...) DL diagnostic aids in healthcare is huge. A recent article in Cell [12] emphasizes the importance of visible (as opposed to black-box) approaches to ML in biomedicine. Thus, the consortium considers that designing, implementing, testing and evaluating by medical professionals and students a small scale understandable digital pathology diagnostic aid could represent a major scientific and technological break-through in the field of software and integration for medical imaging. The importance of this ideas for the quality of life of citizens is also very significant as if pathology image diagnostic aids are up-taken this would lead to quicker and better diagnostics by currently heavy overloaded pathologists which would led to faster interventions and better medical prognosis. Last, but also of great importance, the project would lead to a better position of the industry that develop these solutions in the field of DL based-digital image pathology diagnostic aids in general and more specifically in the new field of understandable DL diagnostic aids.

REFERENCES

[1] W. Knight, "The dark secret at the heart of al," *Technology Review*, vol. 120, no. 3, pp. 54–61, 2017.

[2] A. Lauterbach, "Artificial intelligence and policy: quo vadis?" *Digital Policy, Regulation and Governance*, 2019.

[3] F. Pesapane, C. Volonté, M. Codari, and F. Sardanelli, "Artificial intelligence as a medical device in radiology: ethical and regulatory issues in europe and the united states," *Insights into imaging*, vol. 9, no. 5, pp. 745–753, 2018.

[4] A. Shrikumar, P. Greenside, and A. Kundaje, "Learning important features through propagating activation differences," in *Proceedings of the 34th International Conference on Machine Learning-Volume 70*. JMLR. org, 2017, pp. 3145–3153.

[5] C.-C. J. Kuo, M. Zhang, S. Li, J. Duan, and Y. Chen, "Interpretable convolutional neural networks via feedforward design," *Journal of Visual Communication and Image Representation*, vol. 60, pp. 346–359, 2019.

[6] M. T. Ribeiro, S. Singh, and C. Guestrin, "" why should i trust you?" explaining the predictions of any classifier," in *Proceedings of the 22nd ACM SIGKDD international conference on knowledge discovery and data mining*, 2016, pp. 1135–1144.

[7] A. Holzinger, B. Malle, P. Kieseberg, P. M. Roth, H. Müller, R. Reihs, and K. Zatloukal, "Towards the augmented pathologist: Challenges of explainable-ai in digital pathology," *arXiv preprint arXiv:1712.06657*, 2017.

[8] D. Wang, A. Khosla, R. Gargeya, H. Irshad, and A. H. Beck, "Deep learning for identifying metastatic breast cancer," *arXiv preprint arXiv:1606.05718*, 2016.

[9] B. E. Bejnordi, M. Veta, P. J. Van Diest, B. Van Ginneken, N. Karssemeijer, G. Litjens, J. A. Van Der Laak, M. Hermsen, Q. F. Manson, M. Balkenhol *et al.*, "Diagnostic assessment of deep learning algorithms for detection of lymph node metastases in women with breast cancer," *Jama*, vol. 318, no. 22, pp. 2199–2210, 2017.

[10] A. Cruz-Roa, H. Gilmore, A. Basavanhally, M. Feldman, S. Ganesan, N. N. Shih, J. Tomaszewski, F. A. González, and A. Madabhushi, "Accurate and reproducible invasive breast cancer detection in whole-slide images: A deep learning approach for quantifying tumor extent," *Scientific reports*, vol. 7, p. 46450, 2017.

[11] D. F. Steiner, R. MacDonald, Y. Liu, P. Truszkowski, J. D. Hipp, C. Gammage, F. Thng, L. Peng, and M. C. Stumpe, "Impact of deep learning assistance on the histopathologic review of lymph nodes for metastatic breast cancer," *The American journal of surgical pathology*, vol. 42, no. 12, p. 1636, 2018.

[12] K. Y. Michael, J. Ma, J. Fisher, J. F. Kreisberg, B. J. Raphael, and T. Ideker, "Visible machine learning for biomedicine," *Cell*, vol. 173, no. 7, pp. 1562–1565, 2018.

EMG-Controlled Robotic Prosthetic Arm With Neural Network Training

Javier Civit-Masot, Alfonso Pérez-Rodríguez, Francisco Luna-Perejón, Luis Muñoz-Saavedra, Manuel Domínguez-Morales and Anton Civit
Robotics and Computer Technology Lab
Avda. Reina Mercedes s/n, E.T.S. Ingeniería Informática, Universidad de Sevilla, Sevilla, Spain
Email: {jcivit, fralunper, luimunsaa, mdominguez, civit}@atc.us.es

Abstract—**The project consists of the creation of a robotic arm controlled remotely through a brace developed by Thalmic Labs, able to read the limb muscles' biopotentials. This project aims to create an economic alternative to non-invasive active prostheses that exist today. Our prosthesis can perform the same functions but at a so much affordable price. In order to perform all the functions of a normal joint, the arm has several elements. Strings that simulate tendons and allow the movement of the fingers, gears that allow the rotation of the wrist and motors, which can generate movement based on the data extracted from the bracelet. The bracelet is responsible for transmitting information from the hand to the robotic arm through a wireless module that connects it with the computer, where the signal that extracts the bracelet goes through a filtering process to keep the information that interests us and Transmit it through the USB port to a microcontroller, which will be in charge of moving the engines according to the signals received. To avoid errors in the measurement of the sensors, the information received from the bracelet is trained in the computer using a Neural Network architecture before sending the information to the robotic arm.**

Keywords—**Machine learning, neural network, exoskeleton, prosthetics, EMG.**

I. Introduction

The evolution of prosthetics is long and full of stories, from its primitive beginnings, through the sophisticated present, to the incredible visions of the future. As in the development of any other field, some ideas and inventions have worked and have been explored in more detail, such as the fixed-position foot, while others have been left out or have become obsolete, such as the use of iron in prostheses [1]. The long and complex road to the computerized arm began around 1500 BC and has been in constant evolution ever since. There have been many refinements since the first wooden legs and hand hooks, and the result has been the highly customized fixation and molding found in today's devices [2].

Today, bionics is the application of biological solutions to systems technologies in architecture, design, engineering and modern technology [3]. There is also bionic engineering that covers several disciplines with the aim of concatenating (making biological and electronic systems work together), for example, to create prostheses activated by robots controlled by a biological signal or also to create artificial models of things that only exist in nature, for example, artificial vision and artificial intelligence also called cybernetics [4]. One could say, bionics is that branch of cybernetics that tries to simulate the behavior of living beings by making them better in almost all branches by means of mechanical instruments.

Bionics has had a great development in countries like Germany that has courses qualified in the same way in different schools, Japan that has a great development in Bio robots, United States and United Kingdom. In Latin America and Spain there are also developments of this type. In Mexico, the career of Bionic Engineering was founded at the UPIITA (Professional Interdisciplinary Unit in Engineering and Advanced Technologies) of the IPN (National Polytechnic Institute) in 1996 [5], which has yielded results in the creation of bionic devices [6].

In the United States, 82% of amputations are due to vascular disease, 22% are due to trauma, 4% are congenital, and 4% are tumorous. Approximately 1.6 million people in the United States live with an amputation. 1.5 amputees per 1,000 population in the US and Canada According to the Agency for Healthcare Research and Quality (AHRQ), about 113,000 lower extremity amputations are performed each year [7].

There is a 3:1 ratio between men and women (73.6% vs. 26.4%). Amputations occur predominantly in the lower limbs at 84%, compared to 16% in the upper limbs.

- In the upper limbs the cause is predominantly traumatic with 70.4% followed by congenital cause with 18%.
- In lower limbs the predominant cause is vascular with 69.5% followed by traumatic with 22.5%.

Therefore we are going to try to put a cheap alternative to a big problem that affects millions of people in the world (extrapolating the figures previously seen only in USA and Canada).

The main objective of this work is the creation of cheap non-invasive active prosthesis. Its user will be able to use it thanks to a bracelet, which has several integrated sensors (gyroscope, accelerometer and EMG (electromyographic signal sensors)). This bracelet will be placed on the user's arm and will read the parameters of the resident muscular terminations of the user's residual limb, transmitting them to a microprocessor that will move the specific motor in the prosthesis.

To achieve this purpose, the "Divide and Win" technique has been followed: the main aim of this work has been split into smaller objectives, which can be carried out individually, to finally bring them together and obtain the final goal.

The rest of the manuscript is divided as follows:

II. MATERIALS AND METHODS

The system is divided into three parts: the Myo bracelet (green), the computer (blue) and the microcontroller and arm (red). The first communication (between the bracelet and the computer) is done via a wireless connection and the second communication (between the computer and the microcontroller) is done via a USB port communication. The full system diagram can be observed in Figure 1.

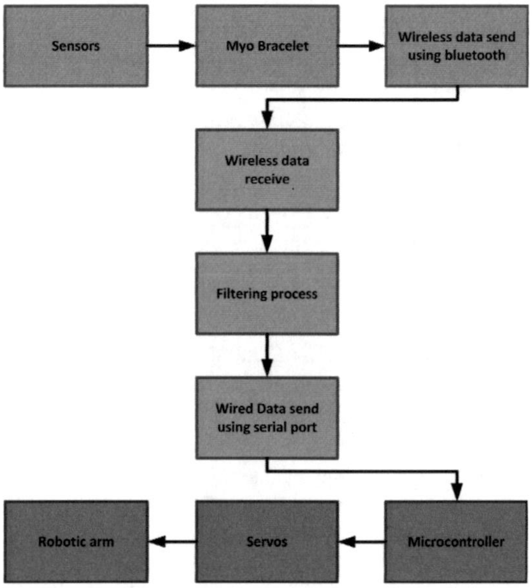

Fig. 1. System block diagram.

The information obtained from the sensors is sent from the bracelet to the computer. In the computer, a signal filtering application is executed to keep the data we are interested in for the arm movement. Once the signal has been filtered and we have the data we want, we transmit the control signal to the microcontroller, which takes care of the movement of the servomotors and therefore the movement of the robotic arm.

The sensors in the wristband encompass the technologies used to measure the different values of the arm:

- Electromyography sensors: These are the sensors placed in each module of the bracelet, they are in charge of measuring the biopotentials of the arm muscles.
- Accelerometer: It is the sensor in charge of taking measurements of arm and wrist accelerations.
- Gyroscope: This is the sensor in charge of taking measurements of changes in position and rotation of the wrist and arm.

In the data transmission of the read values, a USB module is used which provides the bracelet. The device allows data to be sent through it wirelessly. In our case, it is connected to the computer and the bracelet sends the data via Bluetooth.

The computer runs an application created in Visual Studio in C programming language, in which the data from the bracelet is received and training is carried out via a neural network in order to be able to discretize the information to be given to the arm; in this way the system requires a previous training process, as well as a large amount of data to be trained before starting to give an output from the system.

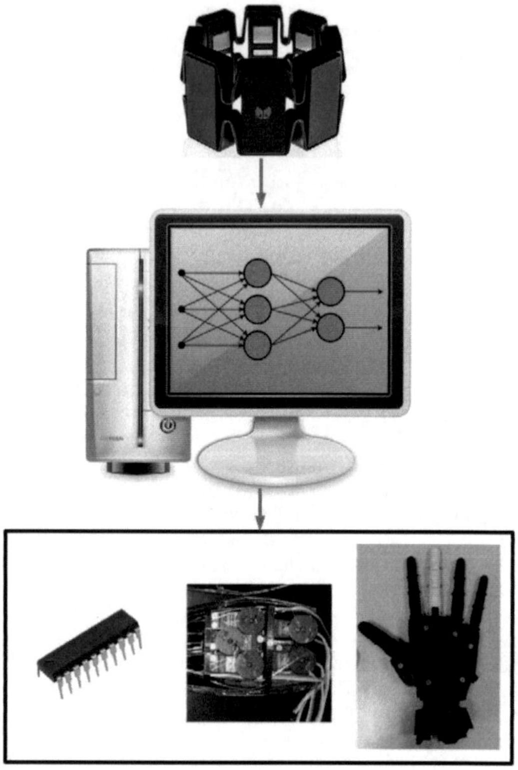

Fig. 2. System block diagram.

The result of the neural network is transformed into programmed movements that are sent to the microcontroller to control the reception of data. On the board, which controls the entire arm, are the necessary connections for the operation of the various devices.

The robotic arm is in charge of imitating the movements that we exercise in the arm that we have placed the bracelet. Servo motors are used on the arm motors. This type of motor is the best alternative for the application that the arm will perform since we can control the movement of the fingers and the rotation of the hand. The servomotors are connected to the microcontroller through the PWM pins on the plate. In this case, the devices are not powered from the microcontroller. An external power supply is used to power the servomotors.

III. RESULTS

Below, we can see the block diagram of the operation of the electromyogram:

The analogical signal is received from the electrodes placed in the bracelet, then the 8 bits analogical digital converter that the bracelet has, converts the analogical signal in a digital range from 0 to 255 bits, if the movements are not calibrated, all the movements are calibrated first, after having all the sensors perfectly calibrated, we make the calibration samples to be trained by the neural network system that we have put in the code so that the program is able to distinguish well the movements automatically. Finally, once the system is trained, the signals are sent to the arm so that it moves the way we send it through the electrical biopotentials of the muscles.

On one hand, the system's neural network structure is composed of an input layer, an output layer and a hidden layer. The entire structure can then be observed:

On the other hand, the application developed, which is in charge of communication and training, can be seen below:

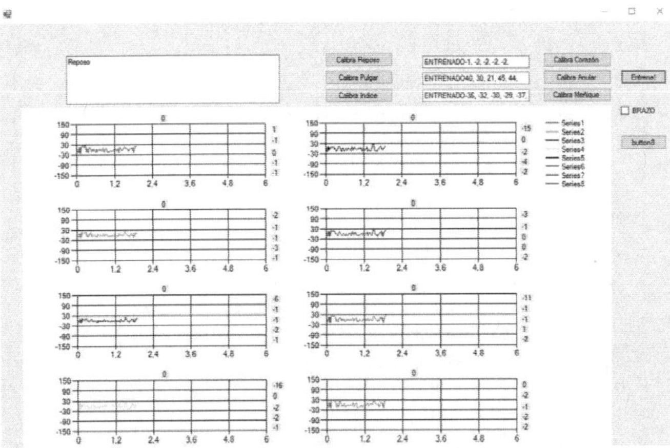

Fig. 3. System block diagram.

As for the tests, three different types were carried out: firstly, system integration tests were carried out to check the correct functioning of all the developed parts; then, for the system training, the success rates were obtained for several subjects and on multiple occasions (the training results did not fall below 85% in any case); and finally, usability tests were carried out with the patients who had served as test subjects for the neural network.

IV. Conclusions

Once the project is fully completed, the project objectives have been fully met. A system based on the movement of a robotic arm through the reading of EMG sensors from a bracelet and the training of the information received by the pc through neural networks has been designed, integrated and tested.

The success rate after training and testing with 8 subjects did not decrease in any case from 85%; being the average above 90% of success in the classification of the neuronal system. The results are satisfactory and users have shown interest in the subject.

Acknowledgment

This work has been developed and funded by the research group TEP-108: Robotics and Computer Technology fron the University of Seville.

References

[1] K. Norton, "A brief history of prosthetics," *InMotion*, vol. 17, no. 7, pp. 11–3, 2007.

[2] C. Lake and J. M. Miguelez, "Evolution of microprocessor based control systems in upper extremity prosthetics," *Technology and Disability*, vol. 15, no. 2, pp. 63–71, 2003.

[3] M. H. Dickinson, "Bionics: Biological insight into mechanical design," *Proceedings of the National Academy of Sciences*, vol. 96, no. 25, pp. 14 208–14 209, 1999.

[4] K. Krippendorff, "The cybernetics of design and the design of cybernetics," in *Design Cybernetics*. Springer, 2019, pp. 119–136.

[5] Z. Lovtchikova Khavrachenko, "Métodos de aprendizaje para los alumnos de ingeniería biónica," 2009.

[6] C. D. Rico Mandujano, "Análisis estructural de una prótesis biónica de brazo para miembro superior," Ph.D. dissertation, 2011.

[7] D. J. Margolis and et al., "Prevalence of diabetes, diabetic foot ulcer, and lower extremity amputation among medicare beneficiaries, 2006 to 2008," in *Data Points Publication Series [Internet]*. Agency for Healthcare Research and Quality (US), 2011.

Proposal of Powered Foot Prosthesis Emulating Motion of Healthy Foot (PEHF)

Yoshitoshi Murata, Haruki Baba
Faculty of Software and Information Science
Iwate Prefectural University
Takizawa, Japan
e-mail: y-murata@iwate-pu.ac.jp, g031q310@s.iwate-pu.ac.jp

Yukihide Nishimura
Department of Rehabilitation Medicine,
Iwate Medical University
Morioka, Japan
e-mail: ynishi@iwate-med.ac.jp

***Abstract*— Several million people around the world live with limb loss. Foot prosthesis is useful to improve their quality of life, and powered foot prosthesis enables them to walk naturally. However, most are too expensive for most amputees to afford. We propose a novel powered foot prosthesis that emulates the motion of a healthy foot with a half cycle delay. When a healthy person is walking, the motions of both feet are basically the same, with a half cycle difference. On the basis of this principle, the angle velocity of the foot part of the proposed prosthesis changes in the same way as the angle velocity of the healthy foot, with a half cycle delay. After introducing the measured motion data for both feet, a prototype of the proposed foot prosthesis is presented. Since this foot prosthesis adopts an industrial cylinder motor to push and pull the foot part of prosthesis, its cost would be dramatically lower than that of existing foot prosthetics.**

Keywords-foot prosthesis; power driven; leg amputee; walking gait; Quality of Life, QOL.

I. INTRODUCTION

There are nearly 2 million people living with limb loss in the United States [1]. LeBlanc [2] estimated the number of amputees is the world was approximately 10 million in 2008, with 30% of them being arm amputees [2]. Therefore, the number of leg amputees was 7 million. Leg amputees use foot prosthesis to improve their quality of life. However, low-priced foot prostheses have rigid ankle parts and no power drive mechanism, which make it difficult to raise the heel and swing the toes up. Therefore, most users need more power to move their foot. Powered foot prostheses enable users to move their foot easily and walk more naturally. However, such prostheses are so expensive for most amputees.

One of the main reasons why the price of introducing an existing powered prosthesis is too expensive may be that the prosthetic market is not open. A few manufacturers provide them as an integrated device, and the components are not compatible between different manufacturers. Another reason is the complex structure.

Most existing powered prostheses measure the change of the ankle angle and torque using sensors mounted on the prosthesis; then, they estimate the state of gait transition and speed, and push or pull the foot part or toe part.

Therefore, we propose a foot prosthesis for the leg amputee whose foot left is healthy. The prosthesis emulates the motion of a healthy foot. When a healthy person is walking, the motions of both feet are basically the same, with a half cycle difference. The basic principle of our prosthesis is that the angle velocity of the foot part of the proposed prosthesis changes in the same as the angle velocity of the healthy foot, with a half cycle delay. The angle velocity of a healthy foot is measured with a gyro sensor mounted on the heel part of the shoe worn on the healthy foot. Since the prosthesis adopts a small industrial cylinder motor that pushes and pulls the foot part of the prosthesis, the rising and falling speed of the foot part are determined on the basis of the pulse speed inputted to a stepping motor. Since the price of small industrial stepping motors is not high and the structure of this foot prosthesis is simple, it would not be too expensive for most amputees to afford.

After introducing existing foot prosthesis in Section II, we describe basic principles of the proposed prosthesis in Section III. Changes of angle velocity and angle for both healthy feet are introduced in Sections IV. A prototype of proposed foot prosthesis is introduced in Section V. Section VI concludes with a summary of key points and future works.

II. EXISTING POWERED FOOT PROSTHETICS

In this section, we introduce existing powered foot prosthesis. Ottobock in Germany and Ösuur in Iceland provide such prosthesis to consumers and the Biomechatronics Group, a research group within MIT Media Lab., has also developed some models.

The powered foot prosthesis developed by the Biomechatronics Group, a research group within MIT Media Lab., is shown in Figure 1 [3]. Its heel part (in-series spring) is pulled up and down by the ball screw driven by the motor through the timing belt. In case of this prosthesis, the gait state transitions are determined using the ankle torque and ankle angle. These data are measured angle sensors and strain gauges mounted on the prosthesis [4]. These prostheses are not a commercial product.

Ottobock provides a power-assist foot prosthesis called "1B1 Meridium [5]." Its mechanism is shown in Figure 2. It adopts a hydraulic pressure mechanism, in which a hydraulic pressure cylinder pushes and pulls a lever on the toe plate, causing the instep to rise and fall, respectively. Unfortunately, we could not find practical kinds of sensors and control method of the toe plate. The control board would estimate the pushing timing and speed from the angle between the shank and floor (= ankle angle) from sensors.

Össur provides a power-assist foot prosthetic called "PROPRIO FOOT® [6]" shown in Figure 3. Due to a lack of relevant material on the product's operation, we assume from observations that an air cylinder positioned in the area of the Achilles' tendon raises and lowers the foot part. We could not find practical kinds of sensors and control method of the foot plate. Since its appearance resembles to the appearance of the prosthesis developed by the Biomechatronics Group, MIT, its control method would be basically same as Biomechatronics Group's control method, its drive system is not different from the Biomechatronics Group's one.

Prices of the PROPRIO FOOT® and 1B1 Meridium are not available to the public, but are assumed to be more than 2 million yen ($18,000) in Japan, which is too expensive for most amputees.

The main purpose of our research is to provide a low price powered prosthetic foot based on a module structure concept.

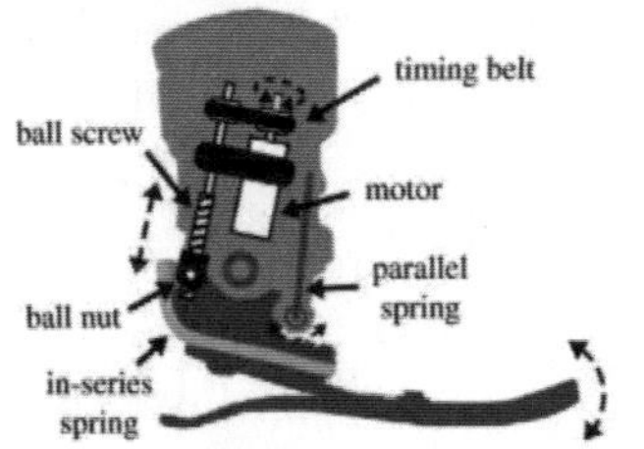

Figure 1. Power-assist foot prosthetic developed by the Biomechatronics Group of MIT Media Laboratory.

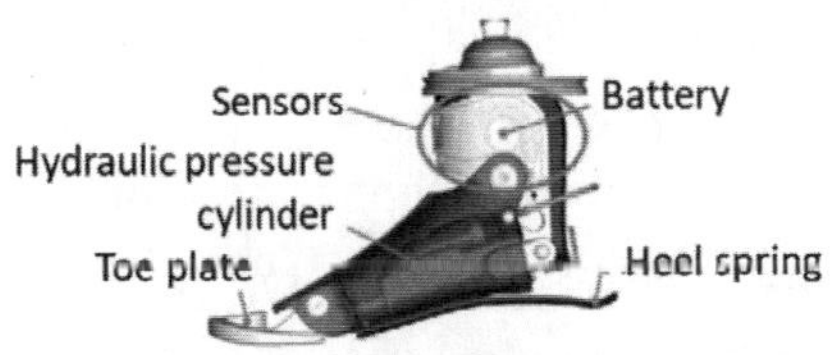

Figure 2. Mechanism of 1B1 Meridium, Ottobock.

Figure 3. PROPRIO FOOT®, Össur.

III. BASIC PRINCIPLE

As the result of measuring the differences between people with a walking disability and healthy people, it was found that the angle velocity when the heel rises and the toe angle when the foot swings for those with a walking disability are lower than healthy people [7]. We are developing walking assist devices that compensate for the shortage of heel raising power. In addition, we noticed that a foot prosthesis was similar to a person whose muscles were very weak and that the gait motion of both feet for healthy people was basically the same with a half cycle difference.

Most existing powered prostheses measure the change of the ankle angle and torque using sensors mounted on the prosthesis; then, they estimate the state of gait transition and speed, and push or pull the foot part or toe part.

However, in our proposed prosthesis, the motion of the foot part of the prosthesis is controlled on the basis of output from a gyro sensor mounted in the heel part of a shoe worn on a healthy foot, not sensors mounted in the prosthesis.

The basic structure of the proposed foot prosthesis is shown in Figure 4. The prosthesis is comprised of the following.

- Socket: a raw lower limb is inserted in a socket.
- Shin rod: a shin rod joins the socket and ankle joint.
- Ankle joint: can rotate foot up/down.
- Foot: one SACH foot is rotated up/down at the ankle joint part.
- Motor cylinder: the cylinder pushes/pulls the foot.
- Link: joins the shin rod to the motor and the motor to the foot.
- Gyro sensor: this is mounted into the heel part of the shoe worn on the healthy foot and the heel part of the foot prosthesis.
- Battery: provides electricity to the motor cylinder.
- Control board: controls the motor cylinder to emulate the gait motion of a healthy foot on the basis of output data from the gyro sensor mounted in the heel part of the shoe for the healthy foot.

The motor cylinder is a stepping motor, and the pulse stream input from the control board is used to change the pulse speed to emulate the motion of a healthy foot. The gyro sensor mounted into the heel part of the prosthesis is used to monitor the motion of the foot prosthesis. The output of both sensors is sent to the control board via Bluetooth. Bluetooth devices are mounted on the gyro sensor and control board.

This dual-link connection was adopted so as to rotate the foot smoothly.

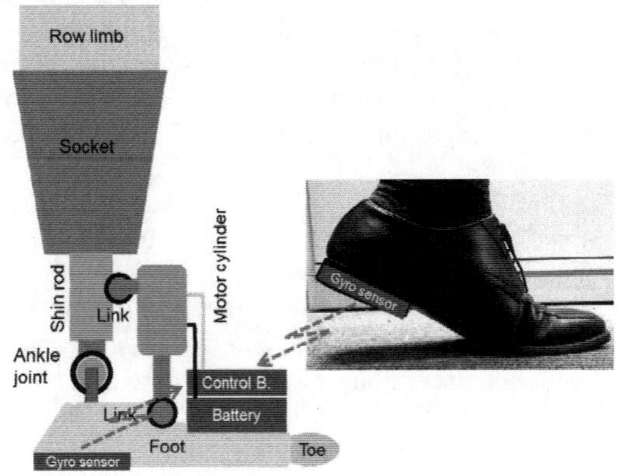

Figure 4. Basic structure of PEHF.

IV. DIFFERENCES BETWEEN THE RIGHT FOOT AND LEFT FOOT WHILE WALKING

Our research is based on the principle that the motion of both feet for healthy people is basically the same and has a half cycle difference. A change of angle velocity and angle of both feet were measured. For this measurement, we used STEVAL-WESU1 by STMicro-electronics (See Figure. 5) [8]. This wearable unit includes four sensors:

- 3D-accelerometer,
- 3D-gyroscope,
- 3D-magnetometer,
- MEMS pressure.

This device is 37 x 40 x 8 mm and weighs 9.6 g.

We inserted STEVAL-WESU1 into the heel of a shoe as shown in Figure. 6.

Figure 5. STEVAL-WESU1 by STMicroelectronics.

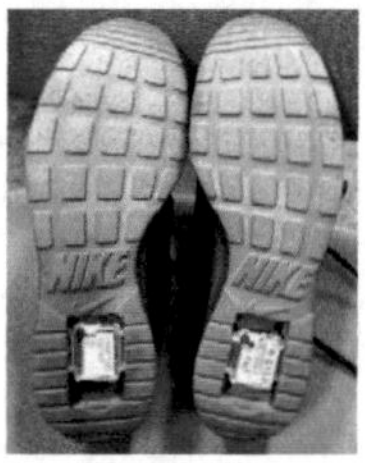

Figure 6. Sensor devices embedded into shoes.

A healthy participant walked on a straight line, 2-m-radius quarter-circle clockwise direction and 2-m-radius quarter-circle counter-clockwise direction.

The measured angle velocity data are shown in Figures 7, 8, and 9. Figure 7 shows the data for walking on a straight line, Figure 8 shows that for walking on the 2-m-radius quarter-circle clockwise direction, and Figure 9 shows that for the 2-m radius quarter-circle counter-clockwise direction. (a) in each figure shows the change in pitch rotation, (b) shows the change in roll rotation, and (c) shows the change in yaw rotation.

The normal walking gait cycle is divided into four phases: (1) entire sole touching, (2) swinging backward (raising heel), (3) swinging forward, and (4) contacting floor. The acceleration is zero for entire sole touching, for the end of swing backward, and for the end of swing forward. The initial entire contact timing, end of the entire contact timing (initial raising heel), end of swing backward, and end of swing forward are shown in Figure 10.

Figure 7 shows that the gait motions of both feet of a healthy person are basically the same and have a half cycle difference. The shape of the angle velocity curve for one foot was very similar to that for the other foot. A foot started to swing forward just after the other foot entirely touched the floor as shown in Figure 7 (a). In addition, the cycle periods of both feet were the same. The roll rotation changed synchronously with the pitch rotation as shown in Figure 7 (a) and (b). However, the ankle joint moved for the pitch rotation, and the change in roll rotation was caused by the motion of the knee. There was little change in yaw rotation because the participant was walking straight.

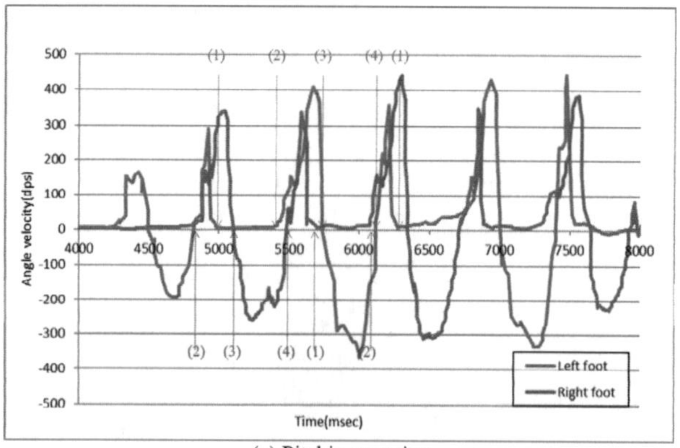

(a) Pitching rotation

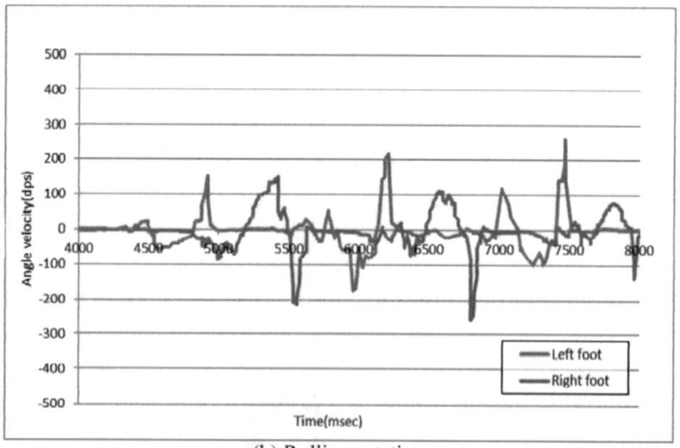

(b) Rolling rotation

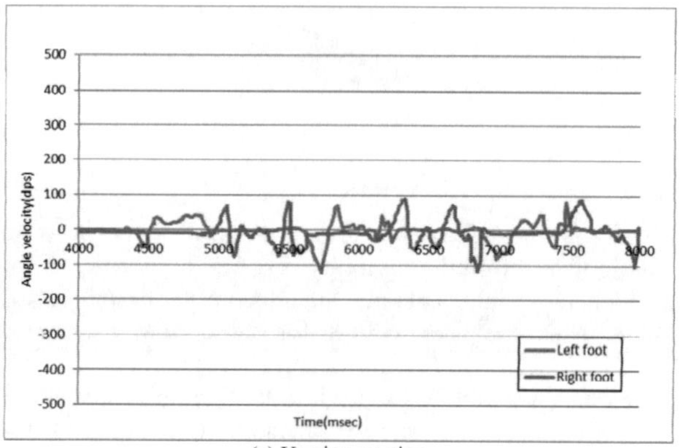

(c) Yawing rotation

Figure 7. Walking on a straight line.
(1) initial entirely contact timing, (2) terminal entirely contact timing, (3) terminal swing backward, and (4) terminal swing forward

The timings of the gait cycle for walking on the 2-m-radius quarter-circle were the same as for walking on the straight line as shown in Figure 7 (a), Figure 8 (a), and Figure 9 (a).

The cycle period of the left foot was the same as that of the right foot in Figure 8 (a) and Figure 9 (a). This characteristic is the same as Figure 7 (a). The yaw rotation in Figure 8 (c) shifted rightward (upward). The yaw rotation in Figure 9 (c) shifted leftward (downward).

Before measuring the differences between both feet for gait motion, we assumed that the cycle period for the outside of the foot would be longer than that of the inside. However, the measured data shows there was no difference in cycle period between both feet. The participant in this case walked on a polygon, not a circle.

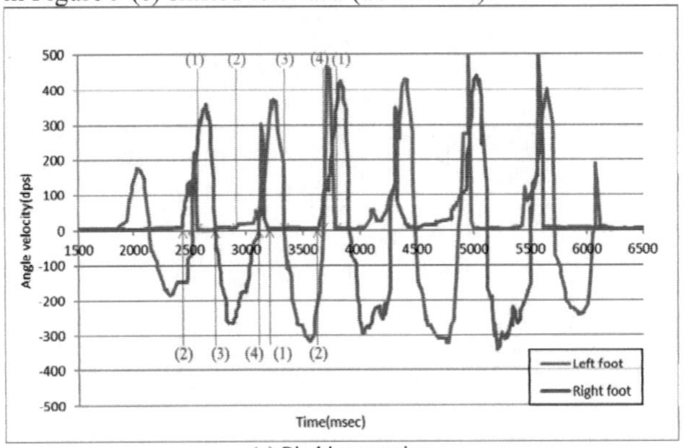

(a) Pitching rotation

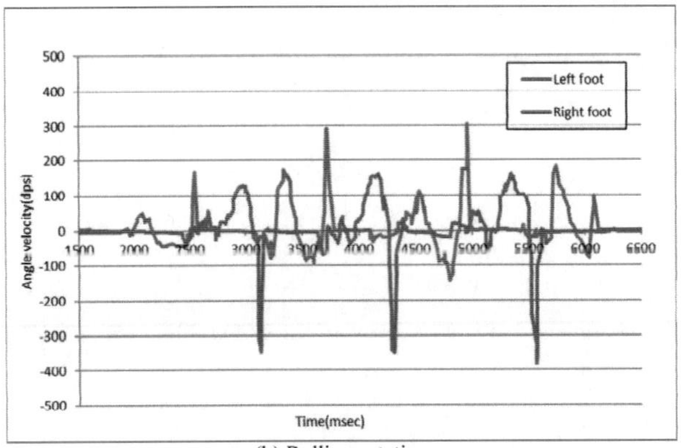

(b) Rolling rotation

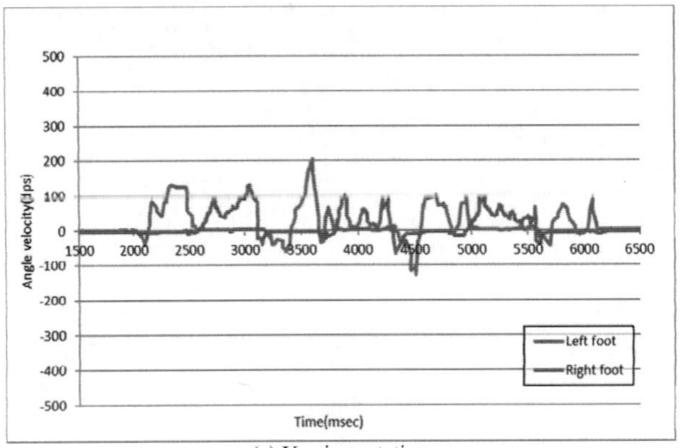

(c) Yawing rotation

Figure 8. Walking on 2 m radius quarter circle clockwise direction. (1) initial entirely contact timing, (2) terminal entirely contact timing, (3) terminal swing backward, and (4) terminal swing forward

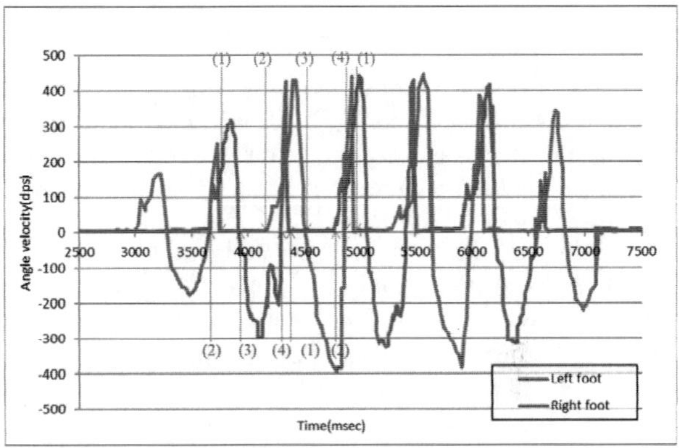

(a) Pitching rotation

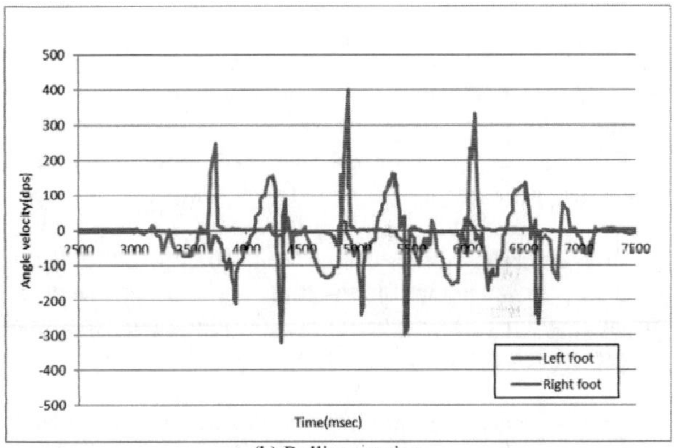

(b) Rolling rotation

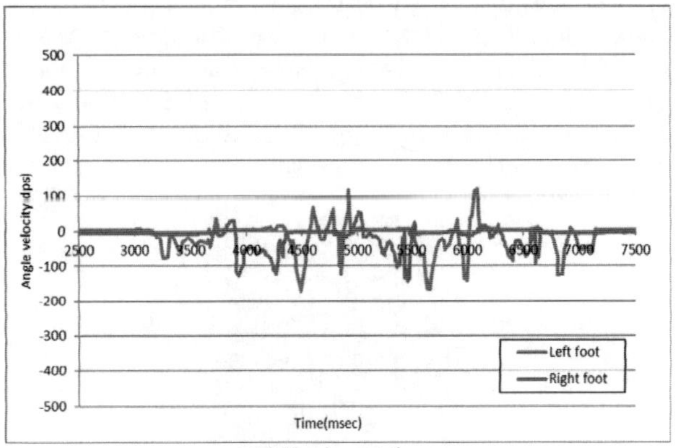

(c) Yawing rotation

Figure 9. walking on 2 m radius quarter circle counter-clockwise direction. (1) initial entirely contact timing, (2) terminal entirely contact timing, (3) terminal swing backward, and (4) terminal swing forward

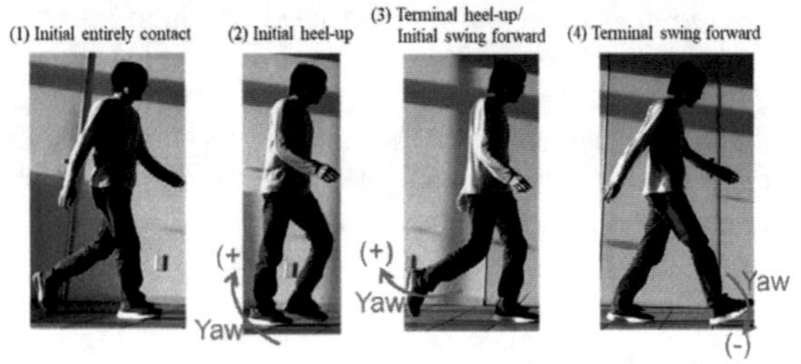

Figure 10. Normal walking gait cycle (See a right foot).

We would like the proposed prosthesis to be used not only for walking but also for jogging. Therefore, jogging motions must be measured to confirm whether there are differences between both feet.

V. PROTOTYPE OF PEHF

We developed a prototype of the foot prosthesis as shown in Figure 11. In case of the powered prosthesis designed by the Biomechatronics group of MIT Media Laboratory and PROPRIO by Össur, the heel part is pushed/pulled in order to rotate the pitch direction of the foot. However, the front part of the foot is pushed/pulled to rotate the foot in the prototype prosthesis because an existing SACH foot is used. A single motor cylinder is adopted to push/pull the foot. The motor is connected to the front of the shin and the foot via two links as shown in Figure 11.

For the single motor cylinder, we used an Oriental Motor DR series with a 30-mm stroke, 4-kg carrying force, 4-Newton thrust, and a 100-mm/sec maximum stroke speed [9]. It is a stepping motor that moves 0.001 mm per pulse. A stream of pulses is input to the motor. The pulse speed and timing are changed according to the output of the gyro sensor mounted on the heel part of the shoe worn on the healthy foot to emulate the motion of the healthy foot.

We examined the characteristics of the prototype when inputting a pulse stream to the motor. As shown in Figure 5, the sensor was attached on the heel part of the foot. The angle velocity and angle data were measured with 30-KHz and 60-KHz pulses. The measured data are shown in Figures 12 and 13. The maximum range of the angle of the prototype was 18 degrees, and the foot rotated 0.0008 degrees per pulse (dpp).

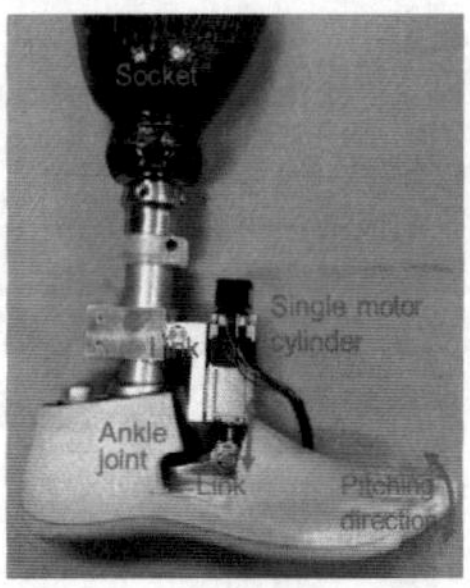

Figure 11. Prototype of PEHF.

The speed of the angle velocity did not stay the same for 30 and 60 Kpps, and the motor stopped at more than 60 Kpps. The reason is that its thrust was not enough. We have to solve this problem, which may involve changing the motor cylinder.

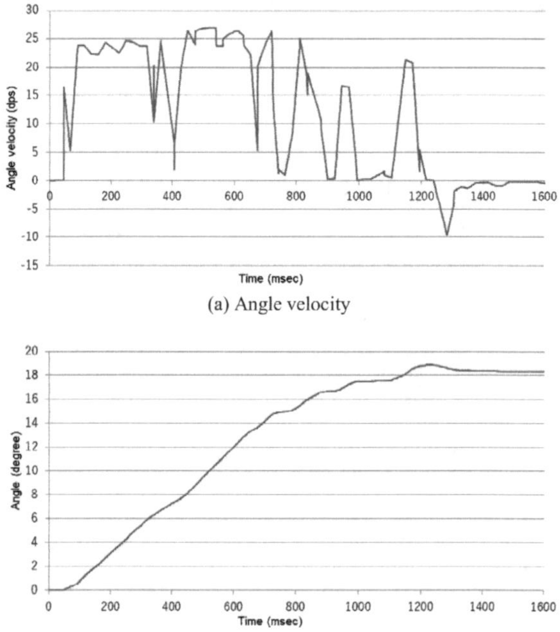

(a) Angle velocity

(b) Angle

Figure 12. Measured angle velocity and angle in 30 KHz pulse.

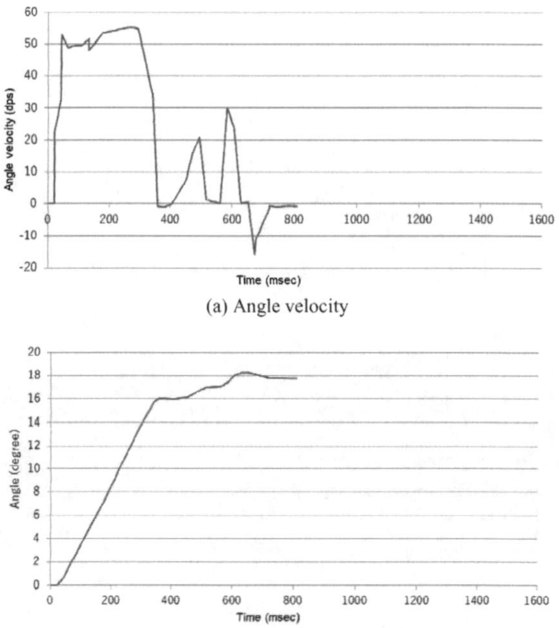

(a) Angle velocity

(b) Angle

Figure 13. Measured angle velocity and angle in 60 KHz pulse.

VI. Conclusion and Future Work

A foot prosthesis that emulates the gait motion of a healthy foot was proposed. After introducing the basic idea, we described the differences in motion between the right and left feet on the basis of the output of gyro sensors mounted in the heel part of the shoes. Basically, there were no differences between both feet except for the half cycle delay. The shape of the curve of angle velocity and the cycle period of the right foot was very similar to those of the left foot. This was the case not only when a participant walking on a straight line but also on a circle line except for the yaw direction. It was possible to recognize whether a person walked straight or turned to the right or left from the yaw data. Thus, it is possible to emulate the motion of a powered foot prosthesis with the gait motion of a healthy foot. Limb amputees could walk smoothly like a healthy person.

However, the research we introduced in this paper is the first step to develop the proposed prosthesis; at least we should work on the followings in the future;

(1) The introduced single motor cylinder could not drive a foot part well. We have to look for a suitable cylinder motor.

(2) Establishing a method for controlling cylinder motors on the basis of the angle velocity of the gyro sensor mounted into the heel part of the shoe worn on the healthy foot.

(3) Examining a prototype with a real limb amputee.

(4) Since the proposed foot prosthesis emulates motion of a healthy foot, leg amputees who use this prosthesis and have healthy foot; must start a walk with their healthy foot. We must investigate whether they could accept this limitation. If not, we have to introduce a new method to solve.

Acknowledgemnt

The authors extend thanks to Mr. Takashi Ushizaki and Mr. Kazuhiro Yoshida for the help they provided in performing this research. This work was supported by JSPS KAKENHI Grant Number 19K11326.

References

[1] K. Ziegler - Graham, E. J. MacKenzie, P. L. Ephraim, T. G. Travison, and R. Brookmeyer, "Estimating the Prevalence of Limb Loss in the United States: 2005 to 2050," Archives of Physical Medicine and Rehabilitation 2008, vol. 89 IS. 3, pp. 422-429, 2008.

[2] M. LeBlanc, ""Give Hope -Give a Hand" The LN-4 Prosthetic Hand", [Online]

https://web.stanford.edu/class/engr110/2011/LeBlanc-03a.pdf,

[retrieved: February, 2020].

[3] M. F. Eilenberg, H. Geyer, and H. Herr, "Control of a Powered Ankle-Foot Prosthesis Based on a Neuromuscular Model," IEEE Trans Neural Syst Rehabil Eng 2010, Vol. 18, pp. 164-173, 2010.

[4] J. Markowitz et al., "Speed adaptation in a powered transtibial prosthesis controlled with a neuromuscular model," Philos Trans R Soc B-Biol Sci 2011, Vo. 366, pp. 1621-1631, 2011.

[5] Ottobock, 1B1 Meridium, [Online] https://professionals.ottobock.com.au/Products/Prosthetics/Prosthetics-Lower-Limb/Feet/1B1-Meridium/p/1B1, [retrieved: February, 2020].

[6] Össur, PROPRIO FOOT®, [Online]

https://www.ossur.com/prosthetic-solutions/products/dynamic-solutions/proprio-foot,

[retrieved: February, 2020].

[7] Y. Murata, S. Yoshida, T. Niinuma, and K. Yoshida, "Comparative Analysis of Walking Gait Cycle between Healthy People and Walking Disabilities to Prevent Tripping Using Wearable Device and KINECT," IARIA, International Journal on Advances in Life Sciences, vol 9 no 3 & 4, pp. 186-187, 2017.

[8] STEVAL-WESU1, STMicroelectronics, [Online]

https://www.st.com/en/evaluation-tools/steval-wesu1.html

[retrieved: February, 2020].

[9] Single motor cylinder, DR series, Oriental Motor Ltd., [Online] https://www.orientalmotor.co.jp/products/new/180611341135dr/

[in Japanese, retrieved: February, 2020].

Proposal of Spring Assist Unit for Walking Disabilities

Haruki Baba, Yoshitoshi Murata
Faculty of Software and Information Science
Iwate Prefectural University
Takizawa, Japan
e-mail: g031q310@s.iwate-pu.ac.jp, y-murata@iwate-pu.ac.jp

Tomoki Yamato
Solution Strategic Department,
DOCOMO Technology, Inc.
Kanagawa, Japan
e-mail: tomoki.yamato.xy@nttdocomo.com

Abstract— **People with diseases, such as hemiplegia, and latter-stage elderly people often have a walking disability, which increases their risk of falling and injuring themselves. The magnitude of the angle velocity during the kicking the floor phase (raising the heel) and swinging the toe forward phase of walking is lower for disabled people than for healthy persons due to lower muscle power. We have developed a spring assist unit that fits in the heel of a shoe and helps disabled people raise their heel when beginning to walk. Experimental results demonstrated that it substantially assists gait walking, that there is a correlation between body weight and optimal spring stiffness, and that the spring assist unit does not affect the person's walking posture.**

Keywords-walking disability; walking assit unit; spring; walking posture; muscle power.

I. INTRODUCTION

As the percentage of elderly people in the world's population is increasing [1], the number of functionally impaired people, such as those with hemiplegia, will also increase. People with such diseases, or latter-stage elderly, often have walking disabilities, which increase their risk of falling, and consequently injuring themselves [2].

Our study of comparing the walking gait between hemiplegia patients and healthy students shows that the magnitude of the angle velocity during the kicking the floor phase (raising the heel) and swinging the toe forward phase of walking is lower for disabled people than for healthy persons; this is because of the lower muscle power for hemiplegia patients compared to the healthy students [3]. This means that assisting with raising the heel and swinging the toe forward while walking could help disabled people and could enable them to have a close to normal gait.

The Solid-Ankle Cushion Heel (SACH) foot (e.g., 1D10, Ottobock, Germany) [4]) and the Energy Storage And Return (ESAR) foot (e.g., Vari-Flex, Össur, Iceland) [5]) are provided for foot amputees to improves their gait so that it is close to a normal gait. They help the wearer raise their heels and take their toes off. Unfortunately, such prostheses cannot assist walking disabilities.

We previously developed a prototype shoe in which a coil spring was built into the heel part of the shoe, and a leaf spring was built into the half backward of sole. Experimental results demonstrated that it reduced the magnitude of muscle power needed to raise the heels and swing the toes forward. They also demonstrated that there would be a correlation between body weight and optimal spring stiffness, and that the spring assist unit would affect walking posture.

We have now developed a spring assist unit that is built into the heel of a shoe. Experimental results using springs with four different degrees of stiffness demonstrated that there was surely a correlation between body weight and optimal spring stiffness, and that the unit did not affect walking posture, unlike the prototype shoe. This means that the assist unit does not have any negative effects.

After introducing two kinds of foot prosthesis, the SACH and ESAR feet, in Section II, we describe in Section III the differences in gait between a hemiplegia patient and a healthy person to clarify the characteristics that need to be addressed. A prototype shoe into which a coil spring and a leaf spring are built and the experimental results are described in Section IV. The structure of the spring assist unit and the experimental results are described in Section V. Section VI concludes with a summary of the key points.

II. WALKING ASSISTANCE MECHANISM IN PASSIVE FOOT PROSTHESIS

Since there are no walking assistance devices for walking disabilities, such as latter-stage elderly people and hemiplegia patients, walking assistance devices for foot prosthesis are introduced in this section. There are two types of walking assistance mechanisms in passive foot prosthesis.

The SACH foot [6], shown in Figure 1, was designed to provide shock absorption and ankle action characteristics close to those of a normal ankle without the use of an articulated ankle joint. The action of the SACH foot is achieved by the use of two functional elements: a properly shaped wedge of cushioning material built into the heel and an internal structural core or keel shaped at the ball of the foot to provide a rocker action. Its primitive form was developed toward the end of the 1800s.

The ESAR foot, shown in Figure 2, has weak push-off power and adequate roll-over shape of the foot, which increases the energy dissipated during the step-to-step transition in gait. Wezenberg et al. reported that the ESAR foot was more effective than the SACH foot in reducing metabolic energy while walking [7], and Houdijk reported that it improved the step length symmetry [8].

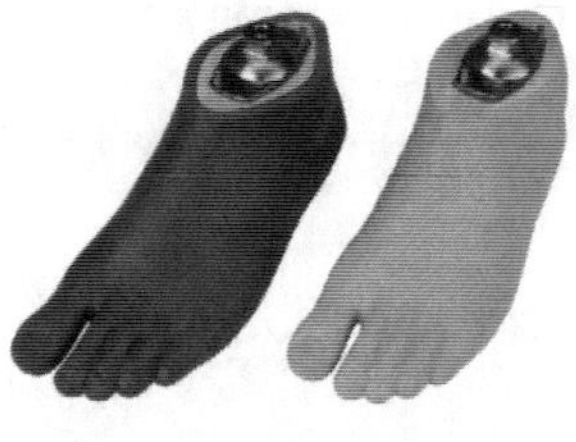

Figure 1. Examples of SACH Foot (1D10, Ottobock, Germany).

Figure 2. Examples of ESAR Foot (Vari-Flex, Össur, Iceland).

III. DIFFERENCES IN GAIT BETWEEN HEMIPLEGIA PATIENT AND HEALTHY PERSON

We analyzed the walking gait cycles of unimpaired people and those with disabilities to walk using a Wearable Device (WD) and a KINECT to detect warning signs of falls [3]. Every walking disability in this experiment had one-side paralysis, and trained periodically at a rehabilitation facility. We experimentally measured the output data of an acceleration sensor and gyroscope sensor in a WD mounted on the front of a shoe to estimate the kicking power and change of angle between a foot and the floor as shown in Figure 3. In this measurement, Smart watch 3, SONY was used as a WD.

Figures 4 and 5 show examples of changes in acceleration, angle velocity, and angle for an unimpaired participant and one with a walking disability, respectively. Data for two steps are plotted. Each flat period (roughly the center period) in these figures represents when the entire shoe sole touched the floor. The maximum angle velocity at timing A indicates the kicking power when raising the heel, and the minimum angle at timing B indicates the angle to the floor at terminal swing.

The lower angle velocity at A in Figure 4 is about 420 deg./sec. On the other hand, the higher angle velocity at A in Figure 5 is about 250 deg./sec. Thus, the participant with a walking disability clearly has a weaker kicking power when raising their heel compared with that of the unimpaired participant, indicating a clear difference in terms of gait.

The higher angle at B in Figure 4 is about -18 deg. On the other hand, the lower angle at B in Figure 5 is about -8 deg. And, the swinging speed of walking disability is slower than that of healthy participant. Thus, the participant with a walking disability expressed difficultly when raising their toe at the terminal swing phase.

(a) WD: SmartWatch 3, Sony (b) WD mounted on foot

Figure 3. Measuring device and WD mounting method.

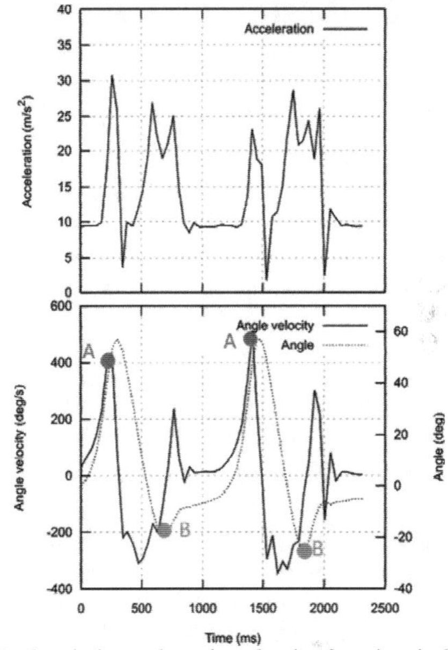

Figure 4. Angle velocity, angle, and acceleration for unimpaired participant.

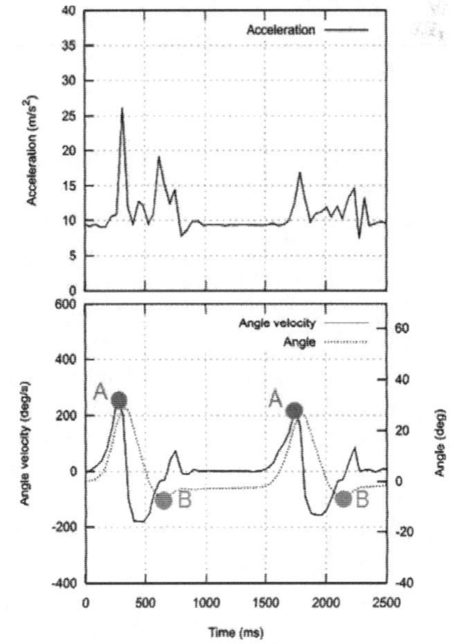

Figure 5. Angle velocity, angle, and acceleration for participant with walking disability.

TABLE I. ANGLE VELOCITY AT THE TERMINAL STANCE

Participant	Average (deg./s)	SD (deg./s)
Unimpaired participant	509.36	18.91
Participant with walking disability	342.06	86.52

TABLE II. ANGLE AT THE TERMINAL SWING

Participant	Average (deg.)	SD (deg.)
Unimpaired Participant	-17.76	8.02
Participant with disability	-7.45	8.02

Tables I and II list the averages and Standard Deviations (SDs) of measured data for angle velocity at timing A and angle at timing B. The angle velocity at timing A is clearly different between unimpaired participants and those with walking disabilities. There is a big difference between them in the angle at timing B; however, this value would have sometimes overlapped each other.

IV. PROTOTYPE OF SHOE TO ASSIST PEOPLE WITH WALKING DISABILITIES

As described in Section III, people with a walking disability, such as those who suffer from hemiplegia, clearly have a weaker kicking power when raising their heel and swing power when swinging their toe forward. We have developed a shoe, shown in Figure 6, that assists with walking disabilities. This shoe has a coil spring and leaf spring to enable a user to easily raise their heel. The spring force of the coil spring is 15 kg. The shoe has a roller to avoid the toe accidentally tripping.

We compare the kicking power (angle velocity) when the heel is raised between a normal shoe and our proposed assist shoe worn by a stroke patient. The data is shown in Figure 7. The kicking power with the assist shoe is lower and more stable than that with a normal shoe.

We then measured a group of 8 students who were asked to walk as if they had a disability while wearing a normal shoe and the assist shoe. Measured data is shown in Figure 8. In every participant except one, their kicking power with the assist shoe was lower and more stable than that with the normal shoe. Authors also examined, and sensed that the shoe compensated to raise their foot slower with weaker power than the normal shoe and its compensation power was stable. Measured data in Figures 7 and 8 indicate the above senses.

We measured the integrated ElectroMyoGram (iEMG) readings for two walking disabilities to confirm the effect of the assist shoe. We used the wireless EMG logger from Logical Product Corporation [9]. The wireless EMG sensors were attached to the gastrocnemius of the right leg as shown in Figure 9. The sampling rate was 500 Hz. Measured data is shown in Figure 10. The results for the assist shoe are lower than those with a normal shoe for both people. The compensation effect of the proposed assist shoe is also confirmed with the iEMG.

Figure 6. Assist shoe prototype.

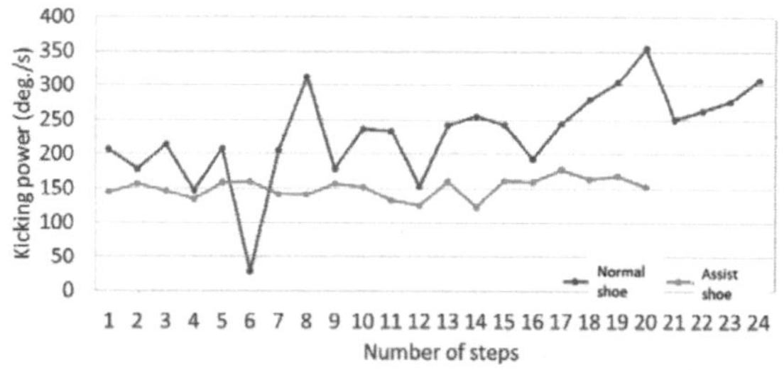

Figure 7. Kicking power when heel is raised with normal and proposed assist shoes for a stroke patient.

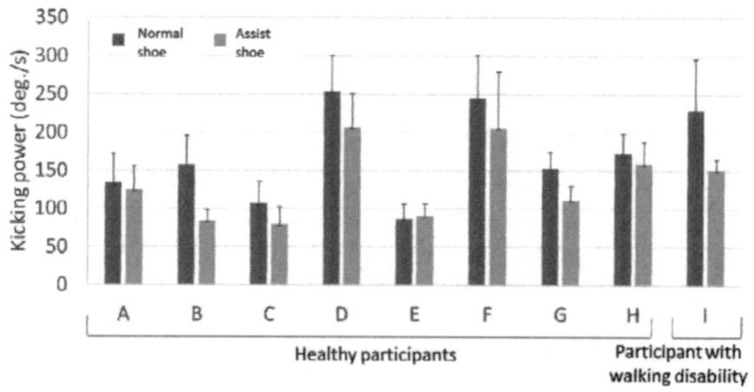

Figure 8. Kicking power when heel is raised with normal and proposed assist shoes.

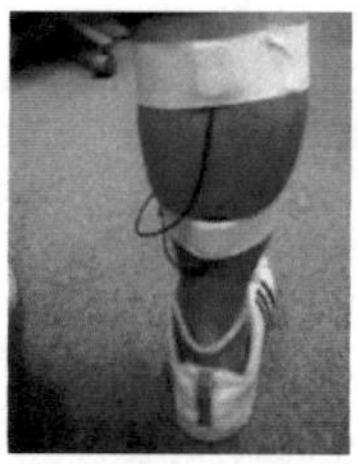

Figure 9. EMG sensors placement.

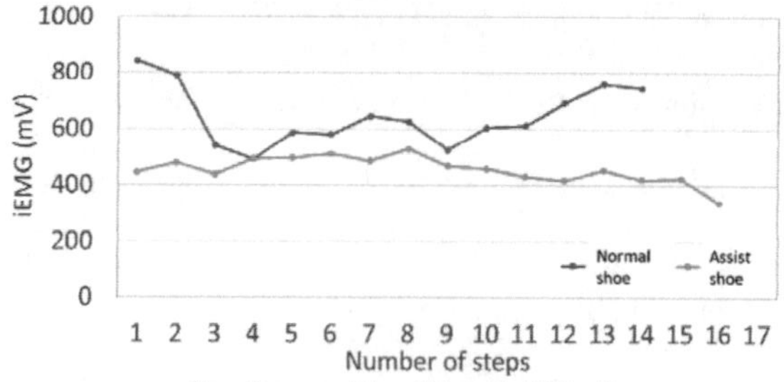

(1) Person with walking disability #1

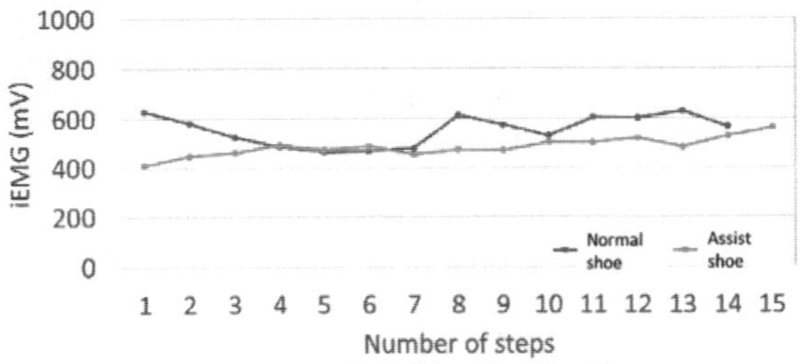

(2) Person with walking disability #2

Figure 10. iEMG data when heel was raised.

It is clear that the proposed shoe compensates for muscle weakness. However, most evaluators including authors felt that the timing to generate a spring reaction force is too early to walk smoothly; the timing at which the knee comes out in front of the ankle is best, and they had to change their gait motion to use a spring power effectively. And, we noticed that there would be a correlation between the body weight and most effective spring power, and would affect the walking posture.

The prototype shoe has a toe roller. However, as it is difficult to have walking disabilities intentionally trip over an obstacle, we could not quantitatively evaluate it.

V. SPRING ASSIST UNIT FOR WALKING DISABLED PEOPLE

A. Structure

As described in Section IV, every participant felt that the timing for generating spring reaction force was too early for walking smoothly. We thus focused on clarifying the correlation between body weight and optimal spring stiffness; the effect on walking posture; and developed the spring assist unit shown in Figure 11. Its mechanism is very simple as it comprises only a conical coil spring and a V-shaped attachment cover. We adopt the conical spring to be thinner when stepping on of which spring power is 3, 5, 9, and 11 Kg. The attachment cover is made of thin stainless steel.

Figure 11. Two views of spring assist unit (heel-up spring).

B. Assistance effect

The prototype assist shoe shown in Figure 6 had a coil spring and a leaf spring and was made for the right foot. In contrast, the spring assist unit shown in Figure 11 was built into the heel part of the right and left shoes, as shown in Figure 12. To measure the assistance effect, we had the participants wear shoes with each of the spring stiffnesses and walk straight for 6 m while we measured the iEMG. We also measured the motions of the head and mid-hip to analyze the effects on walking posture. Wireless EMG sensors were attached to the gastrocnemius of the right leg, as shown in Figure 9. The participant's posture was measured using a MS-Kinect [10]. The participants were ten healthy students. In the near future, we plan to measure the same data for persons with a walking disability.

Examples of the measured iEMG vs. spring stiffness for two participants (A and B weighing 57 and 70 kg) are shown in Figure 13. The iEMG values are lower for every spring stiffness than without the spring assist unit. The value was the lowest at the specified spring stiffness. The value for Participant A was lowest at 5 kg, and that for B was lowest at 9 kg. The spring stiffness at the lowest iEMG vs. participant body weight is shown in Figure 14. The spring stiffness magnitude at the lowest iEMG is linearly bigger, a participant gets more weight.

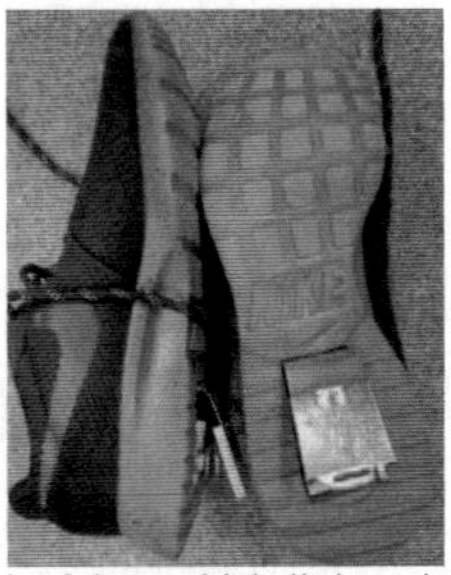

Figure 12. Pair of shoes with built-in spring assist units.

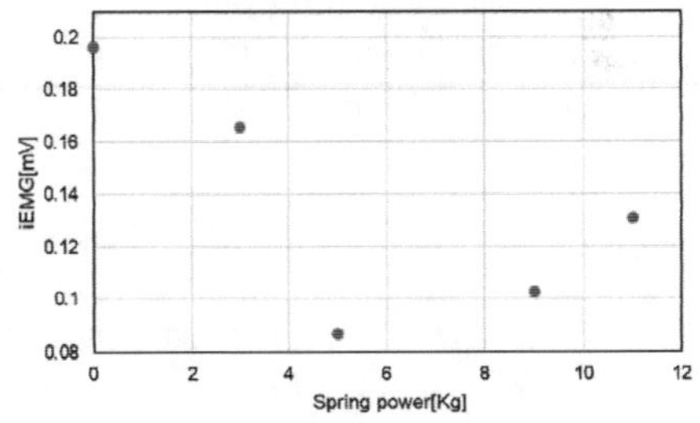

(a) Participant A (weight 57 kg)

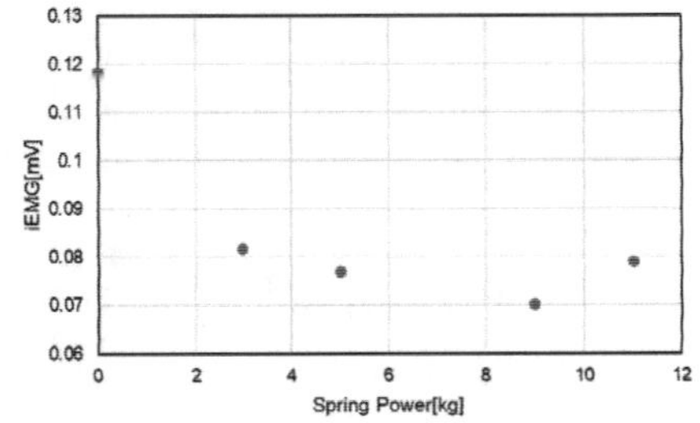

(b) Participant B (weight 70 kg)

Figure 13. Examples of measured iEMG vs. spring stiffness.

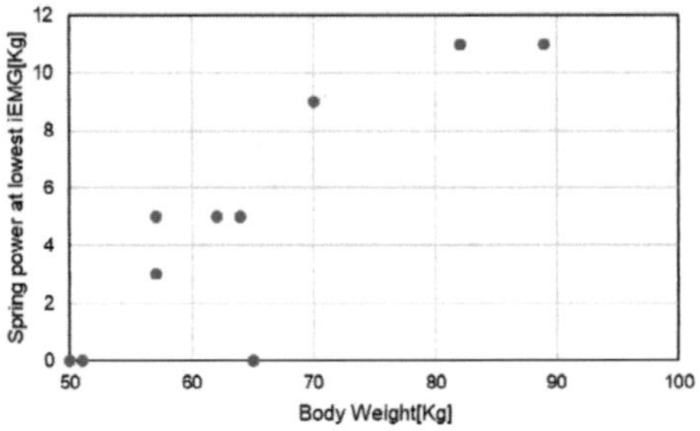

Figure 14. Lowest iEMG vs. participant body weight.

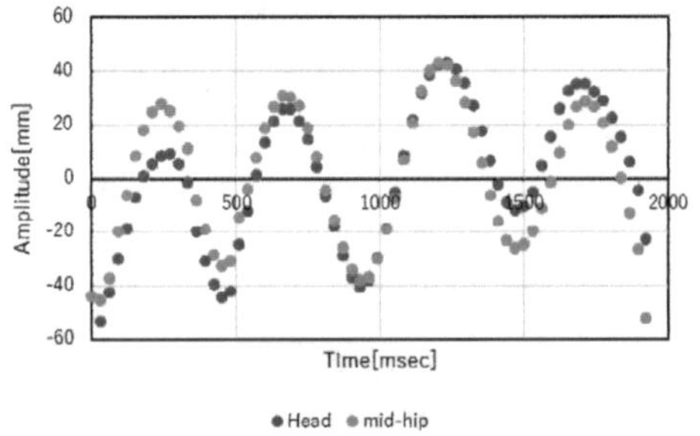

(a) Up and down direction

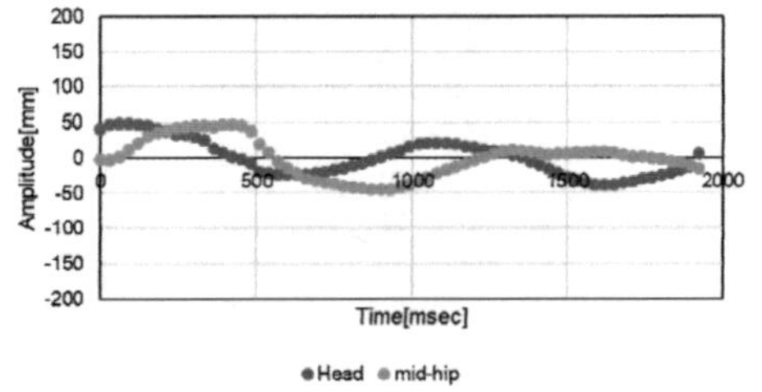

(b) Right and left direction

Figure 15. Motion of Participant C without a spring assist unit.

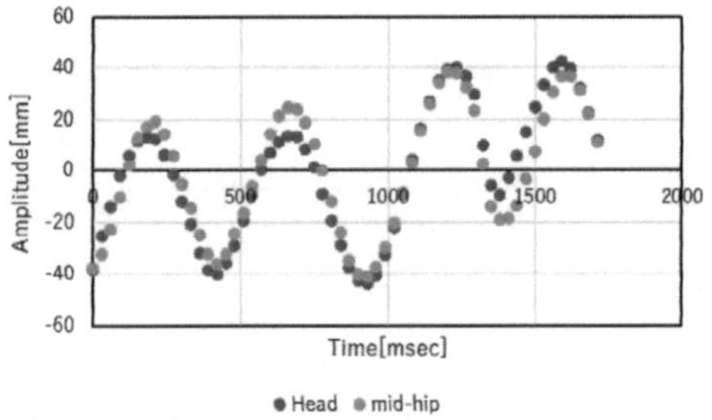

(a) Up and down direction

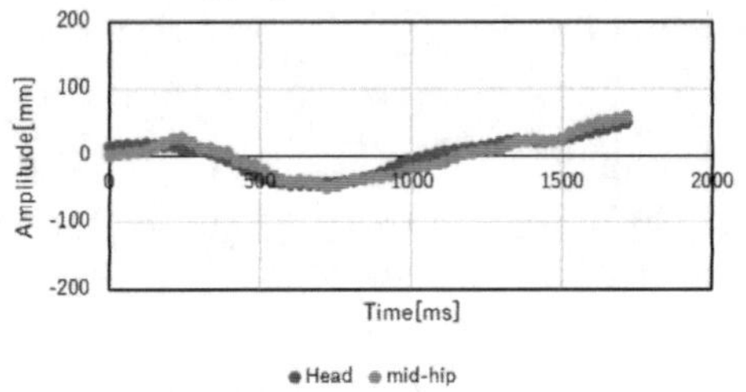

(b) Right and left direction

Figure 16. Motion of Participant C with a spring assist unit.

The measured positions of the head and mid-hip for Participant C are shown in Figure 15 for walking without a spring assist unit and in Figure 16 for walking with the most effective spring assist unit. The changes in the up and down motion are shown in graph (a), and the changes in the right and left motion are shown in graph (b). Without a spring assist unit, the up and down motion of the head and mid-hip clearly changes like a sine wave, as shown in (a) of Figure 15. The right and left motion of the head and mid-hip also changes, but not clearly like a wave as shown in (b) of Figure 15; and the cycle period does not differ from that of the up and down motion. There are not big differences between without a spring assist unit and with a spring assist unit

The average range of each step's peak in the right and left motion (LR) and up and down motion (UD) for each participant is shown in Table III. Although there are differences between participants and spring stiffnesses, the differences are random, with no obvious patterns.

TABLE III. AVERAGE RANGE OF PEAK TO PEAK IN LR AND UD MEASURED OVER TWO STEPS [MM]

Spring power			0[Kg]	3[Kg]	5[Kg]	9[Kg]	11[Kg]
Paticipant A	Head	LR	47.28	34.10	28.54	21.49	30.29
		UD	36.04	33.48	41.07	44.84	42.06
	Mid-hip	LR	32.20	17.11	22.65	17.94	17.44
		UD	29.66	25.80	28.71	22.32	33.61
Paticipant B	Head	LR	47.51	72.87	76.96	71.06	81.35
		UD	13.46	14.68	16.01	18.51	19.18
	Mid-hip	LR	18.83	24.34	22.50	26.19	29.52
		UD	25.87	26.81	28.95	22.72	25.00
Paticipant C	Head	LR	65.83	57.55	56.28	69.76	69.74
		UD	67.89	58.62	62.61	66.90	66.97
	Mid-hip	LR	33.35	33.74	29.23	35.00	32.37
		UD	60.87	67.42	68.07	65.31	54.42
Paticipant D	Head	LR	38.53	41.79	48.00	30.45	51.25
		UD	34.61	37.23	34.01	27.94	30.43
	Mid-hip	LR	31.46	35.64	44.38	23.28	40.88
		UD	38.68	48.60	45.03	45.38	26.01
Paticipant E	Head	LR	34.44	53.31	77.44	63.40	67.88
		UD	30.45	23.43	36.62	21.16	34.64
	Mid-hip	LR	39.86	68.98	71.58	79.92	83.73
		UD	38.71	38.20	34.67	32.90	37.70

A participant who tested the assist shoe shown in Figure 6 and the pair of assist shoes shown in Figure 12 commented that "I had to step on the shoe to walk smoothly in the shoe shown in Figure 6, whereas I did not feel any effect of the spring units when walking with the shoes shown in Figure 12. I could walk smoothly without any additional actions."

We conclude that the spring assist unit does not affect walking posture.

VI. Conclusion

It is difficult for people with walking disabilities to raise their heels because their muscle power is lower. Therefore, most of them shuffle their feet when walking and sometimes stumble over something and fall down. The spring assist unit we developed for walking disabilities enables them to easily raise their heels and walk smoothly. The iEMG was measured to analyze the assistance effect. The iEMG values for every spring stiffness were lower than those without the spring assist unit. The iEMG value was the lowest at the specified spring stiffness; the magnitude of the spring stiffness at the lowest iEMG was linearly bigger and the body weight was greater. These results demonstrate the assistance effect of the spring assist unit and that there is a linear correlation between body weight and the optimal spring stiffness.

We also measured the position of the head and mid-hip with and without the spring assist unit for spring stiffnesses of 3, 5, 9, and 11 kg. There were differences among the participants and among the spring powers, including no spring. However, the differences were random, without any obvious patterns between them. The results also demonstrate that the spring assist unit does not affect walking posture.

We would like to launch the commercial version in a near future.

Acknowledgement

The authors extend thanks to Mr. Kazuhiro Yoshida and to the students in Murata and Suzuki Lab for the help they provided in performing this research. This work was supported by JSPS KAKENHI Grant Number 19K11326.

References

[1] World Population Ageing: 1950-2050, Population Division, Department of Economic and Social Affairs, United Nations, [Online] http://www.un.org/esa/population/publications/worldageing 19502050/ [retrieved: February, 2020].

[2] W. P. Berg, H. M. Alessio, E. M. Mills, and C. Tong, "Circumstances and consequences of falls in independent community-dwelling older adults", Age Ageing, Vol. 26, pp. 261-268, 1997.

[3] Y. Murata, S. Yoshida, T. Niinuma, and K. Yoshida, "Comparative Analysis of Walking Gait Cycle between Healthy People and Walking Disabilities to Prevent Tripping Using Wearable Device and KINECT," IARIA, International Journal on Advances in Life Sciences, vol 9 no 3 & 4, pp. 186-197, 2017.

[4] Ottobock, 1D10 Dynamic foot, [Online] https://professionals.ottobock.com.au/Products/Prosthetics/Prosthetics-Lower-Limb/Feet/1D10-Dynamic-foot/p/1D10, [retrieved: February, 2020].

[5] Össur, Vari-Flex® , [Online] https://www.ossur.ca/prosthetic-solutions/products/dynamic-solutions/vari-flex [retrieved: February, 2020].

[6] A. Staros, "The SACH (Solid-Ankle Cushion-Heel)," Orthopedic & Prosthetic Appliance Journal, pp. 23-31, 1957.

[7] D. Wezenberg, A. G. Cutti, and H. Houdijk, "Differentiation between solid-ankle cushioned heel and energy storage and return prosthetic foot based on step-to-step transition cost," The Journal of Rehabilitation Research and Development, Volume 51, Number 10, pp. 1579-1590, 2014.

[8] H. Houdijk, D. Wezenberg, L. Hak, and A. G. Cutti, "Energy storing and return prosthetic feet improve step length symmetry while preserving margins of stability in persons with transtibial amputation," Journal of NeuroEngineering and Rehabilitation 2018, 15(Suppl 1):76, pp. 41-48, 2018.

[9] Wireless EMG logger, Logical Product Corporation, [Online] http://www.lp-d.co.jp/EMGSensor.html, [in Japanese, retrieved: February, 2020].

[10] Meet Kinect for Windows, [Online] https://developer.microsoft.com/en-us/windows/kinect, [retrieved: February, 2020].

Short-term Changes in Activities of Daily Living and Physical Activity Level of Inpatients Undergoing Rehabilitation Treatment

Yukihide Nishimura, Hiroyuki Tsuboi
Department of Rehabilitation Medicine,
Iwate Medical University
Shiwa, Japan
e-mail: ynishi@iwate-med.ac.jp, hiroyuki.tsuboi@j.iwate-med.ac.jp

Yoshitoshi Murata
Faculty of Software and Information Science,
Iwate Prefectural University,
Takizawa, Japan
e-mail: y-murata@iwate-pu.ac.jp

Abstract—**Though increasing physical activity level is essential for health promotion, inpatients are known to experience a drop in physical activity. However, even short-term rehabilitation treatment is expected to improve Activities of Daily Living (ADL), but accompanying changes in physical activity levels are unknown. This study evaluated change over one week in the Barthel Index (BI) and physical activity level in inpatients undergoing rehabilitation treatment and investigated the correlations between variability in these factors. The BI and physical activity levels were measured twice with a one-week interval between measurements for eight inpatients undergoing rehabilitation treatment at Iwate Medical University Hospital, Japan. The number of steps and time spent walking increased significantly from the first measurement to the second. On the other hand, there was no significant difference in BI score between initial measurement and after one week. However, 4 out of 8 inpatients showed the BI improved, and there was no inpatient showed worsened BI. Also, there were no significant correlations between the BI score variability and variability in the number of steps or time spent walking. The results suggested that rehabilitation treatment improved ADL ability and increased physical activity level among inpatients, but that variability in these factors is not correlated. A future study with an increased sample size divided by medical conditions is necessary.**

Keywords-rehabilitation treatments; Barthel Index; activities of daily living; physical activity level.

I. INTRODUCTION

The Global Recommendations on Physical Activity for Health published by the World Health Organization in 2010 identified physical inactivity as the fourth leading risk factor for global mortality, and lack of activity is known to be involved in the recurrence or exacerbation of various illnesses [1]. Thus, increasing physical activity is essential for maintaining and promoting health (maintaining physical and cognitive functioning, preventing the development of new diseases, etc.) and prolonging lifespan [2][3].

The Japanese Association of Rehabilitation Medicine proposes that "rehabilitation treatment works to return function, overcome disability, and cultivate activity" [4]. In other words, providing rehabilitation treatment (physical therapy, occupational therapy, etc.) to patients with physical disabilities can improve physical functioning, Activities of Daily Living (ADL), and increase the quality of life. However, many inpatients were found to have low physical activity [5]. Therefore, even if the ADL temporarily improve during hospitalization for inpatients undergoing rehabilitation treatment and their physical activity does not simultaneously improve, patients may return to physical inactivity after discharge leading to decreased motor and cognitive functioning, falls, and the development or exacerbation of other illnesses.

Generally, physical activity is evaluated using the International Physical Activity Questionnaire or a similar survey or an activity monitor with an internal triaxial accelerometer [6][7]. Activity monitors are a particularly simple way to measure steps, activity calories, and activity type and are, therefore, used by many researchers and clinicians. The Barthel Index (BI) is one of the most common methods used in clinical settings to evaluate ADL.

In Japan, it is both medically and financially optimal for acute phase hospitals to discharge or transfer patients as quickly as possible. In fact, the average hospital stay at our facility is less than two weeks; however, inpatient care may be continued for patients for whom intensive rehabilitation treatment is found to be effective. In this system, evaluating the short-term ADL and physical activity level is vital to determine the efficacy of rehabilitation treatment. Therefore, this study evaluated the changes over one week in the BI and physical activity level of inpatients undergoing rehabilitation treatment and investigated the correlations between these factors.

In Section II, the method in this study is explained. The results are outlined and discussed in Section III. Conclusion and future work are described in Section IV.

II. METHODS

A. Participants

Participants were eight inpatients undergoing intensive rehabilitation treatment at Iwate Medical University Hospital, Japan. Table I presents the participants' physical characteristics. The main causes of hospitalization were cerebral hemorrhage (n=2), surgery for osteoarthritis of the hip (n=3), surgery for osteoarthritis of the knee (n=2), or surgery for cervical spondylotic myelopathy (n=1). All participants underwent at least two hours of rehabilitation treatment daily, including at least one hour of both physical therapy and occupational therapy. The main exercise therapies conducted in physical and occupational therapy were muscle strengthening exercises, aerobic exercise, joint range of motion exercises, and ADL training. The Ethical Review Board of the Iwate Medical University approved this study. Participants signed a consent form after receiving an oral and written explanation of the research.

B. Measurement

Physical activity levels were measured for all participants using small activity monitors with internal triaxial accelerometers (KSN-200, KISSEI COMTEC) two times with an interval of one week between the measurements. Physical activity was measured continuously for twelve hours from 8:00 a.m. to 8:00 p.m. Activity monitors were placed on the anterior surface of the sternum, right thigh, left thigh, right shin, and left shin (Figure 1). Activity monitors were fixed in place with a transparent adhesive film to prevent chafing. The attending physical therapist evaluated the current BI on measurement days of physical activity. The number of steps, activity calories, and activity type as calculated from the activity monitor were used as indicators of physical activity level. Concerning activity type, the eleven activities configured in the activity monitor were classified into walking, standing, sitting, or lying down, and the duration was calculated for each.

C. Statistical analysis

Data were presented as mean ± standard deviation. The differences in BI score, steps, activity calories and duration for each activity between the initial measurement and after one week was evaluated by the Wilcoxon signed-rank test. Also, correlations between the variability (the difference between initial measurement and measurement after one week) in physical activity and BI were analyzed by using the Spearman's rank correlation coefficient. A p-value of less than 0.05 was considered significant. Statistical analysis was performed using the Statistical Package for Social Sciences software, version 23.0, for Windows (SPSS Inc., Chicago, IL).

TABLE I. PARTICIPANTS' PHYSICAL CHARACTERISTICS

Characteristics	Sample
Sex (male/female)	3/5
Age (years)	66.3±8.1
Height (cm)	158.7±9.8
Weight (kg)	62.1±12.4
Body mass index (kg/m2)	24.8±5.5

TABLE II. DURATION FOR EACH ACTIVITY IN EACH INPATIENT

	Activity Time (min)							
	Walking		Standing		Sitting		Lying down	
	1st	2nd	1st	2nd	1st	2nd	1st	2nd
no. 1	4	50	5	38	546	541	160	78
no. 2	30	55	20	16	286	368	184	52
no. 3	27	39	20	31	366	328	260	291
no. 4	40	57	65	109	450	451	49	11
no. 5	0	50	153	100	341	377	209	189
no. 6	92	80	56	68	374	337	138	196
no. 7	16	62	15	31	313	304	24	131
no. 8	58	56	38	45	300	362	271	231
mean ±SD	33± 30.2	56± 12.1*	46± 47.8	54± 34.1	372± 87.3	383± 77.0	162± 90.2	147± 96.1

1st= initial measurement; 2nd = after one week; *p < .05 vs 1st.

III. Results and Discussion

A. Comparison of the BI, steps, and activity calories

The steps increased significantly after one week compared to the initial measurement (P = .012). However, there was no significant difference after one week for BI score and activity calories (BI score: P = .063; activity calories: P = .208). Figure 2 shows a comparison of the BI, steps, and activity calories at the initial measurement and after one week.

B. Comparison of activity durations

Table II summarizes durations for each activity. Walking time increased significantly after one week compared to the initial measurement (P = .036). However, time spent standing, sitting, and lying down was not significantly different after one week (standing: P = .208; sitting: P = .779; lying down: P = .575).

C. Correlation between BI variability and variability in steps and walking time

There was no significant correlation between BI variability and variability in steps and walking time (Figure 3).

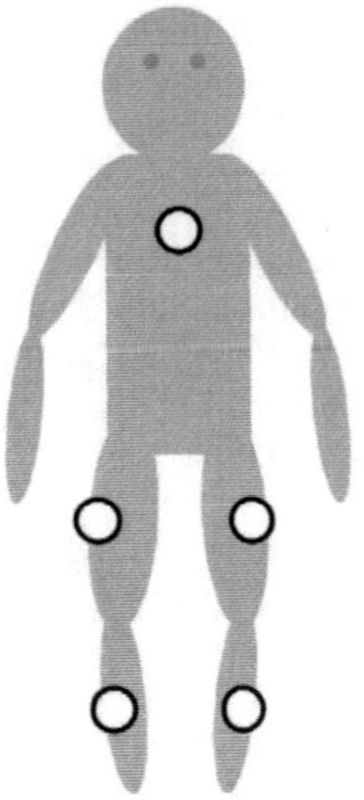

Figure 1. Sensor placement.

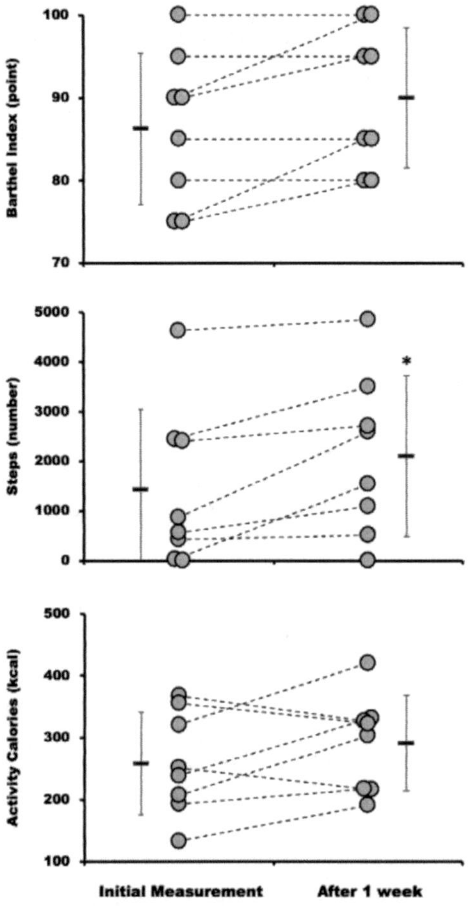

Figure 2. Comparison of values at initial measurement and after one week for Barthel Index, steps, and activity calories. Values are mean ± standard deviation; *p < .05.

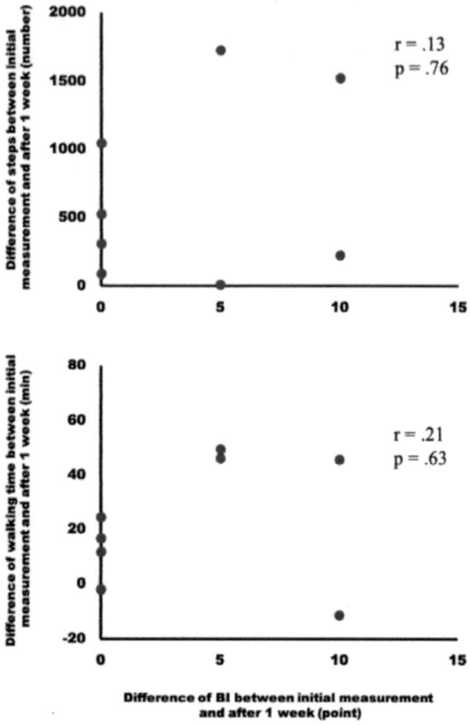

Figure 3. Correlation of increase in steps and walking time with Barthel Index improvement. Values are mean ± standard deviation.

This study found that both the number of steps and walking time increased over one week in inpatients undergoing intensive rehabilitation treatment. On the other hand, there was no significant difference in BI score between initial measurement and after one week. The best possible score of BI is 100 points, and all inpatients had got relatively high BI score at initial measurement. However, 4 out of 8 inpatients showed the BI improved, and no inpatient who showed a worsened BI. Also, the p value of BI between initial measurement and after one week was .063 in spite of small sample size. Therefore, we assume that to increase sample size will significantly improves BI.

Health Japan 21, a strategy for health promotion from the Ministry of Health, Labor and Welfare, proposed an increase in average steps per day. The National Health and Nutrition Survey conducted with community residents in 2010 found that men aged 65 and older took an average of 5628 steps/day while women aged 65 or older took 4585 steps/day. Thus, an increase of approximately 1500 steps/day is the goal for both men and women aged 65 or older [8]. Past research has found that inpatients are inactive [5], and although ADL ability improved and both steps and walking time increased over one week for inpatients in this study, BI variability and variability in steps and walking time were not correlated. This result suggests that improved ADL does not necessarily mean that the physical activity level will increase. In other words, rehabilitation treatment for inpatients should not focus solely on improving physical functioning and ADL, but must increase physical activity level for patients to lead a healthy and fulfilling life in the long-term after discharge. Specifically, besides medically controlling symptoms such as pain that inhibits physical activity, strengthening muscles and cardiopulmonary function, patient education, and environmental adjustments are essential approaches. Further, there are patients for whom ADL improves during hospitalization, but who become inactive after discharge, which leads to a decrease in ADL. Recognizing patients at risk for this and addressing it preemptively, is necessary. However, to the best of our knowledge, there are no studies that report such risk factors.

In our previous work, we presented a data collection system in which movements are analyzed using Google Firebase service and a wearable device equipped with a gyrosensor [9][10]. Popularizing this system will lead to big data, which could potentially establish evidence of many issues from physical activity levels during hospitalization and after discharge. Further, existing activity monitors require professional staff and time to put on and can interfere with inpatient tests and bathing. As such, we believe there is a demand for a device capable of easily measuring physical activity levels.

This study has certain limitations. First, the number of subjects was relatively small. Second, physical activity was measured only once at each measurement, however, the reliability of data is unclear. Therefore, to further improve this study, it is important to measure physical activity several times a week for many patients who are divided according to their medical conditions.

IV. Conclusion and Future Work

In this paper, we evaluated change over one week in BI and physical activity levels in inpatients undergoing rehabilitation treatment and investigated the correlations between variability in these factors. One week of inpatient intensive rehabilitation treatment increased both steps and walking time. However, BI variability and variability in steps and walking time were not correlated. A future study with an increased sample size divided by medical conditions is necessary.

Acknowledgments

We thank Takahito Kondo, Katsunori Takahashi and Fumiaki Yamaguchi for the technical assistance. This work was supported by JSPS KAKENHI Grant Number JP19K11326.

References

[1] World Health Organization. Global Recommendations on Physical Activity for Health. Available from: https://www.who.int/dietphysicalactivity/global-PA-recs-2010.pdf. Accessed January 12, 2020.

[2] US Department of Health and Human Services, Physical Activity Guidelines for Americans. 2nd ed. Available from: https://health.gov/paguidelines/second-edition/pdf/Physical_Activity_Guidelines_2nd_edition.pdf. Accessed January 12, 2020.

[3] K. L. Piercy et al., "The Physical Activity Guidelines for Americans", JAMA, Vol. 320, no. 19, pp. 2020-2028, 2018.

[4] The Japanese Association of Rehabilitation Medicine, Rehabilitation Medicine Core Text, First Edition First Printing, Igaku-Shoin, pp. 3-20, 2018.

[5] C. Baldwin, G. van Kessel, A. Phillips, and K. Johnston, "Accelerometry Shows InpatientsWith Acute Medical or Surgical Conditions Spend Little Time Upright and Are Highly Sedentary: Systematic Review", Phys Ther, Vol. 97, no. 11, pp. 1044-1065, 2017.

[6] C. L. Craig et al., "International physical activity questionnaire: 12-country reliability and validity", Med Sci Sports Exerc, Vol. 35, no. 8, pp. 1381-1395, 2003.

[7] C. E. Matthews, B. E. Ainsworth, R. W. Thompson, and D. R. Jr. Bassett, "Sources of variance in daily physical activity levels as measured by an accelerometer", Med Sci Sports Exerc, Vol. 34, no. 8, pp. 1376-1381, 2002.

[8] M. Miyachi, "Measures of physical activity and exercise for health promotion by the Ministry of Health, Labour and Welfare", J Phys Fitness Sports Med, Vol. 1, no. 3, pp. 467-472,2012.

[9] R. Takahashi, Y. Murata, and O. D. A. Prima, "A System for Collecting Motion Data on Patients' Activities of Daily Living", The Eleventh International Conference on eHealth, Telemedicine, and Social Medicine (eTELEMED 2019) IARIA, Feb. 2019, pp. 7-12, ISSN: 2308-4359, ISBN: 978-1-61208-688-0

[10] Y. Nishimura, H. Tsuboi, Y. Murata, and O. D. A. Prima, "Measurement of Shoulder and Trunk Movements in Hemiplegic Participants Usinga System for Collecting Motion Data", The Eleventh International Conference on eHealth, Telemedicine, and Social Medicine (eTELEMED 2019) IARIA, Feb. 2019, pp. 19-22, ISSN: 2308-4359, ISBN: 978-1-61208-688-0

Multi-action Detection System Using Infrared Omnidirectional Cameras

Takashi Imabuchi
Iwate Monodukuri and Software Integration Technology Center (i-MOS), Office of Research and Regional Cooperation, Iwate Prefectural University
Takizawa, Japan
e-mail: t_ima@ipu-office.iwate-pu.ac.jp

Yoshitoshi Murata, Oky Dicky Ardiansyah Prima
Faculty of Software and Information Science
Iwate Prefectural University
Takizawa, Japan
e-mail: {y-murata, prima}@iwate-pu.ac.jp

Abstract—**Japan is facing a shortage of healthcare workers due to the declining birthrate and aging population. This issue is placing a heavy burden on them to address the growing demand for 24-hour medical services. Information and Communication Technology (ICT) has opened up the possibility of collecting valuable data and providing insights for more control over patients' lives. Early approaches to detect accidental falling and wandering behavior using ICT include the use of invasive and non-invasive sensors. However, in order to put these approaches into practical use, further measures are needed. In this study, we propose a camera monitoring system to automatically detect typical patients' behaviors, such as, rising from the bed, leaving the bed, falling down, and wandering. Our system utilizes infrared omnidirectional cameras that allows a wide range of monitoring actions during day and night. Skeletal information of multiple patients is captured using a computer vision-based pose detection to classify each behavior. Evaluation experiments demonstrated the feasibility of detecting typical patients' behavior using the proposed system.**

Keywords-hospital; patient monitoring; healthcare facility; ommnidirectional camera; human pose estimation.

I. INTRODUCTION

The increasing percentage of elderly people in many national populations [1] is resulting in an increasing number of functionally impaired hospitalized patients, such as, cerebrovascular patients who are paralyzed on one side. Such patients have an increased risk of falling and consequently injuring themselves [2]. Falling is one of the main reasons for them to be hospitalized or placed in residential care. There is also an increasing number of dementia patients, who have a tendency to wander. Inpatient falling and wandering are serious problems for the management of healthcare facilities. Nursing patrols and nighttime monitoring as countermeasures can interfere with patient sleep. The increasing number of patients and the growing shortage of healthcare workers could lead to the stopping of such services. This would result in a lower quality of patient life.

Several measures have been introduced in healthcare settings to detect patient falling and wandering, but their effectiveness is limited. For example, a pressure-sensitive mat sensor placed on or next to a bed can detect the patient leaving the bed but cannot detect wandering or falling down. Moreover, a patient can easily remove or move the mat to prevent sensing, and frequent false detections can results from visitors stepping on the mat. Martinez et al. developed a monitoring system for patients on the bed in a healthcare facility using an infrared camera [3]. It can only detect a patient leaving the bed; it cannot detect wandering or falling. Furthermore, the monitoring area is limited to a bed and its immediate area.

At the research level, Murata et al. developed a multi-action monitoring system for healthcare facilities that uses MS-KINECT sensors [4]. However, they cannot detect a person lying on a bed because they cannot detect differences in the depth between a patient and a bed. Moreover, their coverage is limited to a bed and the surrounding area.

We have developed a monitoring system that uses an infrared omnidirectional camera to automatically detect typical actions by multiple patients in healthcare facilities. Using an infrared omnidirectional camera enables it to monitor patients throughout day and night. It continuously detects and tracks patients on the basis of their skeletal information, estimates the kinds of actions, and notifies the staff if it detects a predefined abnormal action, such as falling.

This paper is organized as follows. Section II describes related work on monitoring systems for the healthcare sector. Section III introduces our proposed system. Sections IV and V describe the experimental setting and results. Section IV discusses the accuracy of estimating a patient's location from an omnidirectional camera image, and Section V describes the evaluation of activity classification based on a patient's skeletal information. Finally, Section VI summarizes the key points and mentions future work.

II. RELATED WORK

Several types of monitoring devices using various kinds of sensors have been introduced in the healthcare sector to detect such patient actions as falling and wandering, as illustrated in Figure 1 [5].

Pressure-sensitive bedside sensor mats (Figure 1 (a)) are commonly used for detecting a patient rising from and/or leaving the bed [6]. Changes in sensor voltage are used to detect rising from the bed, leaving the bed, and standing beside the bed. However, pressure-sensitive mats typically have low durability, can produce false detections due to visitors stepping on them, and can miss detections due to unintentional or intentional movement of the mat.

Clip-type sensors (Figure 1 (b)) have long been used to detect patient leaving the bed. One end of a cord is clipped to the patient's clothing, and the other end is attached to the bed frame with a magnet. If the patient leaves the bed, the cord detaches from the bed, and a notification is sent to the nurse's station. However, they can cause the patient to feel like they are being monitored, can produce false detections due to patient movement in the bed, and can only detect the patient leaving the bed, not falling.

Heat and infrared sensors (Figure 1 (c)) can detect the patient's location and patient wandering, but they can also detect visitors.

Electromagnetic tags (Figure 1 (d)) are useful for tracking patients [7][8], but indoor positioning accuracy is generally poor due to radio interference. Moreover, patients sometimes refuse to wear a tag or deliberately remove them.

A promising alternative to these methods is non-invasive monitoring using optical sensors, such as, web cameras, because patients cannot disable them. Such sensors are well suited for healthcare facilities because they can capture images of multiple patients simultaneously. However, most such systems currently in use do not support 24-hour monitoring, only daytime monitoring. Moreover, their coverage is limited to a bed and the surrounding area.

Depth cameras are commonly used to estimate the human pose in three dimensions, and the depth camera in an MS-Kinect device has shown adequate performance in healthcare imaging applications [9][10]. They can measure not only the changes in body posture, but also pose (skeletal) information for the targeted person. It is possible to estimate a patient's action using this information.

Recent computer vision applications enable the detection of 2D human poses from a single image [11][12]. Furthermore, the 3D human pose can be estimated by using human pose libraries taken from motion capture devices as a reference [13]. Unlike an MS-Kinect sensor, which must be within a certain distance to the target, these applications are more flexible. Skeletal information can be derived for a body located more than 5 m from the camera. Their application in various fields is expected.

III. PATIENT MONITORING SYSTEM

We considered the following requirements to be essential for a patient monitoring system.

- Monitoring both day and night (i.e., 24-hour monitoring).
- Locating and identifying multiple patients.
- Monitoring entire multi-patient room using a minimum amount of easy-to-install equipment.
- Detecting multiple actions, including rising from the bed, leaving the bed, falling down, and wandering.
- Notifying hospital staff of abnormal patient behaviors.
- Protecting privacy.

On the basis of these requirements, we developed a novel patient monitoring system for healthcare facilities. The proposed system detects patient actions using a single infrared omnidirectional camera positioned to face the patient's bed, as illustrated in Figure 2.

An infrared camera is used, which enables 24-hour monitoring. In addition, employing an infrared camera leads to privacy protection because it does not capture clear images compared to a visible-light camera. It has an omnidirectional lens, which enables simultaneously detection of multiple patients in a large patient room. The system continuously and simultaneously detects and tracks multiple patients on the basis of their 2D skeletal information estimated from the camera image. Each patient's actions are classified and labeled in accordance with predefined rules. When an abnormal action is detected, a notification is sent to a hospital staff.

We intend to implement this system on a single-board computer in the near future and complete each image processing as the edge. This will enable easy equipment installation and privacy protection.

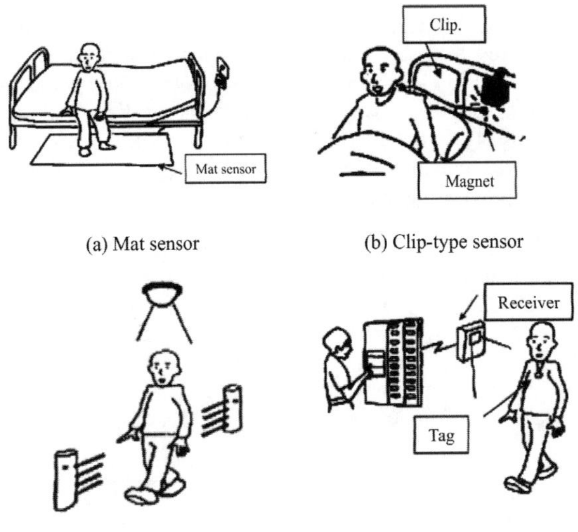

(a) Mat sensor (b) Clip-type sensor

(c) Heat and infrared sensors (d) Tag sensor

Figure 1. Various types of sensors used for detecting wandering.

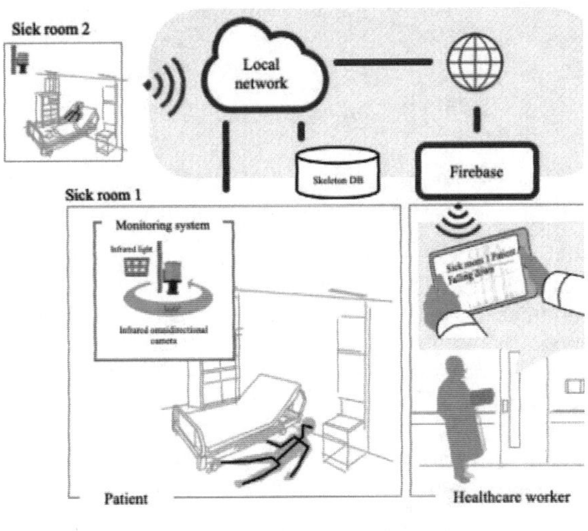

Figure 2. Overview of proposed patient monitoring system.

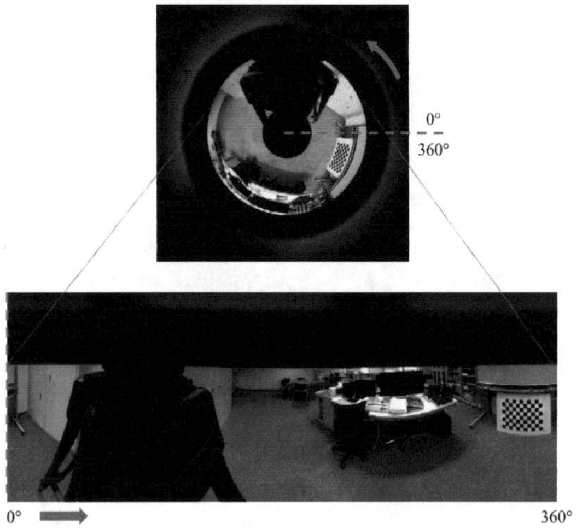

Figure 3. Example expansion of omnidirectional camera image.

In this paper, we describe the monitoring of patients using an omnidirectional camera. In particular, we describe in detail panorama expansion from images captured by the camera, the detection and tracking of patients using 2D skeletal images obtained using the OpenPose library, joint angle calculation, and action classification.

A. Panorama Expansion

The use of a camera with an omnidirectional lens composed of a hyperboloid mirror enables the capture of a 360° image from the projection of the hyperboloid mirror. With this type of lens, image resolution is high on the sides of and below the lens; the area immediately above the lens cannot be captured. Each captured image is expanded to a panorama image by perspective projection transformation. This requires equal division in the circumferential direction from a predefined center point (x_c, y_c) in the omnidirectional image. Four vertex pairs are calculated as parameters using

$$x = -\frac{a^2 fX}{(b^2+c^2)Z-2bc\sqrt{X^2+Y^2+Z^2}} + x_c\ , \qquad (1)$$

$$y = -\frac{a^2 fY}{(b^2+c^2)Z-2bc\sqrt{X^2+Y^2+Z^2}} + y_c\ , \qquad (2)$$

where X, Y, and Z are points in 3D coordinates in the omnidirectional image, x and y are points in the image coordinate system on the panorama image, a, b, and c are parameters for the hyperboloid mirror satisfying $c^2 = a^2 + b^2$, and f is the focal length of the camera. These coordinate pairs and perspective projection transformation are used to calculate the 2D coordinate points in the panorama image. Figure 3 shows an example omnidirectional camera image and the expanded panoramic image.

B. Detecting and Tracking Patient Location

The relative location and orientation of the patient's body from the camera are estimated using the 2D skeletal image for

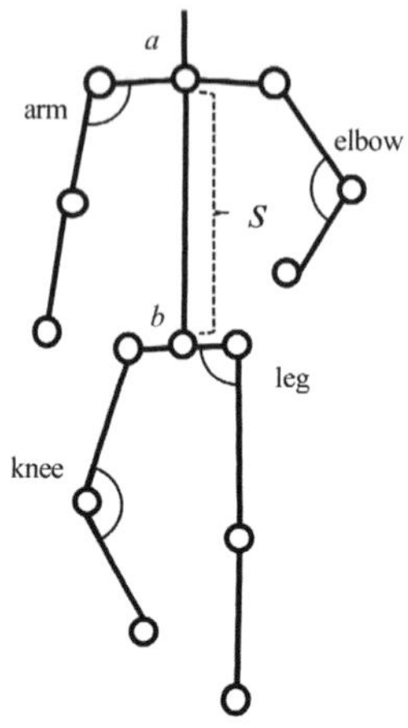

Figure 4. Definition of parameters from skeletal information.

the patient. The location of the patient is defined as a polar coordinate P (θ,R), which is calculated using skeletal coordinate *Joint* ($j_1...j_n$). Figure 4 shows the definitions of the skeletal parameters used in this study. Azimuth θ is centered at the camera and is determined by the ratio indicated by the horizontal coordinate of the Body Center Of Mass (BCOM) under the assumption that horizontal width W of the panoramic image is 360° in all directions. Thus,

$$\theta = 360 \cdot \frac{BCOM_x}{W}\ . \qquad (3)$$

The horizontal coordinate of BCOM is calculated using

$$BCOM_x = \frac{j_{1x}+j_{2x}+j_{3x}\cdots j_{nx}}{n}\ . \qquad (4)$$

Distance R is determined by the scale of body torso s calculated from the distance between joint a and joint b, as shown in Figure 4. Under the assumption that the posture of a person's body is always parallel to the vertical axis of the camera, R can be determined by multiplying s by the weight calculated on the basis of the ratio between and distortion measured using a calibration process.

C. Calculation of Joint Angles and Action Classification

Patient actions are classified using several parameters calculated from the skeletal information: joint coordinates of patient's head, body variation (standard deviation of BCOM), body axis tilt angle, and joint angles from shoulder, elbow, knee, and leg. First, the standard deviation of BCOM indicating the patient's movement is calculated using

$$Var_{bcom} = \sqrt{\frac{1}{m}\sum_{j=1}^{m}\left(BCOM_{j(x,y)} - \overline{BCOM}_{(x,y)}\right)^2}\ . \qquad (6)$$

Then, the tilt angle of the body axis is calculated from the body torso s vector described in Section III-B and the horizontal axis vector.

TABLE I. ACTION CLASSIFICATION RULES

	Inside the bed	Moving	Tilt angle [°]	Joint angles [°]	head position
Rising from the bed	Yes	No	<±30	-	high → low
Leaving the bed	Yes → No	Yes	±30<= → <±30	leg<45 → 150<=leg	-
Falling down	No	Yes → No	±45<=	-	-
Wandering	No	Yes	<±30	160<kne, 150<=elbow	-

The joint angle is calculated as the relative angle between the longitudinal axis of two adjacent segments. For the elbow joint angle, the adjacent segments are the upper arm and forearm. For the arm (shoulder) joint, the adjacent segments are the upper arm and shoulder. For the knee joint angle, the adjacent segments are the upper leg and lower leg. For the leg (hip) joint, the adjacent segments are the upper leg and hip. Let u and v be vectors representing two adjacent segments. Each joint angle between u and v is calculated using

$$\theta_{joint} = 180° - \frac{\overline{u} \cdot \overline{v}}{|\overline{u}| \cdot |\overline{v}|} . \qquad (5)$$

Each joint angle is calculated separately for the left and right sides.

The region of the bed is defined such that the bed is arranged with the long side perpendicular to the camera vertical axis.

Using the parameters described above, we set the rules for the four actions considered in this study (see Table I) and classify the actions on the basis of these rules. The results of action detection are labeled and stored as time series data.

IV. EXPERIMENT I

Experiment I was conducted to measure the accuracy of patient location estimation. Using an omnidirectional camera installed at a height of 2.7 m in a multi-patient room, we captured images spanning an arc of 90°. We measured the position at a total of 35 points in increments of 15° up to 90°, 1 to 5 m in increments of 1 m. The camera was an industrial camera (TXG-50, Baumer [14]) with a resolution of 2840 × 2040 and a speed of 30 fps. It was equipped with a PALNON panoramic lens (elevation 66°, depression 0°). Figure 5 shows an example of the panoramic image captured by an omnidirectional camera and a 2D skeleton coordinates of a participant in the experiment. In Figure 5, the left half of the image is hidden by a tripod.

12 participants participated in the experiment. Figure 6 shows the coordinates of reference points (Black cross marker) and the corresponding points of position estimation (Blue circle marker). In addition, red cross markers show averaged for each reference point. For each participant, we calculated the azimuth and distance using the proposed system. The errors were calculated as the Root Mean Square Error (RMSE),

$$RMSE_\theta = \frac{1}{nm} \sum_{i=1}^{n} \sum_{j=1}^{m} \sqrt{\left(\hat{\theta}_j - \theta_{i,j}\right)^2} , \qquad (6)$$

Figure 5. Example panorama image.

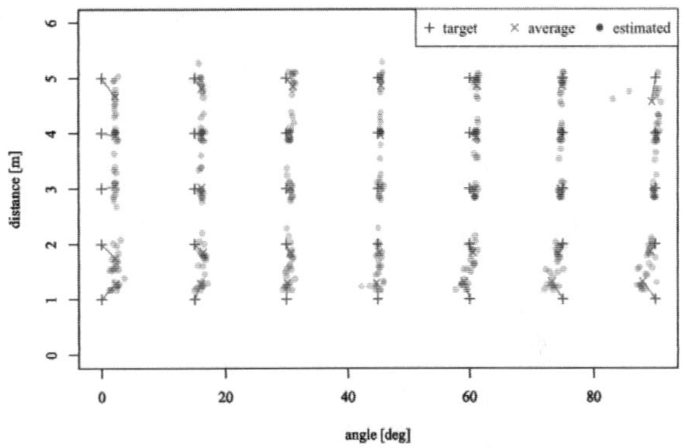

Figure 6. Results of position estimation.

where n is the number of participants and m is the number of reference points. The RMSE for R was calculated in the same way. The RMSE for the azimuth was 1.30°, and the average error for the distance was 0.27 m. These errors are sufficiently small for the proposed system to be used in practical situations because the accuracy is sufficient for estimating the locations of patients between beds in a multi-patient room given that the proposed system enable multiple patients to be simultaneously detected using a single omnidirectional camera. In addition, as shown in Figure 6, no significant error was observed for up to 5 m, indicating that a single omnidirectional camera can effectively cover a large multi-patient room.

V. EXPERIMENT II

Experiment II was conducted to evaluate the accuracy of the proposed action classification method. Patient actions were collected as video data by having the participants perform the four target actions (rising from the bed, leaving the bed, falling down, and wandering), three times each for each participant. For each video frame, we classified the action using the participant's skeletal information on the basis of the rules given in Table I. The results were calculated as a confusion matrix showing how well the actions were correctly estimated for all frames in the video. We calculated the accuracy and precision from the confusion matrix. The correct data was manually annotated while the video was being checked.

Table II shows the results of action classification as a confusion matrix for eight participants. Figure 7 shows an example of each action captured by the omnidirectional camera. The accuracy of estimating each action exceeded 80%,

TABLE II. RESULTS OF ACTION CLASSIFICATION AS CONFUSION MATRIX

[frame]

	Confusion matrix					
	True positive	False positive	False negative	True negative	Accuracy	Precision
Rising from the bed	601	3366	959	22511	84.24%	15.15%
Leaving from the bed	1087	539	828	24983	95.02%	66.85%
Falling down	2216	961	449	23811	94.86%	69.75%
Wandering	3727	2012	2231	19467	84.54%	64.94%

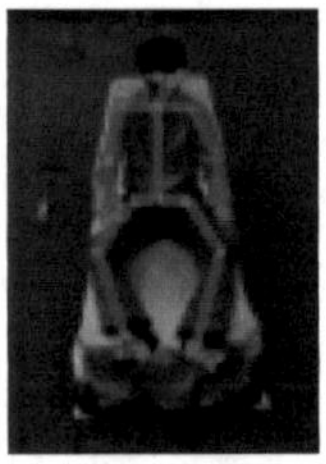

(a) Rising from the bed

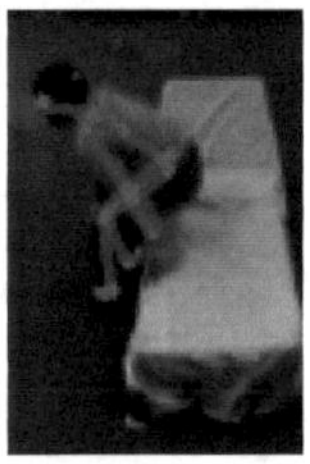

(b) Leaving the bed

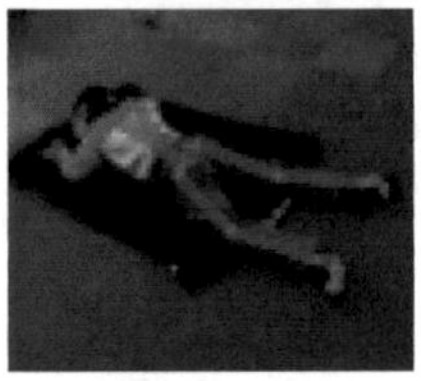

(c) Falling down

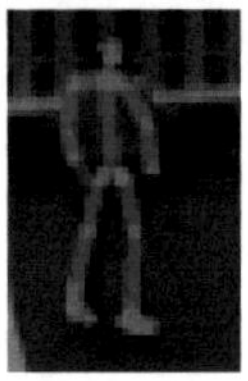

(d) Wandering

Figure 7. Example actions and skeletal information.

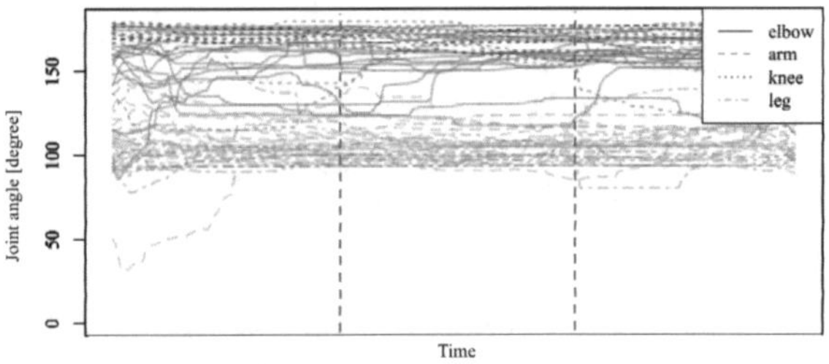

(a) Rising from bed

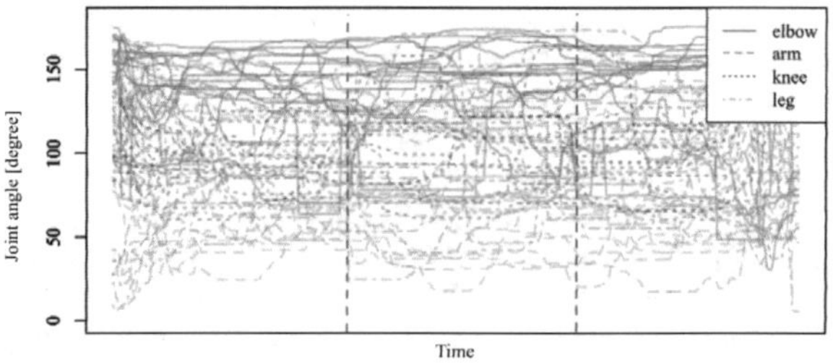

(b) Leaving bed

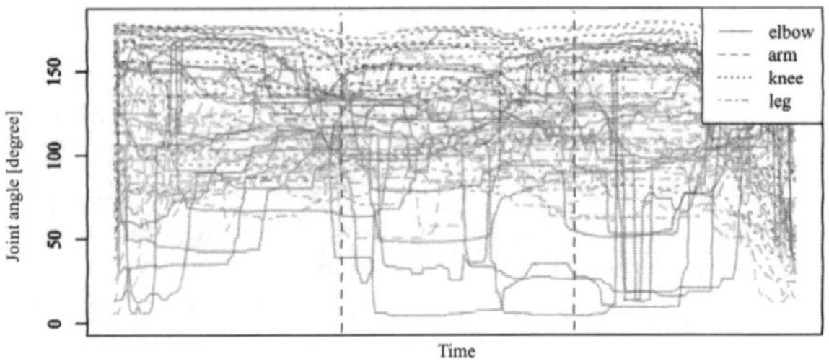

(c) Falling down

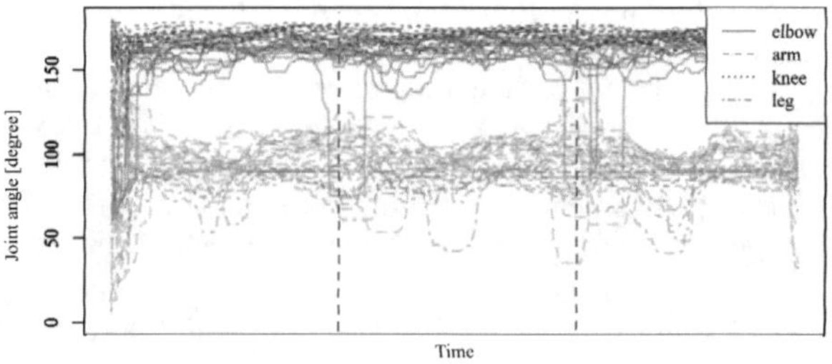

(d) Wandering

Figure 8. Overlaid change in joint angle with time scale aligned across participants.

and the precision was 60-70% except for rising from the bed. The number of false detections was high for each action, and the rising from the bed action could not be estimated for most frames. One reason for this is that the head movement in the vertical direction was not large. It is difficult to measure depth information for the joint coordinate occulted by other body parts; and difficult to estimate the body orientation. We found that our rule-based action classification using 2D skeletal images has limitations. We plan to develop a stereo type omnidirectional camera to get 3D skeleton images.

The change in the joint angle for each action was quite similar for all participants. Figure 8 shows graphs of the overlaid changes in joint angle (elbow, arm, knee, and leg; both sides) with the time scale aligned across participants for the four actions.

VI. CONCLUSION

Our proposed patient monitoring system using an infrared omnidirectional camera for healthcare facilities enables detection and classification of various actions that can be dangerous for patients, such as, falling and wandering. Experimental results demonstrated that this system can accurately estimate the locations of multiple patients, enabling each patient to be identified in a wide area. This system should be applicable not only to healthcare facilities but also to facilities that have wide areas such as, factories and schools for use in detecting dangerous situations.

Given the weakness of action classification, we plan to investigate the effect on classification accuracy of the use of machine learning to estimate the changes in joint angles. In addition, we plan to investigate the effect of using a stereo camera to obtain 3D images. For practical application, we will continue to work on improving the accuracy of action classification, using only joint information for personal authentication to ensure anonymity, and integrating the system with our developed alert notification systems.

REFERENCES

[1] World Population Ageing: 1950-2050, Population Division, Department of Economic and Social Affairs, United Nations, http://www.un.org/esa/population/publications/worldageing19502050/ [retrieved: Febrary, 2020].

[2] W. P. Berg, H. M. Alessio, E. M. Mills, and C. Tong, "Circumstances and consequences of falls in independent community-dwelling older adults," Age Ageing, Vol. 26, pp. 261–268, 1997.

[3] M. Martinez and R. Stiefelhagen, "Automated multi-Camera system for long term behavioral monitoring in intensive care units," 2013 IAPR International Conference on Machine Vision Applications, pp. 97–100, 2013.

[4] Y. Murata, R. Takahashi, T. Yamato, S. Yoshida, and M. Okamura, "Novel Field Oriented Patient Monitoring Platform for Healthcare Facilities," IARIA, International Journal on Advances in Software, Vol. 11 No. 3&4, pp. 368–378, 2018.

[5] I. Kitayama, K. Omori, H. Matsuno, and Y. Sugimoto, "Wandering prevision and monitoring systems for persons with dementia (Part 2) - Usage survey of wandering prevision and monitoring system/machine at nursing homes and other welfare facilities-," Report of the Hyogo Prefectural Town Welfare Research Institute, pp. 98–111, 2003.

[6] Mat sensor, Carecom, https://www.carecom.jp/global/solutions/option/ [retrieved: Febrary, 2020].

[7] T. Gao, D. Greenspan, M. Welsh, R. Juang, and A. Alm, "Vital signs monitoring and patient tracking over a wireless network," Johns Hopkins APL Techincal Digest, Vol. 27, No. 1, pp. 66–74, 2006.

[8] P.Varady, Z. Benyo, and B Benyo, "An open archtectre patient monitoring system using standard technologies," IEEE Tracnsaction on Information Technology in Biomedicine, Vol. 6, No. 1, pp. 95–98, 2002.

[9] E. Gambi, et al., "Heart rate detection using Microsoft Kinect: Validation and comparison to wearable devices," Sensors 2017, Vol. 17, No. 8, pp. 1776, 2017.

[10] S. T. L. Pohlmann, E. F. Harkness, C. J. Taylor, and S. M. Astley, "Evaluation of Kinect 3D sensor for healthcare imaging," Journal of Medical and Biological Engineering, Vol. 36, pp. 857–870, 2016.

[11] A. Newell, K. Yang, and J. Deng, "Stacked Hourglass Networks for Human Pose Estimation," European Conference for Computer Vision (ECCV) , pp .483–499, 2016.

[12] Z. Cao, T. Simon, S.E. Wei, and Y. Sheikh, "Realtime multi-person 2D pose estimation using part affinity fields," Computer Vision andPattern Recognition 2017, pp. 1302–1310, 2017.

[13] H. J. Lee and Z. Chen, "Determination of 3D human body postures from a single view," Computer Vision, Graphics and Image Processing, Vol. 30, pp. 148–168, 1985.

[14] Baummer, https://www.baumer.com/ch/en/ [retrieved: Febrary, 2020]

A Study of Factors Influencing Health Managers' Acceptance of eHealth Services in the Kingdom of Saudi Arabia

Abdullah Alshahrani
Ministry of Health, Kingdom of Saudi Arabia
School of Pharmacy and Life Sciences,
Robert Gordon University, Aberdeen, UK
a.m.h.alshahrani@rgu.ac.uk

Katie MacLure
School of Pharmacy and Life Sciences,
Robert Gordon University, Aberdeen, UK
k.m.maclure@rgu.ac.uk

***Abstract*—Electronic Health (eHealth) offers innovative solutions for improving the access and management of healthcare delivery. This study aims to explore factors that influence health managers' acceptance of eHealth services in the Kingdom of Saudi Arabia (KSA). It has been planned to be carried out in three phases: a Systematic Review (SR), a quantitative survey, and a qualitative interview. Inclusion and exclusion criteria for each phase were applied. Ethical approval to conduct the study had been gained. Thirty-nine factors were identified as influencing health managers' eHealth acceptance in KSA. The top influential factors were: (i) Availability of operational resources, (ii) Trust in confidentiality, security, and data privacy, (iii) Availability of qualified human resources, (iv) Information and Communication Technology (ICT) infrastructure and readiness, and (v) The quality of eHealth systems and applications. Findings from this research have drawn a clearer picture of the key challenges in health managers' acceptance of eHealth services in KSA. Areas for improvement are to be highlighted in the final analysis.**

Keywords - eHealth; acceptance; health managers; Kingdom of Saudi Arabia.

I. INTRODUCTION

Electronic Health (eHealth) is defined as the use of ICT for healthcare [1]. eHealth offers innovative solutions for improving the access to and management of healthcare delivery. The KSA is a country with one of the largest land masses and populations in the Middle East [2]. It has difficult geographical terrain, which makes the delivery of health services challenging. The Ministry of Health (MOH) is the main health provider in KSA, responsible for around 60% of all health services and facilities in the country. Private health sector and other government run health authorities are the providers for the remainder. Many initiatives to embrace technology in healthcare were launched by the MOH to advance the level of health technology acceptance. Despite the growth of eHealth publications in KSA, it is limited to only a few provinces and more research related to the end-users' acceptance of eHealth services is needed. The overall aim of this study is to explore the factors that influence health managers' acceptance of eHealth in KSA.

II. METHODS

In this section, the methods of conducting the three phases of the study are described.

A. 1st phase SR

The Preferred Reporting Items for Systematic review and Meta-Analysis Protocols (PRISMA-P) checklist of 17 items was followed in writing the protocol for the systematic review [3]. The protocol was registered with the prospective register of systematic reviews (CRD Prospero) [4] and a scoping search was conducted in May, 2017.

Five databases, namely Association for Computing Machinery (ACM), Google Scholar, Medline, ScienceDirect, and Web of Science, were searched for articles published between January 1st, 1993 and May 1st, 2017. Studies included were peer-reviewed, full-text primary articles. One reviewer performed the searches; two reviewers independently screened the titles, then abstracts, followed by full articles. Exclusions were recorded, as indicated below. Three tools were applied, based on study design, to assess the quality of the articles that were included:

1. Questionnaire checklist developed by Crombie and adopted by the Centre of Evidence Based Management (CEBMa) [5].
2. Qualitative checklist provided by Critical Appraisal Skills Programme (CASP), Public Health Resource Unit [6].
3. Critical appraisal checklist developed by a group of researchers led by N. Mays for mixed methods studies [7].

B. 2nd phase survey

An online questionnaire was developed based on two sources; first, the findings from the SR, and second, the Unified Theory of Acceptance and Use of Technology (UTAUT) [8]. Professionals in KSA with a health management role were invited to take part. Participation links were distributed across social media platforms.

C. 3rd phase qualitative interviews

For more in depth exploration, semi-structured face-to-face and telephone interviews were conducted with health managers in Aseer province, KSA to address the geographical limitation of studies published to date. A draft interview schedule was developed based on the findings from phase 2 (survey). Invitations to participate in the study were emailed to all potential participants who met the inclusion and exclusion criteria. A study information sheet and consent form were attached to the invitation emails.

Ethical approval for conducting the 2nd and 3rd phases had been gained from:

1. Ethical Review Panel, School of Pharmacy and Life Sciences, Robert Gordon University, Aberdeen, UK.
2. Ethics Committee, Ministry of Health (MOH), Riyadh, Kingdom of Saudi Arabia.

III. RESULTS

A summary of findings from the three phases are presented in this section.

A. 1st phase SR

After duplicates were removed, 110 papers were screened, and 15 studies met the inclusion criteria (Table I). Thirty-nine influential factors were identified from the included studies. Two knowledge gaps were found: lack of eHealth studies from the perspective of health managers and the limitation of studies to few geographical areas in the country [9].

TABLE I. INCLUSION AND EXCLUSION CRITERIA FOR THE SYSTEMATIC REVIEW

Participants	**Inclusion**: • Health professionals (medical doctors, nurses, midwives, pharmacists, dentists, all other allied health professionals e.g. radiology and laboratory technicians). • Health IT professionals. • Health managers. **Exclusion**: • IT professionals who do not have a role in any health facilities and organisations
Interventions	**Inclusion**: The intervention for this study is eHealth. This systematic review aims to include all published articles and literature around eHealth adoption, acceptance, facilitators and barriers in Saudi Arabia from the views of multiple stakeholders. **Exclusion**: Studies that focus on pure technological infrastructure and products without the users views such as: health technology applications and Internet of Things (IoT) for health.
Studies	**Inclusion**: This systematic review focused on peer reviewed primary published articles and literature with all types of study design such as quantitative, qualitative and mixed methods. **Exclusion**: Reviews, blogs, books chapters, and health website contents were excluded.

B. 2nd phase survey

Findings showed the top influential factors to health managers' acceptance of eHealth services were: (i) Availability of operational resources, (ii) Privacy and security of health information, (iii) ICT infrastructure and readiness, (iv) Availability of qualified human resources, and (v) The quality of eHealth systems and applications (Figure 1).

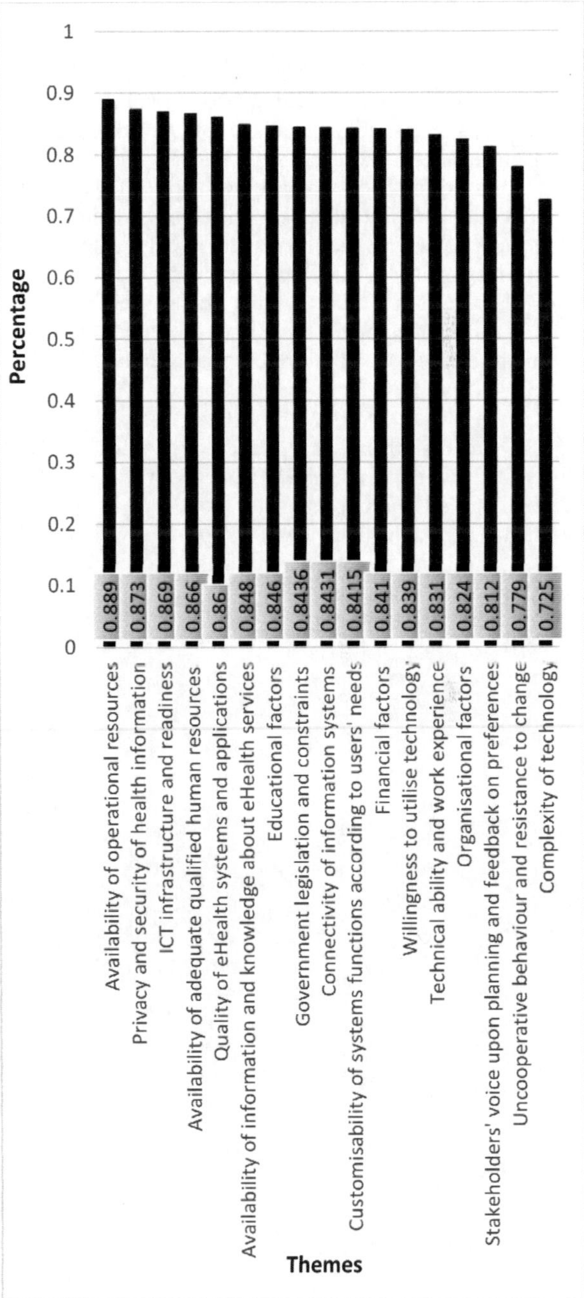

Figure 1. Factors influencing health managers' acceptance of eHealth services in KSA

C. 3rd phase qualitative interviews

Initial analysis confirmed the significance of the factors identified in the 2nd phase of the study. Further work will focus on mapping factors against UTAUT constructs and statistical analysis of the influencing predictors to behavioral intention and actual eHealth use behavior to give a holistic overview of the key challenges in accepting eHealth services in KSA.

IV. CONCLUSION

This study has been conducted by a multidisciplinary team focused on providing a holistic overview of challenges in health managers' acceptance of eHealth services in the KSA. A mixed methods approach was applied to strengthen the findings. It is suggested to apply caution upon interpretation of the results as they only showed the perspectives of health managers. Extending the research to cover the views of other health groups such as health professionals is recommended to draw a clearer picture that may represent KSA health workforce in general.

ACKNOWLEDGMENT

The authors thank the Ministry of Health, KSA represented by the General Directorate for Research and Studies and all participants that took part in both the questionnaire and the interviews.

REFERENCES

[1] World Health Organization. Strategy 2004-2007. eHealth for health care delivery, 2004, Available at: http://www.who.int/ehealth/about/en/ [accessed: March, 2020].

[2] World Atlas. Countries, Saudi Arabia, 2016, Available at: http://www.worldatlas.com/webimage/countrys/asia/sa [accessed: March, 2020]

[3] D. Moher et al., Preferred reporting items for systematic review and meta-analysis protocols (PRISMA-P). Systematic reviews, 4(1), 2015, pp. 1.

[4] A. Alshahrani, K. MacLure, and D. Stewart, Status of eHealth adoption and acceptance in the Kingdom of Saudi Arabia: a systematic review protocol, 2017, Available at: http://www.crd.york.ac.uk/PROSPERO/display_record.php?ID=CRD42017065009 , [accessed: March, 2020]

[5] I. K. Crombie, The pocket guide to critical appraisal: A handbook for health care professionals, 1997.

[6] Critical Appraisal Skills Programme (CASP), CASP systematic review checklist, 2013, Available at: http://www.casp-uk.net/casp-tools-checklists [accessed: March, 2020]

[7] N. Mays, E. Roberts and J. Popay, Synthesising research evidence. Studying the organisation and delivery of health services: Research methods, 2001, pp. 188-220.

[8] V. Venkatesh, M. G. Morris, G. B. Davis, and F. D. Davis, User acceptance of information technology: Toward a unified view. MIS quarterly, 2003.

[9] A. Alshahrani, D. Stewart, and K. MacLure, A systematic review of the adoption and acceptance of eHealth in Saudi Arabia: views of multiple stakeholders. International Journal of Medical Informatics. 2019; 128: pp.7-17.

PROPHECY: Patient Reported Outcomes in Prostate Cancer, a Mobile-Health Experience in Radiotherapy

Bruno Fionda, Anna Rita Alitto, Vincenzo Frascino, Francesco Catucci, Giuditta Chiloiro, Loredana Dinapoli, Ciro Mazzarella, Vito Lanzotti

Radioterapia Oncologica, Dipartimento di Diagnostica per immagini, Radioterapia Oncologica ed Ematologia
Fondazione Policlinico Universitario "A. Gemelli" IRCCS
Rome, Italy
Email: bruno.fionda@policlinicogemelli.it

Antonio Piras, Andrea D'Aviero, Giovanni Palazzoni, Francesco Preziosi, Vincenzo Valentini, Giovanna Mantini

Istituto di Radiologia, Università Cattolica del Sacro Cuore; Radioterapia Oncologica, Dipartimento di Diagnostica per immagini, Radioterapia Oncologica ed Ematologia
Fondazione Policlinico Universitario "A. Gemelli" IRCCS
Rome, Italy

Abstract—**The growing relevance of Patient Reported Outcomes (PROs), the increasing importance of mHeatlh and the objective discrepancy documented between Patient Reported Outcomes and Observer Reported Outcomes are important aspects that need to be addressed when dealing with prostate cancer patients. The aim of this work was to develop an electronic Patient Reported Outcomes tool to systematically assess the impact of radiation therapy on prostate cancer patients' quality of life. We elaborated a four-step process. In the first step, a general literature search was made. The next step was to generate a set of adequate questions and answers. The subsequent step was to identify a reliable scale to report adverse events, namely, the Common Terminology Criteria for Adverse Events version 4.03 scale of toxicity. The last phase was to implement a user-friendly interface. We developed a new and innovative comprehensive list of items for prostate cancer patients receiving radiotherapy whose main characteristic is to link the Patient Reported Outcomes obtained from patients to a well-established scale of toxicity. The conclusive validation of this conceptually innovative tool should allow to provide both patients and physicians with a useful tool to reduce, and possibly prevent, adverse events during and after radiotherapy, with a consequence of improvement in terms of Quality of Life.**

Keywords—Mobile-health; Quality of life; Patient reported outcomes; Prostate cancer.

I. INTRODUCTION

Prostate cancer is the most common cancer in men and the recent developments in therapeutic approaches have allowed to obtain very long overall survival rates [1]. A key aspect that needs to be addressed when dealing with such patients is the Quality of Life (QoL) [2]. There are several reasons which, in our view, should be taken into account when considering to develop a QoL questionnaire for prostate cancer patients undergoing radiotherapy to be used specifically through mobile devices:

1) The growing awareness in the scientific community about the relevance of Patient Reported Outcomes (PROs) [3][4].

2) The increasing importance mHeatlh is rapidly gaining [5][6].

3) The objective discrepancy documented in scientific literature between PROs and Observer Reported Outcomes (OROs) [7][8].

On these premises, we decided to develop a specific Health Related Quality of Life (HRQoL) prostate cancer questionnaire which might be included in a much wider electronic application developed in-house called VALEO+ (VAluation Endorsed by Oncology Patient) with the intent to help all cancer patients undergoing radiotherapy treatment by providing useful tools such as scheduling of appointments, suggestions to improve lifestyle and a specific questionnaire developed to assess toxicity.

The rest of the paper is structured as follows. In Section II, we present the materials and methods used. Section III presents the state of the art. In Section IV, we present the results and we discuss them in Section V. We conclude the work in Section VI.

II. MATERIALS AND METHODS

First of all, we identified two main sources from literature. The first is the guidelines issued by the EORTC for developing Questionnaire Modules. The approach proposed is, in this case, a "modular" one [9]: in particular, the development of modules is specific to tumor site, treatment modality, or a QoL dimension. The second major reference for the questionnaire development was the Food and Drug Administration [10]; for the FDA a PRO instrument needs to capture PRO data used to measure treatment benefit or risk in medical product clinical trials.

The conceptual framework is a straightforward expression of the extracted concepts by the questionnaire and can be represented like a diagram with clear relationships between items, the domain, and concepts (the specific measurement goal) measured. Keeping in mind the framework presented in these guidelines, we proceeded with a four-step process to generate the questionnaire.

In the first phase, in which the relevant QoL issues should be generated, an in-depth literature search was performed [11]-[13]. Since prostate cancer has been widely studied over the years and there is a great amount of literature that extensively covers the issues related to quality of life, we decided to identify the most relevant questionnaires reported in the published reviews. After identifying the most widely reported questionnaires, we started a series of dedicated debates, performed within the urological group of our department, to choose the most relevant items to include in our questionnaire.

In the second phase the list of QoL items was converted into questions in Italian language, keeping in mind the great importance of the major methodological considerations according the guidelines mentioned above in terms of item construction: questions in fact need to be clear, brief and unambiguous.

The third step implied identifying a reliable scale to report adverse events and an accurate literature research was performed to choose a simple and validated system to correlate with the set of items and relative questions generated.

In the last step, a user-friendly interface for patients was realized with the help of graphic experts; this phase was also crucial since this represents a key aspect which differentiates a paper-based from an electronic approach and is a major challenge when dealing with mobile-health PRO [14].

III. STATE OF THE ART

Table I summarizes the results of the relative items from the seven questionnaires included in the analysis [15]-[21].

As shown in Table II, the number of possible choices for patients significantly changes across the different questionnaires already existing and, in some cases, even within the same questionnaire.

IV. RESULTS

After choosing the questionnaires included in the reviews, we created a collection of all the questions presented in the different questionnaires, grouping them according to the relative item of interest.

After identifying the list of the items and of the domains, the list was discussed by prostate experts at our institution and the result was the identification of 3 main domains including urinary symptoms, bowel symptoms and sexual function/hormonal therapy related problems for a total of 14 items, as shown in Figure 1, The first domain (urinary) includes hematuria, urinary incontinence, urinary tract pain and urinary frequency. The second domain (bowel symptoms) includes abdominal pain, diarrhea, rectal hemorrhage and proctitis. The last domain (sexual function and hormonal therapy) includes sexual desire reduction, hot flashes, breast pain, memory or concentration problems, erection problems and ejaculatory problems.

Questions and answers have been formulated while trying to keep the number of words as low as possible, considering the means of delivery that is a smartphone or a tablet.

All questions were subsequently revised by psycho-oncologists with great expertise in cancer patients questionnaires; in this phase, several changes were made in order to make the question not only clear for patients, but also to reduce any possible problem related to the question itself.

TABLE I. NUMBER OF ITEMS IDENTIFIED IN DIFFERENT QUESTIONNAIRES IN LITERATURE.

	U.F.	*U.D.*	*U.I.*	*U.B.*	*I.F.*	*I.D.*
EORTC QLQ - PR25	5	1	2	0	0	0
UCLA - PCI	1	0	3	0	0	1
EPIC	2	2	4	2	3	2
FACT-P	1	1	0	0	0	0
PORPUS	1	0	1	0	1	1
PC-QoL	0	0	5	0	1	1
PCSI - SDS	4	4	4	0	2	2
	A.P.	*I.B.*	*H.F.*	*B.P.*	*Er.P.*	*Ej.P.*
EORTC QLQ - PR25	1	1	1	1	1	1
UCLA - PCI	2	0	0	0	3	0
EPIC	4	2	2	2	5	0
FACT-P	1	0	0	0	1	0
PORPUS	0	0	0	0	1	0
PC-QoL	3	1	0	0	2	0
PCSI - SDS	5	1	0	0	4	1

U.F.=Urinary frequency; U.D.=Urinary tract pain; U.I.= Urinary Incontinence; U.B.= Hematuria; I.F.= Diarrhea; I.D.= Proctitis, A.P.= Abdominal pain; I.B.= Rectal Hemorrhage; H.F.=Hot flashes; B.P.=Breast pain; Er.P.=Erection problems; Ej.P.=Ejaculatory problems

These differences might generate at least two kinds of problems in our view.

The first problem is in patients' perspective because, when they answer questions, patients face diversity, in the range of possible choices, which could, in theory, be a confounding factor in attributing the choice of symptom severity.

The second problem is the physicians' perspective because it is difficult to compare the results from the different questionnaires.

A possible solution to both problems could be found in the third phase of our process since we chose to use the severity scale found in the CTCAE V 4.03, which is a scale going from 1 to 5 from the less severe to the more severe symptoms due to either frequency or entity increase according to the different definitions [22].

TABLE II. DIFFERENT NUMBER OF RESPONSES IDENTIFIED IN QUESTIONNAIRES IN LITERATURE.

EORTC - PR25	1→4 1=no symptom→4=worst
UCLA - PCI	0→6 with a range of 3 to 6 answers and no fixed correlation between severity and number
EPIC	0→5 with a range of 3 to 5 answers and correlation between severity and increasing number
FACT-P	0→4 with correlation between severity and increasing number
PORPUS	No definite number of answers
PC-QoL	1→7 with a range of 3 to 7 answers and correlation between severity and increasing number
PCSI - SDS	1→5 1=no symptom→5=worst

Knowing the severity of the symptoms also includes potentially life-threatening conditions (grade 4) and death (grade 5); the absence of the symptom is not included in the scale, so there is no zero.

We decided to exclude the two higher grades that is to say grade 4 and 5, since they are not compatible with a patient reported symptom and, therefore, we chose a 4 way possible answers with one answer including the absence of the symptom and a growing severity (in frequency or entity, according to the CTCAE definition) for the remaining three answers.

The fact that symptoms were initially derived from the CTCAE, which is not a PRO tool, poses a problem for grade 1 asymptomatic situation such as, for example, for hematuria or rectal bleeding. In these two cases, in order to confirm the real absence of the symptom, we propose to add a urine and stool test.

After choosing the questions and the relative answers, we went on with the fourth phase to generate the graphical interface. We relied on the support from personnel experienced in graphical design and tool creation for cancer patients. The entire graphical interface has been completed and is already fully available, as it can be seen in Figure 2.

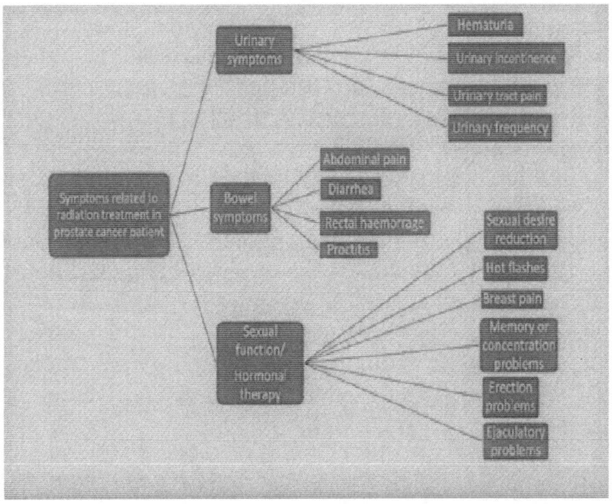

Figure 1. Symptoms domains [27].

Figure 2. VALEO+ graphical interface.

V. DISCUSSION

Assessing PROs has turned out to be a central part of healthcare by measuring the impact of both disease and medical intervention on patients. The first attempts to develop wireless mobile health-related quality of life assessment started more than a decade ago [23]. Few authors have reported about the evaluation of the reliability, usability, and acceptability of point- of-care electronic PRO assessments implemented in prostate cancer clinics [24]-[27] and the available result is that mean scores and standard deviations are similar between the paper-pencil and electronic forms across instrument domains with no assessment bias [28]-[31].

In our work, we followed an item pooling procedure, which was mainly based on previous questionnaires; searching in literature, we found that an item pooling procedure for extracting items based mainly from pre-existing questionnaires is an option that has already been described [32][33]. The advantages of this choice are important because it is possible to obtain a correspondence between the patient reported symptoms and the chance to implement a toxicity record.

The choice of the CTCAE has been used by other groups actively involved in the development of quality of life tools for cancer patients [34]; such choice is a key factor which distinguishes our proposed questionnaire from the existing ones in literature because the CTCAE links the indication of a medical intervention to the severity of the symptom. In this way, it is also possible to generate electronic alerts for the patients who report experiencing certain grades of severity. The chance to generate such alerts has at least two other advantages. The first one is that the alert suggesting to contact a doctor allows the doctor himself to confirm (or not) the severity of the symptom reported (thus implicitly validating the correspondence between the patient reported outcome and the CTCAE). The other important aspect is relative to prevention of severe symptoms: in fact, in case of repeatedly reported low severity symptoms, which by definition require no medical intervention, the system may generate an alert signal to contact the doctor as well so that an in depth analysis can be made of the result to prevent further deterioration of the symptom.

Moreover, CTCAE does not distinguish acute from late side effects, but it is focused on the symptoms themselves so that the same question can actually be used both in the treatment setting and, subsequently, in the follow-up of the patients.

The importance of our choice of integrating a modified version of the CTCAE scoring system in a mobile-health system is, in our view, further strengthened by the very recent release of a PRO- CTCAE item library [35][36].

VI. CONCLUSION

A specific questionnaire for prostate cancer patients undergoing radiotherapy was developed to realize an electronic PRO. A combined approach was used, both traditional and innovative, in order to obtain a HRQOL tool that may help patients, caregivers and physicians to improve the quality of the treatment by focusing on the patient's active role. The subsequent phase will require the testing of the developed questionnaire by patients, in order to fully validate it.

ACKNOWLEDGMENT

The abstract of this paper was presented at the 29th Annual Meeting of the Italian Society of Uro- Oncology (SIUrO) as a poster with interim findings. The poster's abstract was published in "Poster Abstracts" in Anticancer Research 39: 1507-1629 (2019).

REFERENCES

[1] R. L Siegel, K. D. Miller, and A. Jemal, Cancer statistics, 2019. CA Cancer J Clin. 2019 Jan; 69(1), pp.7-34.

[2] C. Acquadro et al., Incorporating the Patient's Perspective into Drug Development and Communication: An Ad Hoc Task Force Report of the Patient-Reported Outcomes (PRO) Harmonization Group Meeting at the Food and Drug Administration, February 16, 2001. Value Health. 2003;6, pp.522–31.

[3] P. R. Deshpande, S. Rajan, B. L. Sudeepthi, and C. P. Abdul Nazir, Patient-reported outcomes: A new era in clinical research. Perspect Clin Res. 2011 Oct, 2(4), pp. 137-44.

[4] L. C. Doward, and S. P. McKenna, Defining patient-reported outcomes. Value Health. 2004 Sep-Oct;7 Suppl 1:S4-8.

[5] R. Miniati, F. Frosini, G. Cecconi, F. Dori, and G. B. Gentili. Development of sustainable models for technology evaluation in hospital. Technol Health Care. 2014, 22(5), pp. 729-39.

[6] G. Nasi, M. Cucciniello, and C. Guerrazzi, The role of mobile technologies in health care processes: the case of cancer supportive care. J Med Internet Res. 2015 Feb 12;17(2):e26.

[7] C. F. Snyder, M. E. Watson, J. D. Jackson, D. Cella, and M.Y. Halyard, Mayo/FDA Patient-Reported Outcomes Consensus Meeting Group. Patient-reported outcome instrument selection: designing a measurement strategy. Value Health. 2007 Nov-Dec;10 Suppl 2:S76-85.

[8] G. A. Sonn, N. Sadetsky, J. C. Presti, and M. S. Litwin, Differing Perceptions of Quality of Life in Patients With Prostate Cancer and Their Doctors. J Urol. 2013 Jan;189(1 Suppl):S59-65.

[9] EORTC QUALITY OF LIFE GROUP – Guidelines for developing Questionnaire Modules, 2011.

[10] U.S. Department of Health and Human Services Food and Drug Administration Guidance for Industry: Patient- Reported Outcome Measures: Use in Medical Product Development to Support Labeling Claims. U.S. FDA, Clinical/Medical, 2009.

[11] J. Bergman and A. Laviana, Quality-of-life assessment tools for men with prostate cancer. Nat Rev Urol. 2014 Jun;11(6):352-9.

[12] E. H. Hamoen, M. De Rooij, J. A. Witjes, J.O. Barentsz, and M. M. Rovers, Measuring health-related quality of life in men with prostate cancer: A systematic review of the most used questionnaires and their validity. Urol Oncol. 2015 Feb;33(2):69.e19-28.

[13] S. Schmidt, O. Garin, Y. Pardo, J. M. Valderas, J. Alonso, and P. Rebollo, Assessing quality of life in patients with prostate cancer: a systematic and standardized comparison of available instruments. Qual Life Res. 2014 Oct;23(8), pp. 2169-81.

[14] E. K. H. Chan, T. C. Edwards, K. Haywood, S. P. Mikles, and L. Newton, Implementing patient-reported outcome measures in clinical practice: a companion guide to the ISOQOL user's guide. Qual Life Res. 2018 Nov 17.

[15] G. van Andel, A. Bottomley, S.D. Fosså, F. Efficace, C. Coens, and S. Guerif, An international field study of the EORTC QLQ-PR25: A questionnaire for assessing the healthrelated quality of life of patients with prostate cancer. European Journal of Cancer. 2008 44, pp. 2418–2424.

[16] P. Esper, F. Mo, G. Chodak, M. Sinner, D. Cella, and K. J. Pienta, Measuring quality of life in men with prostate cancer using the functional assessment of cancer therapy-prostate instrument. Urology. 1997 50, pp. 920–928.

[17] M. Krahn., P. Ritvo, and J. Irvine, Construction of the Patient-Oriented Prostate Utility Scale (PORPUS): A multiattribute health state classification system for prostate cancer. Journal of Clinical Epidemiology. 2000 53, pp. 920–930.

[18] J. T. Wei, R. L. Dunn, M. S. Litwin, H. M. Sandler, and M.G. Sanda, Development and validation of the expanded prostate cancer index composite (EPIC) for comprehensive assessment of health-related quality of life in men with prostate cancer. Urology. 2000 56, pp. 899–905.

[19] R. B. Giesler, B. J. Miles, M. E. Cowen, and M. W. Kattan, Assessing quality of life in men with clinically localized prostate cancer: Development of a new instrument for use in multiple settings. Quality of Life Research. 2000 9, pp. 645–665.

[20] J. A. Clark and J. A. Talcott, Symptom indexes to assess outcomes of treatment for early prostate cancer. Medical Care. 2001, 39, pp. 1118–1130.

[21] M. S. Litwin, R. D. Hays, A. Fink, P.A. Ganz, B. Leake, and R. H. Brook, The UCLA Prostate Cancer Index: Development, reliability, and validity of a health-related quality of life measure. Medical Care. 1998 36, pp. 1002–1012.

[22] T. M. Atkinson, L. J. Rogak, N. Heon, S.J. Ryan, M. Shaw, and L.P. Stark, Exploring differences in adverse symptom event grading thresholds between clinicians and patients in the clinical trial setting. J Cancer Res Clin Oncol. 2017 Apr;143(4) pp. 735-743.

[23] C. Wu, R. J. Volk, J. R. Steinbauer, G. M. Kuo, and J. S. Smith, A wireless mobile health-related quality of life assessment. AMIA Annu Symp Proc. 2003, p. 1054.

[24] P. Sharma, R. L. Dunn, J. T. Wei, J. E. Montie, and S. M. Gilbert, Evaluation of point-of-care PRO assessment in clinic settings: integration, parallel-forms reliability, and patient acceptability of electronic QOL measures during clinic visits.Qual Life Res. 2016 Mar;25(3), pp. 575-83.

[25] M. Spillane, P. Pollock, L. Hedden, E. Wu, L. Goldenberg, and J. M. Jones, Implementation of an electronic patient reported outcome system in a prostate cancer urology clinic. Journal of Clinical Oncology 2018 36:7_suppl, p. 143.

[26] L. Warrington, K. Absolom, P. Holch, A. Gibson, B. Clayton, and G. Velikova, Online tool for monitoring adverse events in patients with cancer during treatment (eRAPID): field testing in a clinical setting. BMJ Open. 2019 Feb 1;9(1):e025185.

[27] B. Fionda, A. D'Aviero, A. R. Alitto, V. Frascino, A. Piras, and F. Catucci, Patient Reported Outcomes in Prostate Cancer, a mobile-health Essay (PROPHECY). Anticancer research. 2019. 39(3) pp. 1507-1629.

[28] J. S. Marcano Belisario, J. Jamsek, K. Huckvale, J. O'Donoghue, C.P. Morrison, and J. Car, Comparison of self-administered survey questionnaire responses collected using mobile apps versus other methods. Cochrane Database Syst Rev. 2015 Jul 27.

[29] A. Rogausch, J. Sigle, A. Seibert, S. Thüring, M. M. Kochen, and W. Himmel, Feasibility and acceptance of electronic quality of life assessment in general practice: an implementation study. Health Qual Life Outcomes. 2009 Jun 3;7:51.

[30] E. Rincon, F. Monteiro-Guerra, O. Rivera-Romero, E. Dorronzoro-Zubiete, C. L. Sanchez-Bocanegra, and E. Gabarron, Mobile Phone Apps for Quality of Life and Well-Being Assessment in Breast and Prostate Cancer Patients: Systematic Review 2017 JMIR vol 5, no 12.

[31] J. Pereira, E. Green, S. Molloy, D. Dudgeon, D. Howell, and M.K. Krzyzanowska, Population-based standardized symptom screening: Cancer Care Ontario's Edmonton Symptom Assessment System and performance status initiatives. J Oncol Pract. 2014 May;10(3), pp. 212-4.

[32] E. Vidhubala, R. R. Kannan, S.C. Mani, K. Karthikesh, R. Muthuvel, and V. Surendran, Validation of quality of life questionnaire for patients with cancer--Indian scenario. Indian J Cancer. 2005 Jul-Sep;42(3):138-44.

[33] M. L. Rothman, P. Beltran, J. C. Cappelleri, J. Lipscomb, and B. Teschendorf, Mayo/FDA Patient-Reported Outcomes Consensus Meeting Group. Patient-reported outcomes: conceptual issues. Value Health. 2007 Nov-Dec;10 Suppl 2:S66-75.

[34] E. Galligioni, E. M. Piras, M. Galvagni, C. Eccher, S. Caramatti, and D. Zanolli, Integrating mHealth in Oncology: Experience in the Province of Trento. J Med Internet Res. 2015 May 13;17(5):e114.

[35] E. Basch, B. B. Reeve, S. A. Mitchell, S. B. Clauser, L. M. Minasian, and A.C. Dueck, Development of the National Cancer Institute's patient-reported outcomes version of the common terminology criteria for adverse events (PRO-CTCAE). J Natl Cancer Inst. 2014 Sep 29;106(9).

[36] J. L. Hay, T. M. Atkinson, B. B. Reeve, S. A. Mitchell, T. R. Mendoza, and G. Willis, Cognitive interviewing of the US National Cancer Institute's Patient-Reported Outcomes version of the Common Terminology Criteria for Adverse Events (PRO-CTCAE). Qual Life Res. 2014 Feb;23(1):257-69.

From Theory to Reality

Health Data Management in a Complex System

Eloria Vigouroux-Zugasti, Christian Bourret
Laboratory of DICEN IDF
University of East-Paris Marne-la-Vallée
Val d'Europe / Serris, France
Email: eloria.vigouroux-zugasti@u-pem.fr,
christian.bourret@u-pem.fr

Sylvie Parrini-Alemanno
CNAM
Paris, France
Email: sylvie.parrini-alemanno@lecnam.net

Abstract— **With the recent promulgation of the General Data Protection Regulations (DGPS), data management is becoming a crucial strategic issue in organizations. The quality of data dissemination is of utmost importance in the healthcare environment. Indeed, medical confidentiality is closely linked to the dissemination of personal information inherent in the patient's record. Yet, how can a complex system, composed of multiple multidisciplinary actors (medical, paramedical, administrative, etc.), deal with the potential disclosure of personal data? What steps can be taken to manage this risk? How to ensure legal compliance with medical confidentiality, while ensuring the interoperability of professionals and the quality of care? To answer these questions, a case study was conducted in the Multidisciplinary Care House of Mimizan (France, Landes, New Aquitaine Region). The goal was to investigate the importance of data management for traceability of care. This medical organization is composed of medical and paramedical professionals, but also a relatively large administrative team for such an institution. Nevertheless, it manages to set up, at the initiative of the professionals, both flexible and structured processes, allowing optimal follow-up of patients, while guaranteeing respect for their personal information.**

Keywords— Health; Data; Patient Records; Quality; Confidential Medical Data; Territory; Organization.

I. Introduction

The management of personal data, with the recent promulgation of the General Data Protection Regulations (GDPR - May 2018), is becoming a crucial and strategic issue for organizations. In the healthcare sector, where respect for medical confidentiality is closely linked to the dissemination of patients' personal records, data have to be managed with great care, in order to limit the risk of unfortunate disclosure or data loss and to tackle their specificity of very sensitive data (privacy challenges).

However, the current French health organization tends to evolve to a collaborative working method. For the past ten years, French health authorities have been witnessing the emergence of some new organizations of health professionals and the exponential growth of groups of (para)medical professionals, attesting the growing complexity of health organizations [1]. This is particularly the case for Multidisciplinary/Multiprofessional Health Houses (MCHs), with 910 establishments active in France in March 2017, compared to only 240 in 2013 [2]. These establishments are made complex by the number of interacting actors they involve. They also tend to rely on the notion of sharing and circulating data, especially health information about patients. By doing so, they tend to improve patients' health care and health monitoring.

Does this new type of establishment pose some risks for private heath data policy? As they are composed of many multidisciplinary actors (medical, paramedical, administrative, etc.), do not they increase the number of possible accidental disclosures or loss of confidential data? Thus, how could legal compliance guarantee medical confidentiality and manage the inherent risks, while ensuring the interoperability of professionals and the quality of care? It seems that the challenge is mainly about the ability of the French health system to increase its level of performance [10].

To answer these questions, a case study was conducted in the Multidisciplinary Care House of Mimizan (MCHM). To do so, an exploratory qualitative approach was conducted via focus groups with the whole team of the institution and via an interview with its two managers. The goal is to establish an inventory of good practices, particularly in terms of data quality management in complex health institutions. The analytical approach presented here is based on the structural level of MCH actions, in which resources (human and material) are mobilized to ensure a good coordination for patients' care and confidential data policy [3]. MCHs are a recent and emergent phenomenon, with heterogeneous ways of working. The goal of this article is to explore one of the biggest and formalized MCH, from a exploratory point of view, in order to understand its organisation and its capacity to be considered as a model for other MCH.

The rest of this article is structured as follows. In Section II, with the overall contextualization of the study. The Section III will present the institution selected for the study, the MCHM, in order to show how it is representative of the new health institutions needed in fragile territory. In Section IV, the article will analyse the quality processes inherent in the management of this institution's data. The Section V will present the specifications of the MCHM. In Section VI, we provide details on the general lack of use of the national digital health record. We conclude the article in Section VII.

II. Study Background

In this section, the study background is explained. The specific context of the current French public health context and policy are also detailed for a better understanding of the study.

A. Challenging health context

In 2014, the population density of Landes fluctuated between 2 and 45 inhabitants/km², with an aging index among the highest in the region [4]. In 2015, 31.5% of Landes' population was over 60 years of age, a great increase compared to 2011 [5]. According to the Regional Health Authority (RHA), in 2016, the rural population represented more than 50% of the whole Landes population. In addition, this region also has a medical demography and a density of specialists lower than regional and national averages, as well as a small number of health establishments. It also has few alternatives to the nursing homes for old people [4]. Therefore, old people's loss of autonomy is more difficult to manage. According to RHA, 1/3 of liberal general practitioners were over 60 years old in 2017 [6]. This is a very problematic issue: combined with the difficulties of attractiveness of the territory, it becomes more and more difficult to maintain the number of practitioners in this area. The RHA demographic patterns of Landes health care show a highly unfavorable public health context:

- White areas (towns located more than fifty kilometers away from a hospital emergency department), combined with an insufficient number of expert services (radiology, rheumatology, gynecology, allergology, pneumology, dermatology, etc.)
- Medical desertification: unattractive territory for young (para)medical professionals (region's remoteness from large cities, low internet coverage, few cultural offers, etc.). Doctors are struggling to find successors, despite administrative provisions and facilitations offered by health authorities.
- Fragile areas: unequal distribution of health professionals in areas with an imbalance between the number of potential patients and the number of doctors, as well as areas where the advanced age of patients (or the doctor) would require urgent decisions.

Facing these difficulties, some political representatives try to shed some light on the issue of medical deserts: in October 2018, the mayor of a small town called Ychoux proposed to prohibit, by municipal decree, his fellow citizens from falling ill, due to the lack of medical care in its surroundings. The study was conducted in this area for its relevance concerning the challenges France will have to face in the coming years. The main problem in France is getting young medical staff to settle in countryside areas, where living standards are less attractive than urban contexts.

B. Management of de-materialized health data in a complex system

Personal data, such as medical data, is defined by the French Data Protection Act, called *Loi informatique et liberté*, (paragraph 2, article 2) as an information relating to a person who is physically identified or who can be identified, directly or indirectly, by reference to an identification number or to one or more elements specific to him/her [7]. In addition, in order to make sense, this data must be processed. This process is defined by the same law (paragraph 3, article 2) as a transaction or set of transactions, whatever the process used [7]: collection, recording, organisation, storage, modification, consultation, communication by transmission, etc. Data are also part of an exchange, characterized by the provision to several professionals, such as health staff, who are entitled to know everything about these data. Their goals are to insure the coordination and continuity of patients' medical care [8].

However, the exploitation of personal data remains a sensitive subject. It directly affects the privacy of each individual [9], especially when it comes to medical data. In complex systems, such as MCH, there are several issues related to the management and to the protection of health data. From an organizational point of view, it is essential to set up procedures to secure access to data. Those procedures tend to limit the structural disorders that can affect the confidential standards of data, by establishing, for example, quality indicators. Measurement and management tools in health establishments are essential [10]. In addition, complex systems have a large number of stakeholders, but they do not have the same level of data access authorization. It increases the risks of fraudulent or accidental access to information.

With the multiplication of MCHs, the French health sector must now face a multitude of risks related to data, which require close scrutiny of each elementary activity [1]. These establishments are the result of a clustering of health professionals who, until now, had been working alone. However, using common resources and administrative staff lead professionals to rethink their working methods, while insisting on control and rigour. The various stakeholders in the project have to develop fundamental procedures for collaborative work. This aims at reducing and optimizing work processes [1], while considering the topics of control and quality as the heart of these processes. Professionals, in this new context, must demonstrate that their services are delivered in a secure environment. This environment helps controlling the risks and meet the expectations/requirements of patients [10].

C. Multidisciplinary health centre: although need for a restructured health policy – from a political point of view

MCHs are a model that catalyzes needs, from health professional, public decision makers and patient care points of view [3]. The creation of this type of institution represents the convergence of three complementary processes, identified by Autès and Dufay [11]:

- Movement initiated by health professionals to gather their activities within MCHs and health centers.
- Reflection of local officials, concerned by the management of health in their districts, involved in logics of prevention, of permanence of care, of first aid and the continuum between outpatient services and hospitals. They also care about offering external and specialized consultations to the people living in areas where there is a shortage of health practitioners;
- Necessary reorganization of the supply of care due, first, to the constraints of modern medicine and pathologies and, second, to the effects induced by the anticipated decline in medical demography.

From a territorial point of view, the MCHM aims at meeting the four standards of public health action: 1) maintaining a local offer, 2) guaranteeing equal access to health for all, 3) ensuring continuity of care between the primary care offer and graduated hospital care 4) and, finally, strengthening health prevention policies. For local officials, the issue is to strengthen weakened health districts and care offer [11], by proposing long-term ways-out to solve the current problems. In addition, since MCHs are subjected to accreditation rules, by responding to quality indicators established by public health authorities [10], they contribute to an increased performance of districts' medical management.

III. Methodological Approach

In this section, the sampling and methodological procedures are presented.

A. *Why Study the MCHM?*

By definition, MCHs depend on specific contexts. It is important to identify the territory's needs, to take into account the needs of its population and its state of supply. MCHM is considered as representative of this movement of territorial restructuring in the field of health, both in its conditions of implementation and in its daily functioning, structured around the interrelationships between territorial stakeholders [3]. Indeed, despite its recent implementation, it manages to meet the whole public health objectives, both mandatory and optional, imposed by the "RHA Inter-professional Agreement" contract, particularly in terms of shared information systems, which are at the heart of the challenges related to data quality. The dynamic of the creation of the MCHM was, in the first place, launched by health practitioners themselves, in reaction to the progressive desertification of their territory and the challenges it involves [11]. The project of creating the MCHM began in 2004. The district's doctors wanted to cluster their activities in a single establishment, to pool their administrative tasks and to offer a better access to care for their patients. This approach is in line with the observations of some researchers, who state that medical desertification in rural areas has been the main motivation for the mobilization of health professionals [11]. One of the main problems lays in Landes' unattractiveness for doctors. The mere proximity of the beach and the "sweetness of life" are not enough to attract young professionals willing to settle down. It is necessary to provide health professionals with some attractive and secure professional conditions of practice.

However, the notion of attractiveness of the project is very important here [3]. Offering a young professional an isolated practice in a small town does not have the same appeal as a long-term position in a multidisciplinary institution, in which he or she could be supervised, advised and supported by administrative services, surrounded by colleagues and supported by financial and material resources. Collegiality and plurality of perspectives make the medical practice both more reassuring and richer [11], especially at the beginning of a career. This is the appeal proposed, in general, by MCHs and, in particular, by the case studied in this article, which is one of the biggest and dynamic MCHs in France. Its professionals have been recently asked to present their institution in the next National French Congress of MCH. This type of organizational dynamism is a movement widely desired and claimed by the younger generations of medical and paramedical professionals [11].

Moreover, Landes is the largest region of France, with a mainly rural territory and offers most of the current and forthcoming public health services presented in the contextualisation part of this article. Having such an innovative MCH in this kind of area is an example of how to deal with public health issues in other regions in France, especially concerning the rural ones.

B. *Methodological Approach and Sampling*

At the beginning of the survey, the goal was to identify the organizational model and rules implemented by the MCHM's team, specially concerning working processes and data management. To do so, a qualitative approach has been chosen This method has been selected for its ability to investigate the practices and interpret the results. It considers that the "confrontation with the corpus is a necessary condition for the perception of social practices" [12]. The goal was also to confront the different points of view concerning the organizational processes. The potential divergences and discordances regarding the positions can be highlighted. Does a secretary think the same thing of the establishment than a doctor or paramedical worker? Concerning data privacy management, the heads of the MHCM in charge have been interviewed. They had to explain their choices in terms of data management policy, of coordination put in place and of emergency plans in case of unfortunate disclosure.

Regarding the questions, all the members of the MHCM staff we asked about two common topics. The first was about the daily-work and its organization, both concerning the inner-group and the relationships with the other members of the MCHM (for example: secretary-secretary, secretary-medical, etc.). The goal was to highlight relational and organizational dysfunctions. Secondly, all the staff members were asked about their own professional uses of patients' health records, in terms of access of use and of transmission. The purpose was to identify good and problematic uses.

The second questions asked for some more specific topics. The goal was to have a better understanding of each specific staff members (medical doctor, paramedical, administrative, etc.), to point out the benefits and the limits of their new work, management and organizational processes since they entered the MCHM. Four focus groups took place in December 2018, with the four specific staff members. Then, interviews were conducted in January 2018, with the head of the administrative staff and the heads of the MCHM. They all were realized in the MCHM, in the meeting room. The goal was to make the people feel comfortable and to prevent conversations from being heard by the patients or the other staffs' members. This approach seemed relevant, as it helped people to speak freely.

TABLE I. DISTRIBUTION OF THE INTERVIEWED SAMPLE

Criteria	Distribution	Number	%
Gender	Male	14	51,9
	Female	13	48,1
Age	25-35	6	22,2
	35-45	7	25,9
	45-55	8	29,7
	55-65	6	22,2
Professional activity	Medical doctors	11	40,7
	Paramedical staff	8	29,7
	Administrative staff	5	18,5
	Executive of administrative staff	1	3,7
	Executive of the MCHM (also doctors)	2	7,4

C. *Health Records Management in the Multidisciplinary Care House of Mimizan*

The MCHM brings together three main crews: doctors (8 + 2 regular substitutes + 1 trainee), paramedics (3 nurses and their collaborators, 2 physiotherapists, 2 podiatrists, 1 psycho-motor therapists, 1 dietician) and administrative staff (5 secretaries, only working for doctors, managed by an executive). Including trainee doctors and nursing staff, the sample is composed of thirty individuals involved in the daily operations of the MCHM. They are all subjected to the institution's collective agreement, in which respect for medical confidentiality is clearly enshrined. Data is managed and stored on a specific medical database software, Weda [19], which is also used by the surrounding external collaborators (pharmacies, specialists, hospitals, etc.). The choice to use the same software aims at facilitating the medical and administrative aspects concerning the caring continuum. Each member of the MCHM has secure access provided by the software. These codes are not stored on the institution's digital devices, in order to limit the risk in case of theft or hacking.

However, not all MCHM members have the same level of access to patients' records. Doctors, as well as their secretaries, have full access to all the information concerning: files, auxiliary session schedules, secure messaging, etc. This is justified by the need for doctor/doctor and doctor/secretary interoperability, for the smooth functioning of the MCHM and good patient care. At the request of a patient or a staff member, restrictions may be applied to limit secretaries' or doctors' access to some records. All patients were asked about this sharing consent.

Then come the paramedics, with a diffusion specific to each specialty. Their access is conditioned by the needs of their activity. A physiotherapist, for example, will have access to the patient's x-rays and related prescriptions; a nurse will have access to history, specialist contacts, blood test results or vaccines, depending on needs. These professionals do not have access to the content of visits, letters, prescriptions, unless some specific case discussed with the doctor. The accreditation of external professionals is aligned with the system applied within the institution, according to the needs of the patient, the activity or the specialty. Collaborative patient follow-up is governed by Multi-Professional Consultation Meetings (MPCMs), planned or impromptu, attended by all the professionals involved in the presented case. Each meeting is documented, stored via Weda and only accessible to the concerned professionals. Doctors and secretaries meet weekly to monitor performance and improve organizational quality processes. The doctors interviewed also associate these meetings with team management (trust, accountability, etc.), to ensure cohesion among all workers and to involve them in the administration of the MCHM.

IV. ANALYSIS OF DATA MANAGEMENT PROCESSES

This section is dedicated to the analysis of MCHM data management concerning patients' records. Thanks to the study, the main protagonists of this type of management tasks have been identified, such as the governance of the establishment, from a managerial point of view.

A. *Data Ethics and Dissemination Quality*

An organization can be defined as a set of recurring transactional programs that constitute transactional flows. They are driven by a set of conventions and rules in a given context [13]. For its proper functioning, given the complexity of its transactions, it is essential to give access to the right information, at the right time, to make a selective transmission of users, in order to fight against misinformation, over-abundance and deviant uses [9]. The principle of data management is based on the ability of actors to select information and analyze it, in such a way that it is only disseminated to its legitimate recipients. The interest of this approach is twofold. On the one hand, it allows a smooth organization of sharing actions, making the institutional processes efficient. On the other hand, it makes it possible to limit the risks linked to the poor dissemination of data, thus guaranteeing respect for confidential medical records.

The ethical processing of information seems to be the starting point of the MCHM's data management strategy. Béranger defines it as a mechanism for the interpretation of data, by a person or an organization, that will lead to give a specific meaning to data [9]. By giving attention to information, by analyzing it, the heads of MCHM tend to give meaning and value to data, as well as to determine the logistics of action to be applied: censorship, global dissemination, limited dissemination, etc. In the medical sector, it is fundamental to establish a reflection on personal health data through an ethical prism "in order to [remove] doubt and control uncertainties" [9] and to manage the risks inherent in the nature of patients' records. It leads to speak of the non-maleficence nature of the MCHM's information strategy: access to data is examined according to the profile and nature of the user [9]. Data sharing is conditioned by the profile of the information receiver, ranging from full sharing to very limited access, depending on activity and needs. This improves the security, confidentiality and protection of such data [9], as well as the performance of the information management system. By analyzing the data, determining the conditions for sharing and clearly identifying the receivers, the quality of access to patients' personal data is guaranteed.

B. *Informational Lean Management: no Unnecessary Information*

The data processing method leads us to analyze the notion of lean in quality management. Lean School is defined as "the search for process optimization by chasing down everything that is inappropriate or superfluous" guaranteeing "performance by eliminating waste" [1]. This method is usually applied to inventory management (0 stock), document management (0 paper), or logistics (0 unnecessary transport, 0 waiting, etc.). This can be relied to information management, in order to analyze the transaction rationalization activities [13].

Indeed, the info-ethical treatment as previously mentioned tends towards a very low entropy, i.e. a degree of almost nil disorder [9]. A system in which information is transmitted without analysis increases the level of confusion, as well as the slowness of decision-making and the risks of accidental dissemination of personal records. On the contrary, in a complex system such as the MCHM, the implementation of a hierarchy in information management (doctors analyse and choose the criteria before disseminating information) makes the actions of all team members easier and more fluid, by sending them only the data that will be useful to them in the exercise of their activity. This is a kind of lean management, applied to information management. This data dissemination method tends towards the goal of "0 useless information", in order to guarantee both respect for medical confidentiality (0 information poorly disseminated), the quality of patient care (0 information missing) and the fluidity of actions (0 dysfunction linked to poor information dissemination). This information management method seems to be perfectly adapted to the performance requirements of MCHM's missions, while benefiting the daily tasks (administrative and patient care).

V. Organizational Processes: Guaranteeing the Quality of Data Management

In this section, the organizational processes observed during the survey will be discussed, which refers to data and patients' records management.

A. *The Human Relations Theory as a Leading Light*

In the MCHM, the mobilization of the whole staff tends to improve the processes' efficiency [1]. The MCHM's management method is based on the involvement of all teams in improving the life of the institution: meetings, taking into account opinions, professional development, empowerment, etc. This method seems to be similar to the collaborative processes set up within the MCHM, although the institution does not claim any particular managerial method: unexpected discussions, weekly team meetings, festive group cohesion events, etc. This team management aims at analyzing defects and dysfunctions, and then seeking solutions [1]. This tends to improve the overall functioning of the establishment, where, according to Zacklad, all persons involved in the transaction are in the position of (co-)director, (co-)beneficiary, (co-)recipient (principal) and (co-)recipient [13]. Emphasis is placed on freedom of speech, professional responsibilities, skills of each individual and, above all, the necessary trust between employees, which is considered essential by all MCHM staff.

As Doucet points out, it is essential that the direction of quality action be collegial. This makes it possible [in particular] to respect responsibilities and involve departments [1]. As a large number of individuals have access to the institution's health data, the use of this collaborative and collegial approach is essential to the MCHM. The increase in performance can only be achieved through the collaboration with the departments involved in this approach. By soliciting and valuing all staff members, the MCHM ensures fine relationship management, but also encourages professionalism and accountability of each individual. They also do so by regularly reminding them of the need for secrecy and rigour (formally and informally), especially concerning the performance of their daily tasks relating to patient health records. Transactional relationships lead to overcome formal/informal oppositions by insisting on their complementarity [3].

B. *Leadership: Team, Quality and Performance Management*

This managerial approach is in line with the objective of promoting confidentiality and trust with producers and suppliers of information, thus contributing to the control of risks and deviations of data [9]. However, for MCHM staff management to be effective, it must also deal with some leadership issues. Leadership and management of the institution must be provided by a person of influence who, thanks to his or her managerial skills is able to guarantee, effective cooperation and coordination, based on mutual trust [3]. The notions of cooperation and trust seem to be interdependent key resources for the management of complex systems. Cooperation relies on a clear commitment of each member of the group and is strengthened by trust and by the working contract [14]. In a complex system, each member of the team contributes to the success of the institution goals, so it is important for all team members to be aware of the values their work involves, such as secrecy, efficiency, empathy, etc. [15] It is up to the leader to make the team understand these fundamental values defended by the institution, from which ethics rules of behaviour flow. To do so, in the MCHM, many meetings are held, with the whole staff or with some subgroups (doctors-doctors, doctors-secretaries, etc.). Managing does not mean dominating. It is rather knowing how to talk to teams and how to get them to work towards a common goal [16]. Even if some members sometimes complain about the high number of meetings, they seem essential for the good management of the institution. Those meetings allow the team to have some feedback on the work and outline all technical or relational issues. It helps the manager to resolve the disagreements before they worsen and help the team to work with fluency [16].

The operational management of the MCHM is based on the involvement of the two doctors, whom will here be called P. and T. The influence they have is based on their legitimacy within the team, gained through their seniority, their involvement in the project and their ability to organize the run of the institution. The team trusts in them. Trust is built over time and in the relationships. It is a capital that the two doctors accumulated through years [15]. They are well known by all the team members, sometimes for more than twenty years. They are also known for their emotional competencies [16], that combines feeling with objective cooperation skills. The long-term trust of the team gives P. et T. the ability to engage and influence each member of the group, which helps the team to solve complex issues [17] and aim for an outstanding performance of their work. P. and T. are complementary, both in terms of relational aspects and in the conduct of data establishment and management. However, for a good cohabitation, the roles must be clear and non-antagonistic [1]. The risk associated is the disappearance of authority representatives. The smooth running of the institution must be based on the clear identification of authority figures, to which the staff can refer. The figure of authority also allows to the control of practices, beyond the "self-control" by the operator himself, in which skills and responsibilities are assumed by himself/herself [1], but not objectified by an external point of view. Each member is a part of the system and they have to work toward the same goal for the institution to reach its goals and insure the quality of care [16].

However, this verification dimension, in order to ensure the quality of the tasks performed, is quite crucial when it comes to such sensitive data as those referred to in this article: respect for medical confidentiality, quality of transmission of information, management of the risks of records leakage, etc. Taking leadership within the MCHM then seems to represent an additional element in the performance of strategies to protect patient medical records. The leaders act here as an element of internal data protection control, which compliance must be assessed and objectified externally by a notational Data Protection Officer [1]. This perspective can also be considered by taking as a model the Zacklad cooperative transaction logic reading grid, framework for analysing action and practice at the *meso* level [13].

VI. General Lack of Use of the National Digital Health Record

The MCHM tends to present organizational processes in accordance with public health requirements, while following effective procedures for the management of patients' personal records. It should, however, be noted that the national digital health folder (*Dossier Medical Partagé - DMP*) is not integrated into any of the care management approaches within this institution. The study case tends to reveal an attitude of rejection of this instrument by MCHM staff members. They themselves state that this folder "is not designed for medical practice", although they admit the promising nature of such a tool.

"In case of emergency, the DMP becomes counterproductive. Of course, all the information about the patient is included, but it is not sorted or classified. It is up to us to find the right information and, in emergency situations, we have something more important to do than sorting information" says E., one of the MCHM doctors. They also confess that they are disturbed by the additional and time-consuming actions required to update the patients' DMP, since this platform does not provide any automatic downloading add-ons for the software they use. It seems that the DMP system is in contradiction with the practices of MCHM professionals, with regard to their quality management processes. This notion is closely linked to the need for procedural rationality of the care action, namely an "orientation of the activity". In this orientation, the action is justified by taking into account the way in which the tools contribute to performance. It is partly defined by the quality of the realization process [13]. Doctors highlight a logic "inherent in our relationship with objects and our environment, which we judge according to their adaptation to our expectations and needs" [1]. Applying this to Doucet reasoning, it would seem that doctors judge the use of the DMP in terms of its field operability and its ability to meet the needs imposed by their profession.

However, according to their statements, data's quality processing on the DMP is incompatible with their needs. Charlotte Maday, in her article [14], uses the image of deep-sea fishing: throwing a net on the ocean floor, collecting information indiscriminately and presenting it to users. This image seems to be applicable to doctors' feeling towards the DMP. By presenting the "raw" data, the system is not in line with their requirements for efficient data management. It does not fit their ethical approaches or the "informational lean", the main data management strategies used within the MCHM. More broadly, DMP raises the need for co-production concerning innovation and the necessary collaboration between producers and users to guarantee the quality of a product or a service. Bringing together the documents and data, for process governance, requires mastering the notion of a system but, above all, acting in a spirit of active collaboration [14]. The main issues related to the DMP concern its digital features, ethical uses of data and the ability of professionals from various trades to collaborate on the same project. It represents a question to investigate, specially concerning the conception and the interoperability of various medical software. The goal is to ensure the performance of software and improve the quality of care.

VII. Conclusion

The sharing procedures introduced at the MCHM raise questions about risk assessment. In this institution, different informational and managerial strategies were put in place to secure the exchange of personal data. However, during the investigation, it seemed that the process of quality of care evaluation and "information crisis" management protocols (accidental or fraudulent disclosure, for example) were relatively minor. Since its launch, the institution does not seem to have been confronted with a crisis of this type. Its youth and its efficient management can explain this situation. Emergency protocols to resolve this type of situation are non-existent in the MCHM. It is problematical, as this implementation is one the most important principles of evaluating the quality of care in France, since the publication of the law of 31 July 1991 [10].

In the field of health, apparently minor errors or failures can have vital consequences [1] or endanger a health care institution and its staff. The management of the quality approach of such an institution cannot be done without a risk management component, nor a more global and formalized evaluation aspect. The measurement principles are inseparable from the principle of quality management [10]. It, therefore, seems necessary to study this institution in greater depth in order to explore more obscure aspects of its management and to consider a global quality approach, in order to find criteria encouraging its external recognition, particularly in the management of personal data.

It would also be interesting to study the management of the MCHM through the notion of emotional intelligence. Quality and data management rely to some rational perspectives. However, the others aspects of the institution depend on human and emotional aspects. The price for the lack of emotional intelligence can end up compromising the existence of the institution [16]. The study highlights some minor problems, especially concerning a doctor, who sometimes awkwardly criticizes his colleagues (doctors and secretaries). Until now, it has been resolved with discretion and fluency. However, this kind of criticism precedes the loss of trust in a team work [16]. The MCHM does not seem to be prepared for this kind of situation. Investigating emotional intelligence as a key resource could insure the stability and the fluency of the MCHM team.

It would also be interesting to investigate, on a larger scale, other MCHs, in order to compare the results of this study with other territorial and technical contexts. The objective is also to consider the development of a single working model for all MCHs, specially concerning data privacy management. As each MCH has its own specifications (socioeconomical context, number of the staff members, equipment, competences, etc.) would it be possible and relevant to propose a single model to all of them?

References

[1] C. Doucet, Quality, University Press of France (Presses Universitaires de France), 2013.

[2] The French Senate. Access to care: health innovation promoted in French territories, 2017. [Online]: http://www.senat.fr/rap/r16-686/r16-68613.htm [Retrieved: January, 2020].

[3] A. Bartoli and J. Sebai, "Multidisciplinary Care Houses facing territorial coordination", Gestion 2000, no. 6, p. 17-37, 2015.

[4] New Aquitaine Regional Health Authority, Population and its characteristics in New Aquitaine, 2018. [Online]: http://www.nouvelleaquitaine.ars.sante.fr/la-population-et-ses-caracteristiques-en-nouvelle-aquitaine [Retrieved: January, 2020].

[5] INSEE, Complete File – Landes (40), 2010. [Online]: https://www.insee.fr/fr/statistiques/2011101?geo=DEP-40#chiffre-cle-1 [Accessed: 23-October-2019].

[6] New Aquitaine Regional Health Authority, Key Metrics, 2017. [Online]: https://www.calameo.com/read/003713143e21e889f9535?page=3 [Accessed: 23-October-2019].

[7] French Data Protection Act. *Loi informatique et liberté,* Law n° 78-17 of January the 6th 1978 (Last modification: July 27th 2019). [Online]: https://www.cnil.fr/fr/la-loi-informatique-et-libertes#article2 [Accessed: 23-October-2019].

[8] L. Miguel-Chamion, "Protection of shared health data, International Journal of Bioethics, vol. 25, no. 3, p. 63-80, 2014.

[9] J. Béranger, "Data ethical value of health big data", Les Carnets du numérique, vol. 12, no. 1, p. 109-32, 2016.

[10] C. Vilcot and H. Leclet, Quality indicators in health: Certification and evaluation of professional practices, Second release, French Association of normalization, 2006.

[11] M. Autès and M-G. Dufay, "Health Care Houses and territories layout: a dynamic to share", Santé Publique, Vol 21, special edition no. 1, p. 17-25, 2009.

[12] L. Monnoyer-Smith, For a complex epistemology of information and communication sciences, presented at the 16th SFSIC Congress, Paris. 2008 [Online]: https://www.sfsic.org/congres_2008/spip.php?article78, [Accessed: 23-October-2019].

[13] M. Zacklad, "Management as a scoping activity in a context of cooperative transactions", Activité, vol. 10, no. 1, p. 192-200, 2013.

[14] G. Le Carnal, "Federating and stimulating energies, building trust to face crisis situations", Inflexions, vol. 2, no°3, p. 153-168, 2006.

[15] G. Le Cardinal, "Trust as a fundamental element of society", Revue Projet, Vol 2, no. 293, p. 65-71, 2006.

[16] D. Goleman, Emotional intelligence, Bantam Books, 1996

[17] D. Goleman, Working with emotional intelligence, Bloomsbury Publishing PLC, 1999.

[18] C. Mayday, "Quality of data and Open data", I2D Information, données documents, vol. 53, no. 4, p. 58-58, 2016.

[19] Software Weda. [On line]: https://www.weda.fr [Accessed: 26-March--2020]

Socio-technical Requirements for Expert Users to Design Structured User-Interfaces for OpenEHR-based Electronic Health Records

Gro-Hilde Severinsen
Norwegian centre for E-health research,
University hospital North Norway,
Tromsø, Norway
Email: Gro-Hilde.Severinsen@ehealthresearch.no

Line Silsand
Norwegian centre for E-health research
University hospital North Norway
Tromsø, Norway
Email: Line.Silsand@ehealthresearch.no

Rune Pedersen
Norwegian centre for E-health research,
University hospital North Norway
UiT Arctic university Tromsø
Tromsø, Norway
Email: Rune.Pedersen@ehealthresearch.no

Gunnar Ellingsen
Norwegian centre for E-health research,
University hospital North Norway
UiT Arctic University Tromsø
NORD University, Bodø, Norway
Tromsø, Norway
Email: Gunnar.Ellingsen@uit.no

Abstract—**The paper addresses socio-technical challenges when expert users design structured user interfaces in openEHR-based Electronic Health Records (EHRs). Structuring healthcare data enables extracting and reusing clinical information for both primary and secondary purposes, contributing to the goal of creating a digital society across multiple social arenas. We use an action research approach to follow an empirical project, developing an openEHR-based registry form, and trying out the initial methodology for designing structured user interfaces in a health organization. The aim of the paper is to describe and discuss socio-technical challenges related to expert users designing openEHR-based forms. The findings relate to 1) the complex design tools 2) the compromise between instant benefit and long-term requirements 3) the need for extensive governance for user-designed software, and 4) the importance of contextualization.**

***Keywords**- structured EHR; openEHR; arcetypes; structured form design; action research.*

I. INTRODUCTION

There are extensive ambitions of reusing data from EHRs, both for clinical use and for secondary purposes, like registries, research, and management. However, one of the main problems when it comes to exchange and reuse of health data is that most of the content in EHRs is recorded as free-text information.

A promising strategy of exploiting health data across different contexts is to structure the clinical information recorded in EHRs through openEHR-based clinical variables (archetypes) [1][2]. This way, the clinical information is recorded as standardized data, in relation to the context it is registered. Hence, such standards can be extracted and reused for both primary and secondary purposes contributing to the goal of creating a digital society across multiple social arenas. The Norwegian vendor DIPS AS is currently developing an openEHR-based EHR system called "DIPS Arena" [3], to comply with the expectations of exchange and reuse of health data in the Norwegian healthcare sector. A core principle of the openEHR specification is that clinical users are in charge of structuring the content of the EHR, i.e., the archetype standards. Furthermore, expert users should be in charge of designing user interfaces, e.g. (forms, templates for notes and documents) where the archetypes are embedded [1]. To be defined as an expert user, a combination of clinical background, technical understanding and knowledge about openEHR is required.

However, previous attempts of healthcare organizations structuring openEHR-based archetypes and forms revealed that this work was more demanding than expected [4]. The main challenges were related to: how to structure clinical variables in terms of high quality archetypes as basis for the structured form, how to make these forms more user friendly than the original paper based ones, and the importance of including a number of clinicians from specific medical specialties' in the standardization work [4].

There exist development methodologies for creating terminologies and ontologies [5]-[7]. In addition, an archetype modelling methodology has recently been released [8]. Some papers describe experiences with using structured data, standardized through the openEHR approach [2][9]-[10]. Still, a comprehensive and formal methodology for modelling structured openEHR-based user interfaces does not exist. It is of great importance to establish a methodology for developing and designing structured user interfaces, since the openEHR ambition gives the expert users a specific role as designers in the process.

However, a methodology is in progress, based on an empirical project in the North Norwegian Health Region. This project is part of the regional program The Future Systems in the Clinic (in Norwegian: Fremtidens Systemer i Klinikken, FRESK) [11]. The program is responsible for implementing a portfolio of new clinical Information and

Communications Technology (ICT) systems in the health region, in which a new openEHR-based EHR system, DIPS Arena, is one of the core systems.

In this paper, we have focused on the development of a structured openEHR-based form, conducted by expert users. The form is supposed to collect and transfer clinical patient data from the EHR to the Norwegian Registry for Spine Surgery (NORspine). During the process of developing the openEHR-based form and the initial framing of the methodology, socio-technical challenges emerged related to the expert users' role in the design process.

This study has followed the empirical project through an action research approach [12][13]. To analyze and discuss the empirical findings in order to answer the research question, we lean on the design principles of Information Infrastructure (II) and Infrastructuring theory [14]-[18].

The aim of the paper is to describe and discuss the socio-technical challenges related to expert users designing openEHR-based forms, in terms of designing clinical variables as archetypes, programming the queries and designing the lay out of the form. Hence, we ask the following research question: Which socio-technical challenges are addressed when expert users design openEHR-based forms?

The rest of the paper is structured as follows: Section 2 presents the action research approach, and data collection. Section 3 describes the design tools, the design process of the registry form, and summarizes the methodology for designing openEHR-based user interfaces. In section 4 we address key socio-technical challenges when dealing with infrastructural complexity in design, this includes working with advanced openEHR design tools, the role of the expert users, and the technical complexity of designing structured forms. Section 5 presents a brief concluding summary of the paper.

II. BACKGROUND

The empirical project started in January 2019, when NORspine approached the FRESK program for collaboration. NORspine set a proposal of transforming their paper-based registry form into an openEHR-based data entry form, and implement it into DIPS Arena. NORspine aims at improving the quality of surgical treatment for degenerative disorders in the cervical and lumbar spine [22]. The purpose of implementing the registry form into the EHR was to raise the registration rate from today's 64% to the national goal of above 80%, and which will raise the quality of the registry's recommendation for the spine surgery service [22].

Today, clinical information within the EHR is mainly recorded as free-text descriptions. Hence, the clinicians have to make double registrations on most of the data reported to registries [23]. This generates a risk that parts of the patient information end up in the registry only, since there is no connection between the registry form and the EHR system. Today surgeons fill out a paper-based form, and then a secretary or a nurse upload the recorded data into an electronic form in the registry's web-portal. Double registration of patient data is a time consuming process taking up an extensive amount of the healthcare personnel's time. Therefore, it is anticipated that automatically extracting and exporting clinical data recorded as part of the EHR documentation process to the registry form, will be a key means to improve the coverage rate, increase the quality of the registry, and reduce the time spent on documentation [22].

The FRESK program accepted the proposal from the spinal surgeons, and the structured EHR team of expert users got the assignment. The request from NORspine implied automatically extracting and exporting clinical data recorded as part of the EHR documentation process to the registry. Accordingly, the clinical information had to be structured as archetypes in accordance to DIPS Arena as an openEHR–based system. The development of the openEHR-based form was done in close collaboration with the vendor DIPS AS, and the National Administration of Archetypes (In Norwegian: Nasjonalt Redaksjons Utvalg for Arketyper NRUA). NRUA is responsible for the library of nationally approved archetypes, in terms of developing new archetypes on request from clinical communities, translating archetypes from international archetype libraries, and running public review rounds with numerous clinicians involved to approve new or translated archetypes [4].

An openEHR-based EHR system is not an 'off-the-shelf' product. In accordance with the openEHR specification, the healthcare organizations and clinicians are expected to take part in the customization process of designing archetypes, templates and forms for their organizations, as well as tailoring clinical process- and decisions support functionalities to their needs [1][8]. The North Norwegian Health Region is committed to the openEHR framework and do their fair share of customization of the EHR system. Several stakeholders are involved: expert users, clinicians, system vendors, ICT consultants, and archetype experts. The expert users have both clinical and technical competence in addition to extensive knowledge on openEHR, which is necessary for using the design tools and understand the capabilities, benefits, and consequences when changes are made in the archetypes, templates and forms.

III. METHOD

We have used a qualitative action research approach to follow the emerging process of developing a methodology for expert users in healthcare organization designing structured user interfaces [12][13][19][20]. Action research requires close collaboration between researchers and clinicians, and it is an iterative research process within a given context [12].

Creswell [21] has defined three elements of action research design. First, the science-theoretical perspective, in where we used II as a theoretical lens, to discuss, understand

and give recommendations to the empirical process. Second, the research strategy, in where we describe action research as the main method of studying practice and organizational development, contributing to a 'co-constructive' learning process for health personnel, developers, and researchers [21]. In this research project, the first and second author participated in the practical work of developing the openEHR-based form, and the outlining of the methodology. In this regard, they had an 'insider' role as expert users and experienced researchers. Working in close collaboration with the empirical program, the preliminary findings were discussed and presented to the project managers, vendors and users involved. In addition, preliminary findings have been used as recommendations for the ongoing process [14]. Third, for data collection and analysis, we used II theory to identify the socio-technical challenges. The last two authors have many years of experience from research on ICT in healthcare in general, and on openEHR research in particular. In this study, these authors had an 'outsider' role, in terms of balancing the insider perspectives and contributing in framing the research.

TABLE I. OVERVIEW OF THE DATA COLLECTION

Participatory observations in the design process	
Participated in:	*Meetings/workshops with:*
• Mapping variables to archetypes • Designing archetypes • Designing templates (OET/OPT) and forms	• The vendor • Clinicians • Project management • Members of Open Q-reg registry • NRUA
In total 320 hours	**In total 50 hours**

We present an overview of the data collection from January - August 2019 in Table I.

IV. CASE – THE DESIGN PROCESS

The findings presented in this section is detected by following the evolving methodology for developing openEHR-based user interfaces.

A. The Design Tools

Two expert users (the first two authors) in FRESK transformed the existing paper-based spinal surgery registry form into an archetype-based registry form for DIPS Arena. To design the archetype-based form, it was necessary to use four different (ICT) programs: Electronic mind map (XMind), Ocean Informatics Archetype Editor, Ocean Informatics Template Designer [24] and a specific Form Designer provided by DIPS AS. The mind map was used to get an overview of the variables in the paper-based form. The Archetype Editor and the Template Designer were used to design archetypes and align and constrain them into a template for this specific use case. Then, the Form Designer was used for further configuration of the template into the registry form as representing the user interface for clinicians. It took long time, extensive training and in-depth knowledge about the openEHR design principles to learn how to use the design tools and how the different design steps related to each other. These steps are described in detail in the following sections.

B. Mapping Variables and Archetypes

The first step was to insert the variables from the existing paper-based registry form into the electronic mind map (XMind). This provided an overview of all the variables in use in relation to e.g. if there was an overlap of variables, and the coherence between them (see Figure 1).

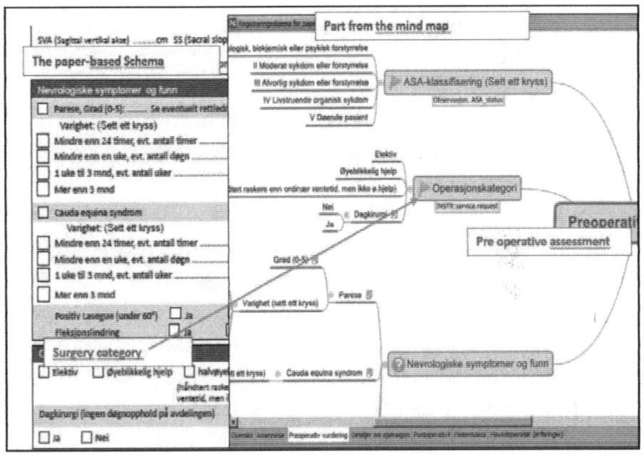

Figure 1. Mapping the variables in XMind

Given the request for automatically extracting and exporting clinical data recorded as part of the EHR documentation process, it was necessary to categorize the variables in relation to where they first emerged in the clinical documentation process (pre, per, or postoperative), and to assess the potential for reuse. In the present documentation process, information was stored in different documents, e.g., outpatient clinic notes, evaluation notes, surgery notes, and discharge notes. In addition, the registry form was not part of the documentation process.

The expert users collaborated closely with clinicians to understand their current use of the clinical documents. For example, when and where did they record the clinical information, what was the coherence between different variables in the registry form, and the relation between clinical needs of specific and unambiguous information.
Then, the expert users made suggestions for which archetype to replace with each of the variable, and discussed the proposal with NRUA and DIPS AS in several meetings.

One important finding from this process was that the variables used in the registry form were aggregated compared to clinical information used for documenting treatment and care. For example, the variable 'other endocrine diseases' used in the register form, covers all kind of endocrine diseases, except from diabetes. However, in clinical practice, 'other endocrine diseases' is too generic to use when documenting assessments, treatments and care.

The different granulation levels will complicate the potential for automatically reusing information about endocrine diseases into the registry form. Hence, differences in the granulation level of registry variables and clinical information was important to consider during the replacement. One suggested way of complying with this problem was to link together the diagnose codes for all endocrine diseases recorded in the EHR and map the linked codes to the specific category in the registry form.

C. Use National Archetypes or Develop Local Ones

After mapping the variables to archetypes, the next step was to decide whether to use archetypes from the national openEHR-library, or design local archetypes for this specific use case. It is preferable to use national approved archetypes to ensure semantic interoperability of clinical information. However, the first version of DIPS Arena did not offer functionality for advanced reuse of clinical information between different documents in the clinical process. In addition, the healthcare organization has not yet started to structure clinical documents as archetype-based notes, e.g., the admission notes, the physicians' daily notes and discharge letters. Accordingly, at this point of the design process, it was not possible to reuse clinical information from different documents to the registry form.

Another factor for using local archetypes was that national approved archetypes are complex standards designed as maximum dataset. This demand for extensive adjustments of the archetypes when using them in the templates - or forms, to comply with specific contextual requirements, e.g., the registry form. For example, the openEHR 'problem diagnosis archetype' is used for recording details about an identified health problem or diagnosis. The registry form asks for the name of the diagnosis only, comparing with the problem/diagnosis archetype that has 13 data elements, e.g., anatomical location, date/time for debut and clinical description to make a comprehensive description of the medical problem and diagnosis. Accordingly, it would be time consuming for the clinicians to fill in information requested in all the 13 data elements when only one data element is required in the form.

In relation to the current limited reuse of archetypes from the clinical process, and the complexity of constraining national archetypes to the registry form, the decision of designing local archetypes to the first version of the registry form was made. This decision was made in agreement with advisors from the vendor and NRUA. In addition, using local archetypes was expected to speed up the design process since they are less complex than the national ones. After a short development process, the first version of the form was presented to the clinicians. They provided feedback about the first version of the electronic form to the expert users, and approved the layout.

A disadvantage of using local archetypes was that some of the archetypes, e.g., (radiological examination) contained several clinical elements with the potential of being reused as autonomous archetypes. However, since the elements were part of a larger archetype, it was possible to reuse the whole archetype only.

D. Coherence between the Design Tools

To design local archetypes, the Ocean Informatics' Archetype Editor was used. The Archetype Editor has a given set of data types used to design clinical variables as data elements in an archetype (Figure 2). There is a coherence between the data type chosen to represent the variables/data elements in an archetype, the possibilities for configuration of the variables/data elements in the DIPS' Form Designer and finally the options of how to display the variables in the form. For example, some data types could generate variables as radio buttons in the Form Designer, while other data types could not.

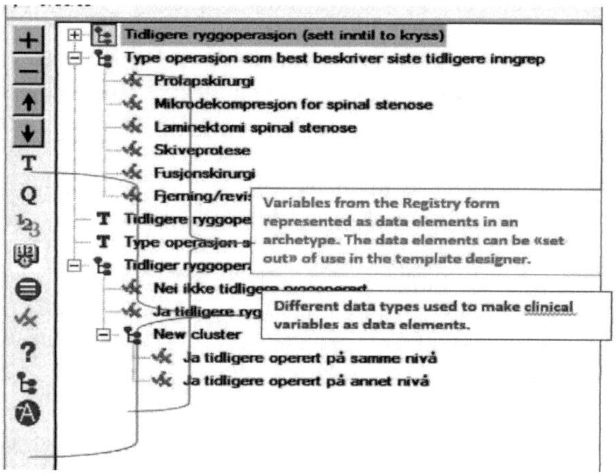

Figure 2. The different data elements for designing variables in an archetype using the Archetype designer.

In addition, if the goal was to create a drop-down menu where users could select one variable from the list only, then the data type 'Text' had to be used. Moreover, the drop-down menu for clinical variables had to be added as a list of 'internal codes' when designing the archetype, or as a 'value set' in the archetype when using the Template Designer. If it was uncertain how the variable would be used in the form, it was possible to design different data elements e.g., both 'Boolean' and 'Text' elements representing the same variable.

E. Assembling the Archetypes to a Template

When every variable in the registry form were represented as archetypes, they were assembled in a template by using Ocean Informatics' Template Designer. In the Template Designer, it was possible to constrain the archetypes in terms of making data elements/variables inaccessible. For instance, the national archetype for American Society of Anesthesiologists risk (ASA) score, containing 11 stages, was used in the form. However, in the registry form, the variable ASA score was defined by five different ASA stages only. Therefore, six stages had to be defined as inaccessible in the template. If archetypes were constrained in the template (Template Designer), and it became necessary to make some of the inaccessible data elements available later, it required creating a new template. After conducting all necessary configurations in the Template Designer, the template was exported to the Form Designer.

The final steps of configuring the form was done in the DIPS' Form Designer, this included adding dependencies between variables to define the relations amongst them, adding calculations, etc. In example, a field in the form addressed medication, and asked if the patient is using anticoagulation regularly. If the answer was 'no', then the option to answer 'yes' disappeared. If this was wrong, you just clicked 'no', and the option 'yes' appeared again. If you answered 'yes', you got more options, e.g., a drop-down list of different kind of anticoagulation medications (made as 'internal codes' in the archetype) to specify the answer. Another example was, if you filled in weight and height, then BMI was automatically calculated. During the process of adding dependencies, it became important to create a system for storing and managing the dependencies, annotations and calculations, to be able to update or check them if something did not work as anticipated, or if changes were made in the template/form. Making change to an archetype required the need for making changes in the calculations, annotations and dependencies where the archetype was used.

When configuring the form in the Form Designer, it was very important to work in close collaboration with the spinal surgeons. The archetype-based form was designed similar to the web-based form, in terms of dependencies amongst variables and options in drop-down menus etc. However, based on feedback from the surgeons, it was necessary to change some of the dependencies and displays in the form. For example, in the paper-based form, the surgeons explained that the list of options related to 'other diseases' was used as a reminder to help them remember to ask for all the different diseases when filling in the form. Therefore, they wanted to see all the options in the electronic form. Dependencies hiding the list of options made it more cumbersome for the surgeons to fill in the form, as 'the reminders' then were hidden.

The vendor had made a technical user manual for the Form Designer. However, the design process addressed a need for expanding the user manual with instructions aimed for the expert user role.

F. The Methodology for Structuring an EHR Form

Table II summarizes the evolving methodology for designing openEHR-based user interfaces.

TABLE II. METHODOLOGY FOR STRUCTURING AN EHR FORM

Tasks	Collaborating partners	Responsible part
Archetype work		
Mapping archetypes and existing variables	Expert users, NRUA, clinicians,	Expert users
Decide to using local or national archetypes	Expert users, NRUA, Vendor	Expert users
Designing local archetypes	Expert users, NRUA, Vendor, clinicians,	Expert users
Template Design		
Design a template from the archetypes	Expert users, Vendor, NRUA	Expert users
Constrain archetypes if needed	Clinicians	Expert users
Forms design		
Upload template to the Form Designer	Expert users	Vendor/ICT department
Define dependencies, annotations, calculations etc. between archetypes	Expert users, clinicians, Vendor and/or ICT consultants in health organizations	Vendor/ICT department

V. DISCUSSION

In the design process, we described the evolving empirical process of designing an openEHR-based EHR form. As mentioned earlier, structuring the spinal surgery form was very useful for uncovering challenges associated with structuring openEHR-based user interfaces in a 'real-life' setting. The design process addressed several socio-technical challenges. We will now discuss four of them; 1) the complex design tools, 2) the tension between instant benefit and long-time requirements, 3) the need for extensive governance for user designed software and 4) the importance of contextualization.

A. The Complex Design Tools

The four design tools available for the expert users are quite complex to use. Everyone can download the mind map tool and the tools from Ocean Informatics. However, behind the scenes, using the tools requires extensive training of the expert users. In an infrastructuring perspective, it is not enough that the expert users have knowledge about using the design tools only. They also need extensive knowledge on the different possibilities and limitations in both designing archetypes and templates, and how designing them affects the design possibilities in the user interfaces. Each step conducted in each of the different tools are interdependent to make the overall infrastructure evolve [19].

In addition, even if the design of archetypes, template and form is of high quality, the final step of the design process needs to end with a user-friendly display of the form. Accordingly, it is necessary to include clinical expertise and technical competence as well. This demand for expert users taking the role as translators between ICT and clinical practice to establish an II were knowledge from these two installed bases are merged.

B. *Tension between Instant Benefit and Long-term Requirements*

There is an inherent tension between 'quick-and-dirty' design, and designing for long-term use. The case describes why local archetypes were used extensively, despite the recommendation of using national approved archetypes. There is a constant negotiation between the importance of an efficient design process, were the clinicians can see early results of a structured form, and the need for robust archetypes as basis for future use. Design for instant usefulness is in accordance with the II design principle of bootstrapping, where users get access to working software as early as possible in the design process, to motivate them to continue in contributing to the work [17]. However, the consequence of 'quick-and-dirty' design can be that long-time usefulness is compromised by for instance extensively constraining the archetypes in the template. Another use-case may demand for briefly constraining an archetype thus, designing a new template is necessary. This demands for extensive work with the form to redesign annotations, calculations and dependencies related to the archetype. Hence, extensively constraining of archetypes in a template decreases the possibility for long-time usefulness.

However, designing for long time use to ensure future needs related to the design principle of adaptability [17]and can generate a risk of designing local archetypes too flexible like in the empirical case where different data elements represented the same clinical variable. To this end, if these archetypes are reused in several EHR systems and one organization decides to use the data element 'Text' and another organization uses use 'Boolean' for the same variable, the consequence is that the clinical variable is not interpreted as the same archetype. Accordingly, the reuse potential is severely compromised.

C. *The Requirement for Extensive Governance for User Designed Software*

User-designed software like variables, forms, and dependencies requires extensive governance. The complex infrastructuring process [18][19] of form design expresses the need for interaction between different actors with dissimilar competences [4]. Structuring user interfaces is a complex infrastructuring process in relation to, e.g., making changes to archetypes, and how that effects the templates and forms were they are used. The infrastructuring process also demands for setting up a shared repository for storing and retrieving both local and national archetypes to be used for structuring the EHR. In addition, governing a structured EHR addresses the need for training the actors involved, to get an overall understanding and knowledge about archetypes, templates and forms, in terms of where archetypes are used and reused, as well as what consequences changes or constraints generates.

Scaling the structured form to a structured EHR will increase the need for control over archetypes, templates and forms implemented in the EHR, and the accompanying dependencies, annotations and calculations. This requires an extensive ICT governance organization unit within the healthcare organization. Establishing an extensive governance organization relates to the design principle of building on the installed base in where existing systems, actors and governance need to be part of the process [17].

D. *Contextualization is a Complex, but Necessary Dealing.*

Nationally approved archetypes are developed through a thorough design process managed by high-qualified expert users in NRUA. These archetypes are developed as maximum datasets to be useful in any clinical context and for different clinical specialties. National approved archetypes constitute the basis for semantic interoperability in openEHR-based systems. This adheres to the II design principle of standardization, and the importance of communicating through defined clinical standards [17]. However, as maximum datasets, these archetypes need to be contextualized, i.e., constrained and tailored to specific use contexts. In line with the openEHR specification, the configurations must be done by local health care organizations. As the case describes, configuration of archetypes, templates and forms is complicated and requires a combination of different skills, such as clinical insight, archetype competence and expertise in ICT/programming. A related contextualization is also found between local archetypes in the EHR form and NORspine. Here some apparently similar data, for instance 'other endocrine diseases' have different granularity. This illustrates that data traveling across settings needs to be contextualized to make the infrastructure grow [17].

VI. CONCLUSION

To summarize, this paper has described and discussed socio-technical requirements, detected when following the design of a registry form and the emerging methodology for designing structured user interfaces by expert users.

From this study, we found:

1) The expert user role demands a thorough understanding of the complex design tools. It also requires extensive training over a longer period of time, in order to create high quality archetypes, templates and forms.
2) It is important to find a balance between an efficient design process and the need for robust archetypes as a basis for the infrastructuring process.
3) Structured user interfaces demand for establishing extensive governance at different organizational levels.
4) It is essential to contextualize national high quality complex archetypes to different local contexts, in order to make them useful for clinical practice. This must be done in collaboration with clinicians and local governance organizations.

The infrastructuring process of designing regional openEHR-based user interfaces is a complex process, which demands for several socio-technical requirements to succeed. Hence, it is important to establish a 'scaffolding' organization, in terms of governance and extensive training programs to make it possible for expert users to be in charge of the infrastructuring processes of designing user interfaces. This is an ongoing process in the health region and, accordingly, it is necessary to do more research in this field.

REFERENCES

[1] T. Beale and S. Heard, "OpenEHR architecture: Architecture overview," The openEHR foundation, 2008.

[2] R. Chen, G. O. Klein, E. Sundvall, D. Karlsson, and H. Ahlfeldt, "Archetype-Based conversion of EHR content models: Pilot experience with a regional EHR system." BMC Medical Informatics and Decision Making, vol. 9, no. 1, pp. 1-33, 2009.

[3] DIPS. *Our solutions*. Available from: https://www.dips.com/uk/solutions

[4] G-H. Ulriksen, R. Pedersen, and G. Ellingsen, "Infrastructuring in healthcare through the openEHR architecture," Computer Supported Cooperative Work (CSCW), vol. 26, no. 1-2, pp. 33-69, 2017.

[5] A. De Nicola, M. Missikoff, and R. Navigli, "A software engineering approach to ontology building." Information systems, vol. 34, no. 2, pp. 258-275, 2009.

[6] Y. Sure, S. Staab, and R. Studer, On-to-knowledge methodology (OTKM), in Handbook on ontologies. Springer. pp. 117-132, 2004.

[7] M. Fernández-López, A. Gómez-Pérez, and N. Juristo, Methontology: from ontological art towards ontological engineering, 1997.

[8] D. Moner, J. A. Maldonaldo, and M. Robles, "Archetype modelling methodology." Journal of biomedical informatics, vol. 79, pp.71-81, 2018.

[9] J. Buck, S. Garde, C. D. Kohl, and P. Knaup-Gregori, "Towards a comprehensive electronic patient record to support an innovative individual care concept for premature infants using the openEHR approach." International journal of medical informatics, vol. 78, no. 8, pp. 521-531, 2009.

[10] C. Rinner et al., Creating ISO/EN 13606 archetypes based on clinical information needs, in EFMI-STC, 2011.

[11] FRESK-HelseNord. *FRESK*. Available from: https://helse-nord.no/fresk

[12] R. Baskerville and M.D. Myers, "Special issue on action research in information systems: Making IS research relevant to practice." MIS Quarterly, vol. 28, no. 3, pp. 329–335, 2004.

[13] M. S. Granlien and M. Hertzum, "Implementing New ways of working: Interventions and their effect on the use of an electronic medication record", in International Conference on Supporting Group Work. Florida: ACM -Digital library, pp. 321-330, 2009.

[14] O. Hanseth, E. Monteiro, and M. Hatling, "Developing information infrastructure: The tension between standardization and flexibility." Science, Technology, & Human Values, vol. 21, no. 4, pp. 407-426, 1996.

[15] S. L. Star and K. Ruhleder, "Steps toward an ecology of infrastructure: Design and access for large information spaces." Information systems research, vol. 7, no.1, pp. 111-134, 1996.

[16] O. Hanseth and N. Lundberg, "Designing work oriented infrastructures". Computer Supported Cooperative Work, vol. 10, no. 3-4, pp. 347-372, 2001.

[17] C. Bossen and R. Markussen, "Infrastructuring and ordering devices in health care: Medication plans and practices on a hospital ward." Computer Supported Cooperative Work (CSCW), vol. 19, no. 6, pp. 615–637, 2010.

[18] H. Karasti and K. Baker, "Infrastructuring for the long-term: Ecological information management." Proceedings of the 37th Annual Hawaii International Conference on System Sciences pp. 1-10, 2004.

[19] E. Monteiro, N. Pollock, O. Hanseth, and R. William, "From artefacts to infrastructures." Computer supported cooperative work (CSCW), vol. 22, no. 4-6, pp. 575-607, 2013.

[20] C. Bossen, "Accounting and co-constructing: the development of a standard for electronic health records." Computer Supported Cooperative Work (CSCW), vol. 20, no. 6, pp. 473, 2011.

[21] J. W. A. Creswell, Framework for design - in research design, qualitative, quantitative and mixed methods approaches. Sage: Thousand Oaks, CA, 2003.

[22] T. K. Solberg and L. R. Olsen, Norwegian registry for spine surgery (NORspine), Annual report 7 with improvement plan, SKDE. 2018.

[23] L. Silsand, G-H. Severinsen, G. Ellingsen, and B. Christensen, "Structuring electronic patient record data, a smart way to extract registry information?" in Infrahealth 2019-Proceedings of the 7th International Workshop on Infrastructure in Healthcare. Reports of the European Society for Socially Embedded Technologies: vol. 3, no. 4, 2019.

[24] S. Bakke, *Verktøystøtte*. Available from: https://kilden.sykehusene.no/pages/viewpage.action?pageId=36406914

Barriers and Enablers to Implementation of mHealth Programmes

Robyn Whittaker, Rosie Dobson, Gayl Humphrey
National Institute for Health Innovation
University of Auckland
Auckland, New Zealand
Email:{r.whittaker, r.dobson, g.humphrey}@auckland.ac.nz

Leah Friedman
NHS Lothian
NHS Scotland
United Kingdom
Email: Leah.friedman@nhs.net

Abstract—**Despite developing several successful mHealth interventions, researchers have struggled with implementation in practice. Key stakeholders in New Zealand were interviewed for their perspectives on barriers and enablers. Their feedback was mapped to the Consolidated Framework for Implementation Research (CFIR), and, from there, to the Expert Recommendations for Implementing Change (ERIC) framework. In this way, twenty recommended implementation strategies were identified. Some of these may be beyond the ability of researchers to influence, however, many can be employed during the development and research phases and may increase the likelihood of translation into real world implementation.**

Keywords-mHealth; implementation.

I. INTRODUCTION

mHealth describes health interventions and health information that is provided to people using their mobile phones. Although the mHealth field is progressing at a rapid pace, there continues to be a lack of significant large-scale implementations.

There have been various studies regarding the success or failure of implementing mHealth innovations in practice. Often, barriers to implementation concern the structural and cultural aspects of the system rather than the intervention or technology itself [1]. In fact, a WHO study from 2011 found that competing priorities in an overworked health system were the main barrier to implementing mHealth, followed by a lack of knowledge about its applications, a need for policy that recognises mHealth as a legitimate approach for addressing health, and cost-effectiveness [2]. Other studies of barriers have included usability of the intervention, integration of the tool into existing systems, data security and privacy, resistance to change, and a lack of planning, funding, capacity, training, and support [3]-[6].

We set out to examine the perspectives of key stakeholders in New Zealand to the enablers and barriers impacting mHealth implementation. Our group (the National Institute for Health Innovation (NIHI) at the University of Auckland) has developed many mHealth programmes. Some were successful in large randomised controlled trials [7]-[11], or pilots [12][13], and some were not proven effective in the research phase [14][15]. One was developed as an ongoing service rather than out of a research project [16]. To date, only one programme has successfully moved from proven research evidence to implementation [7][8], being run for many years as a national smoking cessation service and stimulating many other such cessation services internationally [17]. Considering other successful programmes languishing between publication and implementation, we wished to learn if there was more that we could do to prepare our planned future research developments for eventual implementation. Therefore, the purpose of this study was to obtain input from decision makers in the health sector to inform our current and future mHealth research programmes for greater translation into practice.

This paper describes our original research with Section II outlining the methods used, Section III describing the key findings from interviews and the results of our mapping to constructs and implementation strategies, and in Section IV we discuss how we could use these strategies and our next steps.

II. METHODS

Key senior stakeholders in New Zealand were interviewed as part of several ongoing mobile health development and implementation projects being conducted by NIHI at the University of Auckland. These ten stakeholders covered the spectrum from the Ministry of Health to local funding organisations, primary health care organisations, a University, and an academic/industry/health service research partnership. Their roles included funding, contracting, health service improvement, clinical leadership, research and innovation. Stakeholders were identified by the lead mHealth researcher (RW) as key people familiar with the implementation of previous, existing or planned NIHI mHealth programmes. These programmes focus on various health topics such as smoking cessation, diabetes self-management support, cardiac rehabilitation and pulmonary rehabilitation. Potential interviewees were asked to participate by the lead mHealth researcher. All those approached consented and were interviewed.

Interviewees were asked about their role in mHealth implementation and both barriers and enablers that they have faced in the past or would anticipate for future projects. Semi-structured interview guidelines were developed by the entire team. All interviews were conducted by an independent student (LF) during an international internship at NIHI. The interviewer also conducted all analyses independently of the researchers at NIHI. Interview notes were summarised and main ideas were identified and mapped.

As this study was focused on enablers and barriers, the Consolidated Framework for Implementation Research (CFIR), developed by Damschroder et al. [18], was chosen as a means to frame interview responses and recommendations. This model provides a comprehensive structure for identifying what has/has not worked in the past and what might work/not work in future mHealth projects. It brings together the various existing implementation theories and key constructs. The goal of the CFIR is to look at the context of an intervention and assess possible barriers and enablers to its implementation. CFIR consists of five domains, and a total of 39 specific constructs within these:

- *The intervention domain* comprises the flexibility, complexity, adaptability, and other characteristics of the intervention itself.
- *The inner setting* involves organisational elements such as culture, structure, leadership, and readiness for change.
- *The outer setting* involves the economic, social, and political context of the organisation. The outer setting also includes patient needs.
- *The individual level* looks at choice and behaviour of those involved in the implementation and is driven by their personalities, mind-sets, and so on.
- *The process level* examines the actual change process surrounding the intervention—this includes engagement, planning, executing, and reflecting/evaluating.

The Expert Recommendations for Implementing Change (ERIC) framework developed by Powell et al. [19] proposes 73 different implementation strategies that can be used in isolation or conjunction. These have been categorised under nine domain headings by Waltz et al. [20]:

- Use evaluative and iterative strategies
- Provide interactive assistance
- Adapt and tailor to context
- Develop stakeholder interrelationships
- Train and educate stakeholders
- Support clinicians
- Engage consumers
- Utilise financial strategies
- Change infrastructure

These strategies were then mapped to the CFIR (framed as barriers to implementation) based on respondents choosing the most appropriate ones for each construct [18]. A matching tool is available at [21].

III. RESULTS

All of the interviewees were enthusiastic about the potential for mHealth to positively impact health outcomes. However, interviewees indicated that despite evidence of positive health outcomes, there is a lack of funds and other resources in the system to implement and maintain the use of these tools. There was a general tone of frustration around inertia and a feeling that tools that are shown to be effective in supporting and improving patient wellbeing cannot get implemented.

Overall the tone was pessimistic, with a general feeling that systemic changes were needed to successfully implement mHealth and that these changes were a long way off. While technology advances, the challenges of its implementation remain constant.

The most popular responses were:

- Tools/interventions are often viewed as additive, rather than substitutive, and therefore compete with other demands and priorities
- There is difficulty working across health service organisations (eg. primary and secondary care) to implement interventions - in terms of competing priorities, disconnect in data and information, many different stakeholders, and varying structures
- It is incredibly important to manage relationships between clinical and management staff and ensure clinical engagement with the intervention
- The innovation and its impact should be aligned with wider organisational (and national) strategies, goals, and priorities however, this can be difficult as priorities change
- It is difficult to integrate a new technology into current processes where old legacy systems exist and there is a lack of interoperability across units, departments, hospitals, and organisations
- There is a general culture of risk aversion that is resistant to change
- Using intensive on-site training, ongoing support, and multi-disciplinary teams to plan and implement the tool is helpful
- Having consumer champions in addition to clinical champions is important
- Funding needs to be committed and secured early so that work can continue after a successful pilot finishes – agree outcomes to be demonstrated to release funds

Key points brought up by interviewees were categorised as either enablers or barriers and mapped to the specific CFIR constructs (Table I). This shows that planning, engaging, networks and communications, and external policies and incentives were the most commonly cited CFIR constructs.

TABLE I. KEY BARRIERS AND ENABLERS

CFIR Domain	Main Ideas from Interviews (Barrier (B) or Enabler (E))	No. (n=9)	Specific CFIR Construct
Intervention Characteristics	Easily integrated into existing systems and work processes (E)	2	Adaptability
	Generic interventions more likely than disease-specific to get funding (B/E)	1	Relative Advantage
	Convenient and functional for clinicians (E)	1	Complexity
	Robust process for approving apps, based on clinical and privacy issues (E)	1	Evidence Strength and Quality
	Design with end-user in mind (E)	1	Evidence Strength and Quality
	Private PHOs are able to get things done if commercial value can be demonstrated (E)	1	Relative Advantage
	Strong evidence demonstrated over reasonable length of time (E)	2	Evidence Strength and Quality
Individual Level	Tools/interventions often viewed as additive rather than substitutive. Competing demands (B)	3	Knowledge and beliefs about the intervention
	Culture of fear/risk-aversion (B)	3	Other Personal Attributes
	GPs operate in commercial environment and may not value public health projects (B)	2	Knowledge and beliefs about the intervention
	Find early adopters for the intervention (E)	2	Individual Stage of Change
Inner Setting	Alignment with organisational strategy/goals/priorities (E)	3	Compatibility
	Securing executive leadership and multiple sign-offs (B)	1	Leadership engagement
	Difficulty working across DHBs and PHOs (B)	6	Networks and Communication
	Disconnect of data and information sharing across organisations and primary/acute care (B)	1	Networks and Communication
	Culture of fear/risk-aversion (B)	3	Culture, Implementation Climate
	Old legacy systems, lack of interoperability (B)	3	Compatibility
	Lack of time and resources dedicated to operationalising tools (B)	1	Available Resources
	Broad promotion and board engagement (E)	1	Networks and Communication Leadership Engagement
	Incentivise use of tool for patients and staff (E)	1	Organisational Incentives and Rewards
	Managing clinical relationships and clinical engagement (B)	4	Networks and Communication
	Board priorities can change quickly (B)	2	Relative Priority
	No place in Allied Health/nursing budget for technology (B)	1	Available Resources
Outer Setting	Different patient engagement than with traditional care system (B)	1	Patient needs and resources
	Politics and relationships get involved when choosing projects to fund (B)	1	Networks and Communication
	No framework to help prioritisation process (B)	1	External Policies and incentives
	No framework for measuring and evaluating innovations (like what exists for medicines) (B)	1	External Policies and incentives
	Issues with patient data—security/privacy (B)	1	External Policies and incentives
	Patients with multiple comorbidities may need a suite of tools (B)	1	Patient needs and resources
	Competition exists amongst big DHBs (B)	1	Peer Pressure
	National priorities can change quickly (B)	1	External Policies and incentives
	Poor health literacy and non-compliance of patients (B)	1	Patient needs and resources
	Fit mHealth into accreditation, ongoing education, training, medical council guidance, etc. (E)	2	External Policies and incentives
Process	Poor management of control and adoption phases, translating to implementation (B)	2	Executing
	Use of MDTs (E)	3	Engaging
	Both clinical and consumer champions (E)	3	Champions
	Design for implementation from the start (E)	1	Planning
	Difficult to scale projects from local to national level (B)	3	Executing
	No framework to help prioritisation process (B)	1	Planning
	No framework for measuring and evaluating innovations (like what exists for medicines) (B)	1	Reflecting and Evaluating
	Find early adopters for the intervention (E)	2	Opinion Leaders, Champions
	Use MBIE sourcing rules early in process to create plan post-pilot (E)	1	Planning
	Secure funding for continuation of intervention after pilot finishes (E)	3	Planning
	Change the timing of funding—agree outcomes before that must be demonstrated to release funds; payments contingent on milestone reporting (E)	3	Planning
	Using expanded health teams—not just GPs—to deliver intervention (E)	2	Engaging
	Intensive on-site training and support available (E)	1	Executing
	Secure early buy in, socialise people to the idea early on (E)	2	Engaging
	Need to see pathway to commercialisation from beginning (E)	1	Planning
	Using expanded health teams—not just GPs—to deliver intervention (E)	2	Engaging
	Need a group to enable the bureaucratic process (E)	2	Formally appointed implementation leaders

Using the mapping tool provided by the CFIR website [21], the most relevant ERIC strategies for these CFIR constructs were:

1. Identify and prepare champions
2. Assess for readiness and identify barriers and facilitators
3. Conduct local consensus discussions—to discuss whether the chosen problem is important and the tool is appropriate
4. Inform local opinion leaders—about the innovation, so that they can influence others
5. Build a coalition—recruit and cultivate relationships with partners in effort to implement
6. Capture and share local knowledge—from implementation sites on how others made it work
7. Conduct educational meetings—targeted at different stakeholder groups to teach about the innovation
8. Alter incentive/allowance structures—to Incentivise adoption and implementation

9. Conduct local needs assessment—regarding the need for the innovation
10. Create a learning collaborative—groups of providers to learn and improve implementation
11. Facilitation
12. Identify early adopters
13. Promote adaptability—tailor to meet local needs
14. Develop a formal implementation blueprint—to include all goals and strategies, scope of change, timeframe, milestones, and progress measures
15. Tailor strategies—in order to address barriers and leverage facilitators
16. Organise clinician implementation team meetings—protected time to reflect, learn, and support each other during implementation
17. Involve executive boards
18. Recruit, designate, and train for leadership—for the change effort
19. Use advisory boards and workgroups
20. Conduct cyclical small tests of change

IV. CONCLUSION

Our key stakeholders believe in the potential for mhealth to have positive impacts on helping our patients, however, they feel these programmes are unlikely to be implemented due to lack of funds and other resources to implement and maintain the use of these tools. Our findings are not dissimilar to those previously identified - lack of policy or national standards around the provision of mHealth, a need for compatibility with current work systems and processes, and insufficient resources and funding [22].

We mapped identified enablers and barriers to implementation framework constructs and, from there, to recommended implementation strategies. These fall into three groups. First are those strategies that are outside our control, such as altering incentive structures to promote adoption and implementation, and organising clinician implementation team meetings. Developing a formal implementation blueprint is something that we could perhaps conduct with a willing implementing organization, but in our experience is unlikely to be supported until there is actually approved funding for an implementation.

Second is the group of strategies that NIHI already uses in the development and research phases of our mHealth programmes [23]. From the start we try to involve clinical champions, advisory boards and workgroups, and build a coalition of relevant local organisations. In this way we build on local knowledge and networks and tailor for local needs. We work with the target audience in focus groups, surveys and other formative research methods to determine their needs, whether an mHealth initiative could be helpful, and their preferred tools. Recruitment methods are tailored according to how we think the programme will be implemented locally.

The third group of identified strategies included areas that NIHI could focus more on in our current and future mHealth developments. Our existing methods tend to focus on particular levels in the system – that is, the consumers/end users, clinicians, champions and local services – but not at the executive board or funder level. We could spend more time engaging at this higher level, assessing readiness and identifying barriers and facilitators upfront. This could have a greater focus on the longer term implementation strategy to fit with national and regional priorities and programmes, and therefore increase the likelihood of committing funding.

This may also align with the He Pikinga Waiora Implementation Framework developed in New Zealand that has indigenous self-determination at its core [24]. Under four over-arching elements (cultural centredness, community engagement, systems thinking and integrated knowledge translation) seven components can be scored as high/medium/low/negative in order to assess the likely effectiveness of proposed interventions. One of these components includes the degree to which different levels of change (macro, meso and micro) are taken into account, with rationale and context for each level.

The next step will be to deliberately consider the identified strategies and how they apply at each of these levels, in the early stages of our mHealth programme development in New Zealand, and to evaluate whether these strategies have an impact on implementation.

ACKNOWLEDGMENTS

We would like to thank all the interviewees and the NHS Scotland for supporting LF.

REFERENCES

[1] E. H. Shortliffe, "Strategic action in health information technology: Why the obvious has taken so long," Health Affairs, vol. 24(5), pp. 1222-1233, 2005.

[2] WHO Global Observatory for eHealth, mHealth: new horizons for health through mobile technologies: second global survey on eHealth. Geneva: World Health Organization. http://www.who.int/iris/handle/10665/44607, 2011.

[3] V. P. Gurupur and T. Wan, "Challenges in implementing mHealth interventions: a technical perspective," mHealth, vol. 3, pp. 32, 2017.

[4] V. M. Kiberu, M. Mars, and R. E. Scott, R, "Barriers and opportunities to implementation of sustainable e-Health programmes in Uganda: A literature review," African journal of primary health care & family medicine, vol. 9(1), pp. e1-e10, 2017.

[5] D. J. Meyers et al., "Management challenges in mHealth: failures of a mobile community health worker surveillance programme in rural Nepal," BMJ Innovations, vol. 3, pp. 19-25, 2017.

[6] L. Wallis et al., "A roadmap for the implementation of mHealth innovations for image-based diagnostic support in clinical and public-health settings: a focus on front-line health workers and health-system organizations," Global Health Action, vol. 10 (sup3), 1340254, 2017.

[7] A. Rodgers et al., "Do u smoke after txt? Results of a randomised trial of smoking cessation using mobile phone

text messaging," Tobacco Control, vol. 14(4), pp. 255-61, Aug. 1, 2005.

[8] D. Bramley et al., "Smoking cessation using mobile phone text messaging is as effective in Maori as non-Maori," New Zealand Medical Journal, vol. 118, pp. 1216, 2005.

[9] C. Free et al., A randomised controlled trial of mobile (cell) phone text messaging smoking cessation support: txt2stop. Lancet 2011; 378(9785), pp. 49-55.

[10] R. Maddison et al., A Mobile Phone Intervention Increases Physical Activity in People with Cardiovascular Disease: Results from the HEART Randomised Controlled Trial. European Journal of Preventive Cardiology, 2014.

[11] R. Dobson, R. Whittaker, L. Pfaeffli Dale, and R. Maddison, The effectiveness of text message-based self-management interventions for poorly-controlled diabetes: A systematic review. Digital Health 2017, 3, pp. 1-12.

[12] L. Pfaeffli-Dale et al., Text message and Internet support for coronary heart disease self-management: results from the Text4Heart randomised controlled trial. JMIR 2015; 17 (10).

[13] C. Ni Mhurchu et al., Feasibility, acceptability and potential efficacy of a mobile health (mHealth) weight management programme for ethnically diverse adults. BMC Obesity 2014, 1:10.

[14] R. Whittaker et al., STUB IT: randomized controlled trial of a theory-based video messaging mobile phone smoking cessation intervention. J Med Internet Res 2011: 13(1); e10.

[15] R. Whittaker et al., MEMO: an mHealth intervention to prevent the onset of depression in adolescents. A double blind randomised placebo-controlled trial. J Child Psychol & Psych June 2, 2017.

[16] R. Dobson et al., Development of a culturally tailored text message maternal health program (TextMATCH). JMIR mHealth & uHealth 2017;5(4):e49.

[17] WHO Be Healthy Be Mobile global initiative https://www.who.int/ncds/prevention/be-healthy-be-mobile/en/

[18] L. J. Damschroder et al., Fostering implementation of health services research findings into practice: a consolidated framework for advancing implementation science. Implement Sci. 4(50), 2009.

[19] B. J. Powell et al., A refined compilation of implementation strategies: results from the Expert Recommendations for Implementing Change (ERIC) project. Implementation Science. 10(1), 2015.

[20] T. J. Waltz et al., Use of concept mapping to characterize relationships among implementation strategies and assess their feasibility and importance: results from the Expert Recommendations for Implementing Change (ERIC) study. Implementation Science. 10(1), 2015.

[21] CFIR Research Team. The Consolidated Framework for Implementation Research. Retrieved from https://cfirguide.org/, 2018.

[22] J. Ross, F. Stevenson, R. Lau, and E. Murray, Factors that influence the implementation of e-health: a systematic review of systematic reviews (an update). Implementation Science.11 (1), 2016, pp. 146- 153.

[23] R. Whittaker, S. Merry, E. Dorey, and R. Maddison, A development and evaluation process for mHealth interventions: examples from New Zealand. Journal of Health Communication 2012, 17, sup.1, pp. 11–21.

[24] J. Oetzel et al., He Pikinga Waiora Implementation Framework: A tool for chronic disease intervention effectiveness in Māori and other indigenous communities. International Journal of Integrated Care. 2018 Mar 12;18(s1).

Communication between Mentor and Mentee Using Videoconferencing in Surgical Training

Line Lundvoll Warth
Norwegian Centre for E-health Research
Tromsø, Norway
University of Tromsø, The Arctic University of Norway
Tromsø, Norway
email: line.lundvoll.warth@ehealthresearch.no

Abstract—**In surgical training, mentors and mentees communicate to expand mentees' technical skills. However, access to mentors for education in surgical subspecialties is a challenge in many hospitals. Videoconferencing (VC), which enables real-time communication between mentors and mentees in different geographical locations, can overcome this challenge. This study examines a practice in Norway in which VC is used to provide education on a specific laparoscopic surgical procedure. Specifically, the study explores the characteristics of communication between a mentor and mentee using VC and how it affected communication. The empirical material consists of video recordings of an educational trajectory that included eight patient cases and focus group meetings. The communication reveals knowledge gaps and their closure through establishment of a shared understanding. In this way, VC supported the learning of technical skills while enabling feedback on non-technical elements. Both the mentor and mentee could reach their full potentials—expanding their own communicative skills and reflecting on their own abilities. VC also affected the relationship between the mentor and mentee, who were peers and colleagues rather than participants in the traditional mentee–mentor relationship. The results of this study will enable the development of an activity for non-technical skills to become relevant using VC.**

Keywords—non-technical skills; surgical training; communication; videoconferencing; qualitative study.

I. INTRODUCTION

Although technical skills in surgery are obviously important, communication in the operating room (OR) plays an important role in patient safety, as operations are social situations in which tasks are accomplished through communication between team members. Communication errors within surgical teams have been studied in terms of communication failures [1], and studies have attempted to explain how surgical procedures are influenced by the quality and efficiency of teamwork. The results have shown that deficiencies in teamwork in the OR contribute significantly to adverse events and patient harm [2], as there is a strong relationship between teamwork failure and technical errors [3]. In other words, a good surgeon is more than just a good "pair of hands" [4]; he or she must be a good team player, must listen and communicate with colleagues, and must empower colleagues to reach their full potential [4]. These qualities are related to collective and cognitive competence, defined as non-technical skills.

Non-technical skills are gaining importance in surgery and surgical training [4]. The Royal College of Surgeons of Edinburgh defines non-technical skills as skills and behaviors related to situational awareness, decision-making, communication, teamwork, and leadership [5]. Others have defined non-technical skills as interpersonal (e.g., communication, teamwork), cognitive (e.g., decision-making, situational awareness), and personal resource skills (e.g., coping with stress and fatigue) [6]. Communication and teamwork related to decision-making are also important non-technical skills. All of these skills are essential for surgeons to operate safely in the OR, and, although they are developed in an informal and tacit manner [5], they need to be explicitly addressed in training.

Surgical training involves individual work and guidance from an expert mentor. Mentees gain significant skills and experience by participating in simulated environments with virtual simulators and models prior to performing procedures on patients in the OR. Work in the OR involves collaboration; each team member has his or her own tasks to perform. Although each team member's individual technical skills are important, good collaboration is necessary for a good surgical outcome [7][8]. Hence, both mentors and mentees need to develop non-technical skills in surgical training. A recent review of surgical telementoring reported limited understanding of VC in surgical practice; the review concluded that little attention has been paid to the educational and non-technical elements but has instead been focused on piloting the technology [9][10].

Surgical training is an educational process in which the mentor's and mentee's competence and work serve as parts of a collective activity and communicative process. Both communication and teamwork are important for modern surgical education and practice; indeed, a review of the role of non-technical skills in surgery stated that the key root cause of surgical errors worldwide is a lack of non-technical skills [3]. The review also provided evidence that non-technical skills have an effect on technical performance and suggested that training that is focused on improving non-technical skills can improve teamwork, performance, and safety in the OR and thereby positively contribute to patient outcomes [3]. This indicates that there is a need to focus on the development of non-technical skills in surgical training.

In surgical training, access to local mentors for surgical subspecialties is a challenge in many hospitals. Videoconferencing (VC) is a technology that enables real-time communication between mentors and mentees, even if they are in different geographical locations. It can thus help to overcome the issue of lack of access to local experts. This study focuses on a practice in Norway during which VC was used as a tool for surgical education concerning a specific hernia procedure.

Research on VC has stressed its educational benefits [9] and described VC for mentoring as an effective way to develop surgical skills [11]. However, within this field, special focus on communication and team performance is needed to better understand the factors that influence surgical outcomes [12]. Specifically, a systematic literature review of communication in the OR concluded that further detailed observational research that provides detailed transcripts and analyses of communication patterns is needed to gain a better understanding of non-technical skills [13].

This study explores communication and teamwork between a mentor and mentee using VC as well as the knowledge needed to complete a surgery. Even though it is important to gather information about the outcomes of work in the OR, it is also necessary to gain a detailed understanding of the processes and communication patterns that lead to those outcomes, and these are often overlooked in favor of technical skills. This study therefore aims to provide insight into how mentors and mentees organize and accomplish collaborative work through VC in the OR.

This study expands upon previous work by investigating knowledge-sharing between a mentor and mentee—specifically, the manner in which individual knowledge is shared and constructed—to ensure that mentees apply best practices. The use of VC and communication between a mentor and mentee were followed in real-time surgical training regarding a laparoscopic hernia procedure. The characteristics of communication between the mentor and mentee through VC and how VC affects communication were explored.

The rest of this paper is organized as follows: Section II describes the framework and scenario of the study, Section III describes the methods used, Section IV presents the results, and Section V presents the discussion. The article ends with Conclusions and Acknowledgments.

II. Scenario and Framework

Laparoscopy is a visual technique that uses several small ports in the abdomen with an instrument inserted through each. The procedure is visual, as a small camera is inserted into the patient`s abdomen. The images obtained from the camera are transmitted to a monitor in the OR and enable communication with participants outside the OR. In the case examined in this article, the mentee in the surgical team used VC to communicate with a geographically distant mentor. The mentee was experienced with laparoscopy; before practicing this procedure on patients, he underwent the traditional education pathway for a new procedure (i.e., simulations using models and videos of the procedure).

Communication in the OR was framed with an activity theoretical perspective [14], focusing on the complex interactions between individual subjects and their wider context [15]. The mentor and mentee were part of a collaborative educational and communicative process mediated by a cultural tool (i.e., VC). This approach enabled expansion of the unit of analysis of education and learning beyond the individual [16]. The collaborative activity happened between the activity system of the mentor and mentee, enabling the use of VC in practice.

III. Methods

This is an ethnographic study [17] that explores the use of VC for communication between a mentor and a mentee within an educational process. The study was carried out from 2014–2016 in Norway and involved observations, interviews, focus groups, and field notes. Five semi-structured interviews, which lasted a total of six hours, took place from 2015–2016, and all were transcribed and analyzed. For three months in 2014 and 2015, surgical training of a mentee on the specific hernia procedure was observed and videotaped. The dataset covers the entire educational trajectory, which included eight cases and six hours of video observations. The whole dataset was transcribed. All participants participated in two focus group meetings to discuss the procedure. These meetings were also videotaped and transcribed. The data protection officer at the selected hospital approved the study, and the study participants provided informed consent.

The analysis focused on the interactions between the mentor and the mentee, particularly when tensions appeared [17] and knowledge gaps needed to be closed, which directed the opportunities for expanding verbal decision-making and non-technical skills [18]. The observations allowed communication and team performance (as opposed to individuals) to be studied. The eight sessions revealed communication patterns and non-technical skills (but not individual deficiencies) in a series of operations that utilized VC for educational purposes.

IV. Results

The surgical training examined in this study was organized into eight sessions. The first three sessions occurred onsite in the OR and involved preparation for the VC, and the next five sessions used VC. After the eighth session, the mentee was considered an expert in this procedure and the VC sessions ceased [19].

A. *Communication using VC*

The characteristics of communication using VC are illustrated in Figure 1. This extract includes data starting from the four-minute mark of the seventh session, which was videotaped for about 25 minutes.

The mentee referred to a history of communication by suggesting a course of action for the day based on previous sessions. Specifically, he suggested cauterization and pulling the sack into the abdominal cavity. He then asked the mentor what he thought about the suggestion (utterance 1). The mentor supported the proposal but had a hunch that simply pulling the sack was not adequate, based on his own practice with stitches (utterance 2).

Extract from the 7th session (A: mentee, B: mentor)

1	A:	I thought maybe today we could try just to cauterize it, if it's possible to, uh, pull the sack out into the abdominal cavity. Or, what do you think?
2	B:	Yeah. You can see. You can try. Um, it depends. You can try. I always start by turning and, and then if it seems like it's not adequate, then I put a stitch in.
3	A:	It's quite deep, you see....
4	B:	Yeah, it might be hard to do with just cauterizing.
5	A:	Yeah, I think so too. Because it goes into the....
6	B:	All the way down.
7	A:	Labia majora. Yeah. Okay, I think we will go for....
8	B:	Yeah.
9	A:	I don't think it's even necessary to try. Do you agree?
10	B:	No, you [...] but the good thing is, you could do a lot of cauterizing, you don't have to worry about, uh, injuring it.
11	A:	That's good. Okay.

Figure 1. Communication using VC.

The mentee referred to the hernia as deep and acknowledged the suggestion to use stitches (utterance 3). The mentor then confirmed that it might be hard to just cauterize (utterance 4). The mentee considered going deep with the instrument (utterance 5), and the mentor elaborated on the depth (utterance 6). After this reflection, the mentee decided to use stiches (utterance 7), a decision that was supported by the mentor (utterance 8). The mentee reconsidered his decision to try to pull the sack into the abdominal cavity and asked the mentor to support this decision (utterance 9). The mentor did support the decision and elaborated on the opportunity to perform cauterization without injuring the patient (utterance 10). The mentee confirmed the shared understanding (utterance 11).

The characteristics of the communication involved skills related to choosing an appropriate course of action and a shared understanding. First, the mentee presented a knowledge gap (i.e., whether to use stiches). This tension between the mentor's and mentee's knowledge provided an opportunity to close the knowledge gap, thereby expanding the collective activity of decision-making. The mentor supported the suggestion while mentioning the tension between the possible actions (i.e., pulling the sack or using stiches). Drawing upon the mentor's experience and knowledge, the mentor and mentee communicated, closing the knowledge gap by establishing a shared understanding. This shared understanding was based on a collective activity in which the participants were able to bridge the gap and perform a successful procedure.

B. Reviewing the procedure

After two training sessions, focus group meetings were held to review the sessions and allow the mentor and mentee to discuss the content and how VC affected communication. Figure 2 illustrates how this meeting progressed. The mentor asked the mentee about his experience in one of the sessions and how he could improve as a mentor (utterance 1). The mentee pointed out the tension between anticipated and "comfortable" knowledge, referring to the fact that the mentee had watched the training videos of the procedure (utterance 2). The mentor asked if the mentee felt that he provided too little instruction during the session (utterance 3). As the session went well, he was not sure whether there was a gap in knowledge between the mentor and mentee, but guidance would have made the mentee feel "safer" during decision-making (utterance 4). The mentor then reflected on the communication between the mentor and mentee, illustrating the tension between the traditional way of locally training mentees (in which the expert mentor holds a more powerful position) and the use of VC as a pre-planned tool for collaborative distributed work (in which the mentor and mentee act as colleagues; utterance 5).

Reviewing a session (A: mentee, B: mentor)

1	B:	What was not good? Don't be polite [...] What could I have done better as a mentor?
2	A:	We just assumed that I had seen the video that I knew [...] You just let me do it, and then you corrected me....
3	B:	I didn't give enough instructions [...]? You wish I had given more instructions?
4	A:	I don't know if it was necessary, but maybe it would [...] feel more safe in a way.
5	B:	This is a problem that [...] Not feeling comfortable as a mentor, knowing not to say too much. When I have a relationship with a resident, I say whatever I want. He is my resident. But, when it is a colleague, I am a little bit more shy about being too talkative. Does that make sense? So, it is that different relationships exist between an attending and a trainee, a resident, than between an attending and another surgeon. I don't want them to be annoyed too much....

Figure 2. Reviewing the procedure.

Overall, the extract shows the mentor's and mentee's reflections on their own communicative skills, including how the mentor related to those around him. By exchanging reflections after the surgical procedure, the mentor could better understand his performance as a mentor. This learning activity led to a shared understanding between the activity systems of the mentee and mentor.

V. Discussion

The purpose of this study was to explore the characteristics of communication between a mentor and mentee using VC and how VC affected communication. Observing communication using VC (Figure 1) enabled identification of successful communication and teamwork. This educational process was a collective activity mediated by VC as a cultural tool. Tensions in the work illustrated the limitations of the mentee's individual knowledge, opening up opportunities to bridge the knowledge gap between the expert mentor and mentee. Collective decision-making led to learning opportunities, which allowed the mentee to become an expert in the procedure. Hence, communication using VC supported the learning of technical skills.

VC also has the capacity to support collaborative (i.e., non-technical) skills. The communication examined here refers to prior sessions (a history) and the progress made in expanding the mentee's knowledge. The mentor reflected on his prior actions and modified his teaching according to the mentee's needs.

The emphasis on decision-making skills in the training allowed the mentee to develop skills related to assessing situations and agreeing on an appropriate course of action within the team. Even though there was a gap in the mentee's knowledge that the mentor had to bridge, the mentor and mentee discussed options in a balanced way, considering the consequences and benefits of each option, and maintained flexibility while making the shared decision. Afterwards, the mentor explained why he recommended a specific course of action.

The communication built upon traditional problem-solving in the OR. Laparoscopy is a visual procedure in which a small camera is inserted into the patient's abdomen and the image is transmitted to a monitor in the OR. In this case, VC was used to show the mentee the same images seen by the mentor. Unlike in traditional training, where both the mentor and mentee are in the OR, this training occurred using VC. This created tension between the traditional method of local training, in which the mentor and mentee are both at the patient's bedside and are aware of all activity in the OR, and remote guidance, in which the mentor has expert knowledge of the procedure but not complete knowledge of all the activity in the OR.

The problem-solving process and the technical skills are both based on the same information, using the monitor. Nevertheless, there is teamwork in the OR that cannot be experienced by the mentor using VC. Both mentor and mentee develop awareness of the situation, which includes all activities in the OR and the pre- and post-operative situations of the patient. Hence, as both have responsibilities, the mentee is more of a colleague than a resident.

When reviewing the procedure (Figure 2), the mentor and mentee discussed both the technical skills and dynamics of communication patterns (i.e., non-technical skills). This allowed the mentor to support the mentee while improving his own communicative skills through reflection. This activity also supported the mentee in reflecting on his own communicative skills.

VC was used because the mentee had experience, but not in this specific procedure; the competences of the mentor and mentee were therefore unequal. However, the VC tool seemed to reformulate this inequality between the mentor and mentee and redefine the traditional mentee/resident–mentor/expert relationship as one between peers and colleagues. Based on the division of labor, the mentee held the leadership in the OR, but the mentor was the expert on the procedure (defined as the picture on the monitor). This allowed non-technical skills, rather than just technical skills, to be developed, and enabled the participants to reflect on how teamwork could be improved.

Communication is shaped by organizational culture and historical activities, which play an important role in how work is performed. Communication problems are attributed to a lack of clarity regarding roles and power relationships [14]. Implementing VC for collaboration in surgical education challenges traditional surgical training and communication patterns between mentors and mentees. Specifically, the results of this study illustrate that VC promotes effective reasoning and good communication between mentors and mentees. Communication and teamwork related to decision-making are characterized by reflection on performed work, which leads to the development of non-technical skills and the ability to emphasize non-technical skills as important in surgical training.

VI. Conclusion

In the case studied, collaboration occurred between the activity systems of the mentor and mentee, and their communication and use of VC in practice were examined. It was found that VC allowed knowledge exchange during surgical training, resulting in the mentee becoming an expert on this procedure. The results provide insights into the way in which surgical training and practice are performed, the use of VC within surgical training in the OR, and how VC facilitates the development of communication skills.

Incorporating VC into surgical training within the current training paradigm would allow both technical and non-technical elements to be included in the feedback provided to mentees. The use of VC in surgical training could be a step toward raising awareness of non-technical skills, facilitating changes in the workplace, and emphasizing collaborative skills (i.e., communication and teamwork) among both mentors and mentees in the educational process (and, later, in daily work). In this way, VC could help produce a new generation of surgeons who are competent in all the skills required for knowledge expansion and safe, high-quality patient care.

Acknowledgments

Thanks to the Northern Regional Health Authority for funding this project (HST-1181-14) and to all the surgeons who participated in the study.

REFERENCES

[1] L. Lingard et al., "Communication failures in the operating room: an observational classification of recurrent types and effects," J. Qual. Saf. Health Care, vol. 13, pp. 330–334, 2004.

[2] N. Sevdalis et al., "Quantitative analysis of intraoperative communication in open and laparoscopic surgery," Surg. Endosc., vol. 26, pp. 2931–2938, Oct. 2012.

[3] L. Hull et al., "The impact of nontechnical skills on technical performance in surgery: a systematic review," J. Am. Coll. Surg., vol. 214, pp. 214–230, 2012, doi:10.1016/j.jamcollsurg. 2011.10.016.

[4] R. A. Agha, A. J. Fowler, and N. Sevdaliscet, "The role of non-technical skills in surgery" Ann. Med. Surg., vol. 4, pp. 422–427.

[5] The Royal College of Surgeons of Edinburgh. *NOTSS: Non-technical Skills for Surgeons*. [Online]. Available from: http://www.abdn.ac.uk/iprc/notss/. 2020.01.21.

[6] R. Flin, P. O'Connor, and M. Crichton, Safety at the Sharp End: A Guide to Nontechnical Skills, Aldershot: Ashgate, 2008.

[7] B. S. Nedrebo et al., "Survival effect of implementing national treatment strategies for curatively resected colonic and rectal cancer," Br. J. Surg., vol. 98, pp. 716–723, 2011.

[8] N. J. Birkmeyer et al., "Safety culture and complications after bariatric surgery," Ann. Surg., vol. 257, pp. 260–265, 2013.

[9] K. M. Augestad et al., Surgical telementoring in knowledge translation – clinical outcomes and educational benefits: A comprehensive review. Surg. Innov., vol. 20, no. 3, pp. 273–281, 2013.

[10] K. M. Augestad et al., Educational implications for surgical telementoring: a current review with recommendations for future practice, policy, and research. Surg. Endosc., vol. 31, pp. 3836–3846, 2017.

[11] P. Lamba, "Teleconferencing in medical education: a useful tool," Australas. Med. J., vol. 4, pp. 442–447, 2011.

[12] L. Panait et al., "Telementoring versus on-site mentoring in virtual reality-based surgical training," Surg. Endosc., vol. 20, pp. 113–118, 2006.

[13] C. Vincent, K. Moorthy, S. Sarker, A. Chang, and A. W. Darzi, "Systems approaches to surgical quality and safety. From concept to measurement," Ann. Surg., vol. 239, 475–482, 2004.

[14] Y. Engeström, "Expansive learning at work: towards an activity theory reconceptualization," J. Educ. Work, vol. 14, pp. 133–156, 2001.

[15] S. M. Weldon, T. Korkiakangas, J. Bezemer, and R. Kneebone, "Communication in the operating theatre," Br. J. Surg., vol. 100, pp. 1677–1688, Dec. 2013, doi:10.1002/bjs.9332.

[16] Y. Engeström and H. Kerosuo, "From workplace learning to inter-organizational learning and back: the contribution of activity theory. Guest editorial" J. Workplace Learn., vol. 19, pp. 336–342, 2007.

[17] D. Silverman, Interpreting Qualitative Data: Methods for Analysing Talk, Text and Interaction, London: Sage Publications, 2001.

[18] P. Linell, Approaching Dialogue: Talk, Interaction and Contexts in Dialogical Perspective, Amsterdam: John Benjamins, 1998.

[19] L. L. Warth, "Creating learning opportunities by using videoconferencing in surgical education," Stud. Health Technol. Inform., vol. 262, pp. 15–18, 2019, doi:10.3233/SHTI190005.

Can Increased Patient Involvement Reduce the Number of Surgery Cancellations?

Lessons learned from a research and development project in Norway

Kari Dyb, Conceição Granja
Norwegian Centre for E-health Research
Tromsø, Norway
e-mail: {kari.dyb, conceicao.granja}@ehealthresearch.no

Abstract—**Increased patient involvement is the new mantra in modern healthcare, indicating that patients can play an important role in improving the quality of their health and care services. In this paper, we study the reality of increased patient involvement within the current organizational structure of hospitals by revisiting the findings from a recent research and development project at a university hospital in Norway. Through the development and implementation of a tool for electronic two-way communication between patients and hospitals, the overarching goal of the eTeam-Surgery project (2013–2019) was to better prepare the hospital and the patient for surgical procedures and thereby reduce the number of surgery cancellations. To approach this goal, researchers from the field of health informatics, medicine and sociology explored a quality improvement initiative at the hospital, made field observations and interviewed patients and health professionals. eTeam-Surgery demonstrated that establishing a tool for active involvement of patients in the pre-operative planning process was not an easy task. The research revealed (1) diverse reasons for surgery cancellations, (2) a lack of uniformity in the pre-operative workflow and (3) a hospital practice of marking a surgery as cancelled when patients tried to rebook for a different date. Focusing on increased patient involvement, this paper demonstrates how the main reasons for cancellation at the hospital and the lack of uniformity of the pre-operative workflow both originate from hospital organizational issues and, consequently, will not be solved with increased patient involvement. On the other hand, increased patient involvement can improve the current hospital practice of marking surgeries as cancellations when patients are actually trying to rebook. It is our conclusion that hospitals have a long way to go before the new mantra of active patient involvement can be a reality. However, we still believe that active patient involvement can positively contribute to the problem of elective surgery cancellation. The first step would be for hospitals to offer patients the option of choosing their own date of surgery, similar to booking an airline ticket online.**

Keywords-elective surgery; e-health; cancellations; planning; increased patient involvement.

I. Introduction

Increased patient involvement is the new mantra in modern health and care services, indicating that patients can play an important role in improving the quality of care by becoming actively involved in their own health [1]. Empowering patients to take an active role in their own health has been nationally and internationally identified as a key initiative to improve health and care services [2][3]. A number of trends and policy shifts can be identified that are promoting greater patient involvement in health and care delivery through consultations, treatments and continuous care [4].

Health Information Technology (health IT) is frequently used to increase patient involvement. In Norway, the Norwegian Ministry of Health and Care Services and the Directorate of eHealth have defined providing patients' digital access to important health information as one of the three main priorities areas for IT development and implementation [5][6]. Health IT is seen as an important tool to increase patient involvement and improve the quality of health and care services.

In this paper, we will study the reality of increased patient involvement in hospitals by answering the following research question: "Can increased patient involvement and digitalization reduce the number of elective surgery cancellations at the University Hospital in North Norway (UNN, from the Norwegian: Universitetssykehuset Nord-Norge)?" We approach the research question by revisiting the findings from a recent research and development project, the eTeam-Surgery (2013–2019) [7].

A. Elective surgery cancellations

For many patients, undergoing surgery is a major life event involving a high level of anxiety before admission to the hospital [8]-[11]. Surgery cancellations can be troublesome for both patients and their families [12][13]. Patients may also suffer psychological stress and/or financial hardships due to surgery cancellations [14]. Thus, for patients, cancellations are stressful and costly, with a high level of emotional involvement before surgery [15]. However, the research literature illustrates that surgical patients who adopt a more active role have a higher general patient satisfaction compared with patients who did not participate in decision making [16], including being involved in choosing the date of surgery [17].

Furthermore, in most hospitals, surgical departments are both a major area of investment and the greatest source of revenue [15][18]. Nonetheless, hospitals, in particularly public hospitals, regularly cancel elective surgeries, evidencing cancellation rates of 10–40% [10][11][15][19]. However, it has been shown that 50% of elective surgery cancellations might be avoided [14][15][20], and that patients can contribute to the reduction [17]. Avoidable cancellations include, for instance, cancellations due to lacking information; predominantly these were situations in which information existed prior to surgery but was not available when required [14][21]-[24].

At our site of research, UNN, 10% of elective surgeries are cancelled on the day of surgery, adding up to a total of 1847 cancellations per year [25]. In 2008, UNN reported that more than 50% of their elective surgery cancellations were related to inadequate pre-operative planning and that a substantial number of these cancellations could have been avoided [25].

B. The eTeam-Surgery project

The eTeam-Surgery project (2013–2019) was a research and development project at UNN funded by the Northern Norway Regional Health Authorities (Helse Nord). The project departed from the internal report and the dominant narrative among local health professionals at that time, revealing that a considerable number of the elective surgery cancellations at the hospital could relate to lacking information during the pre-operative planning process.

The health professionals described the lacking information as information that could affect the surgery, for example if the patients had a cold or other virus infections, allergies or chronic conditions; if they were taking pharmaceuticals; if they were prone to anaesthetic reactions; if they had surgery-related anxiety or distress and other reasons. The common reasoning at the hospital was that the patient could provide much of this information, and that cancellations could be avoided by simply asking the patient these questions.

By developing and implementing a new tool for electronic two-way communication between patients and hospitals, the overarching goal of the eTeam-Surgery project was to provide the lacking information and thereby better prepare the hospital and the patients for surgical procedures, hence reducing the number of elective surgery cancellations at the hospital.

To support the development of a technical prototype of a tool for two-way communication between surgical patients and health professionals, the project established an interdisciplinary research group. The research group consisted of a professor of medical informatics and three researchers from the field of health informatics, medicine and sociology.

To determine how part of the pre-operative planning could be moved from the hospital to the patient at home, through electronic collaboration prior to hospitalization, they studied the reported reasons for cancellations and the pre-operative planning processes at the hospital from technical, medical and organizational perspectives.

The interdisciplinary research group soon learned that creating tools for active patient involvement within the current hospital organizational structure for pre-operative planning was not an easy task. Therefore, a revisit of the empirical complexity and heterogeneity found in the eTeam project from an increased patient involvement focus is appropriate.

In this paper, two of the involved researchers approach the potential of increased patient involvement within the current organizational structure of hospitals by revisiting the findings from the eTeam-Surgery project.

This paper is divided in four sections. In the first section, we present the objective of the study, a brief review of the state of the art on elective surgery cancellations and the eTeam-Surgery project. Data collection methodologies, with which the results were obtained, are presented and explained in the second section. The results are disclosed and interpreted in section three. In the last section, we discuss the results and conclusions about the results are drawn, some indicators of future work in the area are foreseen.

II. METHODS

In the methods section, both the methods used in the eTeam-Surgery project and the methods used for revisiting these findings are described, focusing on the reality of increased patient involvement within a hospital context.

A. Quality improvement initiative

The management at UNN, our site of research, was determined to reduce the cancellation rate at the hospital. Therefore, in 2012, they allocated resources to a quality improvement initiative—that is, a Lean project for elective surgical patient pathways at the Operation and Intensive Care Clinic. Lean projects are commonly used to implement improvements in patient care in healthcare organizations through the development of a quality driven culture [26]. The aim of the quality improvement initiative at UNN was to determine new and effective surgical pathways by identifying and reducing waste in current practices.

Prior to the start of the eTeam-Surgery project, two members of the eTeam-Surgery research team followed the quality improvement initiative from the initial group meeting in April 2012. One had an active role and contributed as an anaesthesiologist in the quality improvement initiative. The other participated solely as a researcher by conducting observations during project meetings (2014–2015). In total, the two researchers observed and participated in more than 20 meetings.

One task of the quality improvement initiative was to identify the reasons for elective surgery cancellation. From the reasons reported in the Electronic Health Record (EHR), they identified 17 different reasons for elective surgery cancellation at UNN, illustrated in Figure 1 [19].

The data, as presented by the quality improvement initiative, shown in Figure 1, demonstrates 17 different reasons for elective surgery cancellations at UNN, and that insufficient surgery indication was believed to be the main reason for elective surgery cancellation at the hospital.

In this paper, we performed a thematic analysis of the data on the reasons for elective surgery cancellations. Focusing on increased patient involvement, we reanalysed, compared and translated the 17 different reasons for elective surgery cancellation, aiming for a new understanding of the reported reasons for surgery cancellations at the hospital.

B. Fieldwork at the hospital

In addition to following the quality improvement initiative, the eTeam-Surgery research group completed three weeks (approximately 100 hours) of fieldwork at UNN, mainly at the Operation and Intensive Care Clinic. During the fieldwork, the researchers conducted observations and unstructured interviews while following an anaesthesiologist and an anaesthetist nurse in their daily work. To complement the observations, the researchers interviewed 13 physicians, nurses and administrative personnel from four different surgical departments. They conducted the interviews at the workplace of the informants, and the interviews lasted between 30 and 60 minutes each.

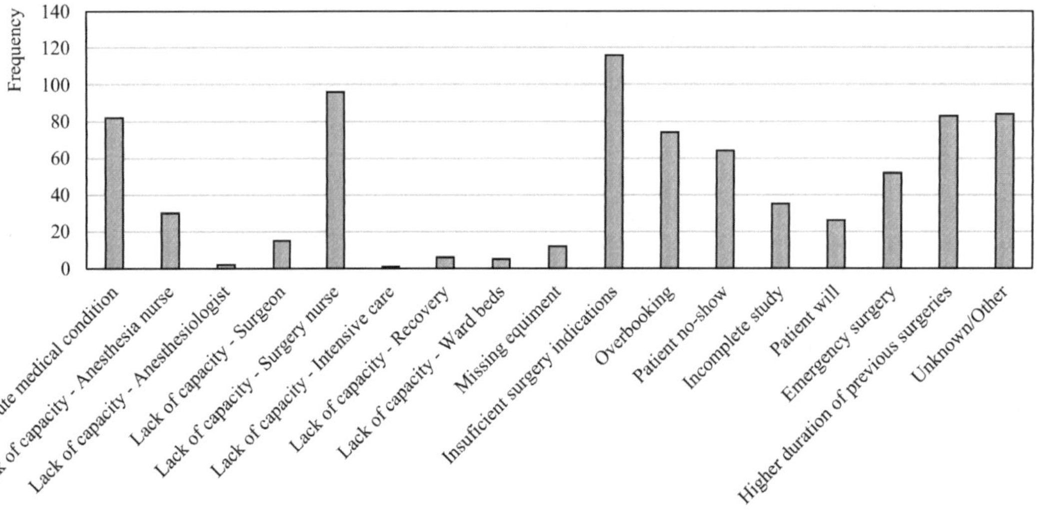

Figure 1. Reasons for day of surgery cancellations distribution at UNN from January to June 2011 [19].

The fieldwork demonstrated internal variation between the different departments in who plans the surgery and how and when the pre-operative planning takes place [27]. Additionally, it revealed how the different departments and, to some extent, even individual physicians, had developed their own local, pre-operative planning practice [28]. In some departments, senior surgeons did the pre-operative planning. In other departments, interdisciplinary teams involving senior and junior physicians, nurses and secretaries did the planning together. The data from the fieldwork demonstrated that in order to complete the daily schedule, health professionals used personal and empirical knowledge. These findings illustrate why the eTeam-Surgery research group could not describe a standard model of the pre-operative planning process needed for establishing two-way electronic communication between the hospital and the patient.

In this paper, we made a qualitative analysis of the challenges of the lack of standard pre-operative work practices, focusing on how increased patient involvement can influence the way that hospitals plan and organize elective surgeries. We reanalysed the heterogeneity of the workflow models, aiming for new knowledge on the patient role in the pre-operative planning process.

C. Patient interviews

The "Patient will" and "Patient no-show" categories (Figure 1) indicate that the responsibility for a significant amount of elective surgery cancellations at the hospital resided with the patients. In the Norwegian healthcare system, cancellations have consequences for patient priority in accessing surgery, hence, to cancel or not showing up for a surgery have direct consequences for the patient. Therefore, the research group approached the patients' perspectives on elective surgery cancellations.

To approach patients, a health professional at UNN assisted the researchers by identifying the patients who had cancelled an elective surgery within the last three weeks. To maintain patient anonymity, the health professional mailed letters of invitation to the identified patients, requesting their participation in the eTeam-Surgery project through qualitative telephone interviews.

Of the 48 letters of invitation, 11 patients stated that they were available for an interview. These 11 patients became informants in the eTeam-Surgery project. Prior to being enrolled as informants, they signed and returned the letter of consent and provided a telephone number for the researcher to contact them. One of the researchers called all the informants to make an appointment for the interview. The informants were adults ages 32 to 70 and included six men and five women. The phone interviews were structured and lasted 10 to 60 minutes each. At the beginning of each phone interview, the researcher introduced herself as a researcher in the field of e-health, with an interest in surgery cancellations. She emphasized that all information provided during the interview would be anonymized and that participation in the research project would not interfere, in any way, with current or future treatment at UNN or any other health or care institution.

After the introduction, the researcher communicated the data from UNN stating that the patient had cancelled an elective surgery at the hospital. She asked if this information was correct. If verified, the researcher asked why the patient had cancelled the surgery. If the patient contradicted the information, she asked the patient to tell his/her story about the situation.

The admittance letter is the first piece of information that patients receive from the hospital regarding their surgery, and it already has a fixed date for surgery. The interviews revealed that these patients had called the hospital to reschedule the date of surgery immediately after receiving the admittance letter. For the informants, the surgery date provided by the hospital was inconvenient and ill-timed for them or their families. The data from the patient interviews illustrates that, contrary to the hospital's assumption, these patients wanted their surgery. The findings demonstrate that surgeries are cancelled when patients try to rebook. This is a situation that the patients described as frustrating, outdated and not appropriate for the digital era. The eTeam-Surgery project revealed that the patient-reported reasons for surgery cancellations did not correspond with the hospital's representation of the same problem [29].

In this paper, we revisit the findings from the patient interviews, by conducting a thematic analysis focusing on the patients' representation of increased patient involvement in pre-operative planning at the hospital. We reanalysed the patients' reports to gather new, in-depth knowledge on the patients' expectations to be involved in the decision of the date of surgery.

III. Results

In this section, we present new knowledge on increased patient involvement and digitalization.

A. Quality improvement initiative

Focusing on the potential of increased patient involvement in the pre-operative planning process at UNN, the 17 different reasons for elective surgery cancellation were reanalysed, compared and translated. This analysis revealed four new general categories for the reasons for elective surgery cancellations:

1) Maladjusted resource planning, 67%
2) Patient will and no-show, 12%
3) Other, 11%
4) Medical, 10%

The first category, "Maladjusted resource planning", comprises 11 of the original categories and refers to cancellations due to a lack of available nurses or doctors, theatres, ward beds or equipment; overbooking; and that the provisos surgery had a higher duration than expected, and that there were not enough time for the next (longer duration of previous surgery). This category presents itself as the main reason for elective surgery cancellation (67%) and relates to maladjusted resources planning within the hospital.

The second category, "Patient will and no-show" (12%), comprises two of the original categories and refers to surgery cancellations due to patient decisions. The category includes, inter alia, the surgeries patients cancel themselves, and when the patient does not show up for an elective surgery at the hospital.

The quality improvement initiative used the category "Unknown/Other" (11%) when the reason for cancellation was not documented in the system or not available for selection in the system. This category was directly transferred into the new categorization.

The fourth category, "Medical", constitutes 10% of all surgery cancellations. It refers to cancellations due to medical issues with the patient such as insufficient surgery indications and other acute medical conditions.

The revisit demonstrates that the main reasons for surgery cancellation at UNN is maladjusted resource planning (67%).

B. Fieldwork at the hospital

The re-analysis of the workflow models at the hospital explored the potential of increased patient involvement in the existing pre-operative workflows at the hospital. It exposed a very low level of formal structure of the pre-operative planning process at UNN, hindering any attempts of re-engineering the workflow models to include a patient focus. Even when attempting to build a workflow model that only included activities involving the patient, we still found variation and heterogeneity, both at the department and individual level. The reality of the situation was illustrated in the observations and interviews completed during the fieldwork:

1) the patient data requested by the hospital, prior to surgery, differed between departments;
2) if data were returned by the patients (by the postal system) to the hospital, it had different meanings for different departments and individuals; and
3) different individuals within the same department interacted with the patient in different moments of the pre-operative planning process.

The revisit revealed that increased patient involvement is severally limited by the lack of uniformity of the pre-operative planning process at UNN.

C. Patients interviews

Revisiting the patient interviews demonstrated that patients wanted to be involved in choosing the date of their own surgery. The patient reports mentioned that

1) they were assigned a date of surgery without being involved or asked their availability;
2) the surgery was cancelled when they tried to reschedule it to another day (in the Norwegian healthcare system, cancellations have consequences for patient priority in accessing surgery); and
3) some patients expressed their wish to be digitally involved in deciding the date of surgery.

The revisit illustrates that an improved scheduling system and increased patient involvement might contribute to the surgery cancellation problem at the hospital.

IV. Discussion and Conclusions

In this section, we will consider the reality of increased patient involvement in the current hospital practice of pre-operative planning at a university hospital in Norway. We will do so by answering the following research question:

"Can increased patient involvement and digitalization reduce the number of elective surgery cancellations at the University Hospital in North Norway (UNN)?"

We have addressed the research question by revisiting the findings from a research and innovation project, particularly suitable for embracing the complexity of increased patient involvement in a hospital context, the eTeam-Surgery project (2013–2019). The lesson learned from this project was that creating digital tools for two-way electronic communication between the hospital and the patient within the current hospital organizational structure for pre-operative planning is not an easy task. For example, the quality improvement initiative at the hospital identified 17 different reasons for elective surgery cancellation, the fieldwork at the hospital exposed the heterogeneity of the pre-operative planning process and the findings from patient interviews demonstrated that surgeries were cancelled when the patient was actually trying to rebook to a different date.

By revisiting these findings focusing on the potential of increased patient involvement, we demonstrate how 67% of elective surgery cancellations at UNN relate to maladjusted resource planning. Maladjusted resource planning is a reflection of internal hospital organizational issues and is not solvable with increased patient involvement.

The revisit also shows that, in 2016, the hospital had not adopted a formal pathway to support the pre-operative workflow. To the best of our knowledge, this is still the situation in 2019. Lack of a formal pathway of the pre-operative planning process means that in order to complete the daily schedule, health care workers rely on personal and empirical knowledge. Lack of structures in everyday planning practices relate to internal organizational issues and is not an ideal starting point for developing e-health interventions for increased patient involvement.

The patient interviews demonstrated that at UNN, elective surgeries were cancelled when patients were trying to rebook the date of their surgery. Despite growing recognition internationally that patients can help promote their own wellbeing, little evidence exists on how willing patients are to take on an active role. It is the authors' understanding that these surgical patients wanted increased involvement and that the cancellations were due to the hospital's poorly functioning scheduling system or lack of knowledge on how to operate the system, not because the patient did not want the surgery. The revisit illustrates that an improved scheduling system and increased patient involvement might reduce the surgery cancellation problem at the hospital.

Can increased patient involvement contribute to the reduction of surgery cancellations? Five years later, after revisiting the eTeam-Surgery project, we still believe that active patient involvement can help solve the elective surgery cancellation problem. However, as this paper demonstrates, in 2013-2019, the eTeam-Surgery project was ahead of its time. We are concerned that similar e-health interventions today might suffer from similar prematurity. To achieve patient involvement, the first step for hospitals might be to allow patients to schedule their own surgery date, in a similar way to how they book an airline ticket online.

It is our conclusion that increased patient involvement is wanted and needed, but hospitals, as organizations, are not ready for it yet. Hospitals have a long way to go before the new mantra of increased patient involvement is a useful asset in improving health and care services.

ACKNOWLEDGMENT

The authors would like to thank the regional health authority Helse Nord for funding the research projects HST 1119-13 and HST 1125-13. We will also thank the personnel and patients at UNN for their participation in the eTeam-Surgey project, and Professor Gunnar Hartvigsen and MD, PhD, Stein Roald Bolle for research collaboration.

REFERENCES

[1] R. E. Davis, R. Jacklin, N. Sevdalis, and C. A. Vincent, "Patient Involvement in Patient Safety: What Factors Influence Patient Participation and Engagement?," Health expectations, vol. 10, pp. 259-267, 2007.

[2] L. T. Kohn, J. M. Corrigan, and M. S. Donaldson, To Err Is Human: Building a Safer Health System. Washington (DC): National Academies Press (US), 2000.

[3] Department of Health/NHS, "Creating a Patient-Led Nhs – Delivering the NHS Improvement Plan," NHS, Available.

[4] A. G. Thompson, "The Meaning of Patient Involvement and Participation in Health Care Consultations: A Taxonomy," Social science & medicine, vol. 64, pp. 1297-1310, 2007.

[5] *Én Innbygger – Én Journal,* Ministry of Health and Care Services Meld.St. 9 (2012-2013), 2012.

[6] *The Coordination Reform: Proper Treatment - at the Right Place and Right Time,* Norwegian Ministry of Health and Care Services Stortingsmelding 47 (2008- 2009), 2009.

[7] Helse Nord. Moving Pre-Surgical Planning from the Hospital to the Patient at Home through Electronic Collaboration (eTeam-Surgery) [Online]. Available: https://forskningsprosjekter.ihelse.net/prosjekt/HST1116-13/ 2019.09.12.

[8] M. Johnston, "Anxiety in Surgical Patients," Psychological medicine, vol. 10, pp. 145-152, 1980.

[9] M. J. Pritchard, "Identifying and Assessing Anxiety in Pre-Operative Patients," Nursing standard, vol. 23, p. 35, 2009.

[10] B. Ivarsson, P. O. Kimblad, T. Sjoberg, and S. Larsson, "Patient Reactions to Cancelled or Postponed Heart Operations," J Nurs Manag, vol. 10, pp. 75-81, 2002.

[11] M. J. Lacqua and J. T. Evans, "Cancelled Elective Surgery: An Evaluation," The American Surgeon, vol. 60, pp. 809-811, 1994.

[12] M. Perroca, M. C. Jericó, and S. Facundin, "Surgery Cancelling at a Teaching Hospital: Implications for Cost Management," Revista latino-americana de enfermagem, vol. 15, p. 1018, 2007.

[13] A. R. Tait, T. Voepel-Lewis, H. M. Munro, H. B. Gutstein, and P. I. Reynolds, "Cancellation of Pediatric Outpatient Surgery: Economic

and Emotional Implications for Patients and Their Families," Journal of clinical anesthesia, vol. 9, pp. 213-219, 1997.

[14] T. L. Trentman, J. T. Mueller, S. L. Fassett, C. L. Dormer, and K. P. Weinmeister, "Day of Surgery Cancellations in a Tertiary Care Hospital: A One Year Review," Journal of Anesthesia & Clinical Research, vol. 1, 2010.

[15] W. N. Schofield, G. L. Rubin, M. Piza, Y. Y. Lai, D. Sindhusake, M. R. Fearnside, and P. L. Klineberg, "Cancellation of Operations on the Day of Intended Surgery at a Major Australian Referral Hospital," Medical Journal of Australia, vol. 182, pp. 612-615, 2005.

[16] A. A. Ashraf, S. Colakoglu, J. T. Nguyen, A. J. Anastasopulos, A. M. S. Ibrahim, J. H. Yueh, S. J. Lin, A. M. Tobias, and B. T. Lee, "Patient Involvement in the Decision Making Process Improves Satisfaction and Quality of Life in Postmastectomy Breast Reconstruction," Journal of Surgical Research, vol. 184, pp. 665-670, 2013.

[17] E. Hovlid, C. von Plessen, K. Haug, A. B. Aslaksen, and O. Bukve, "Patient Experiences with Interventions to Reduce Surgery Cancellations: A Qualitative Study," BMC surgery, vol. 13, p. 30, 2013.

[18] B. Denton, J. Viapiano, and A. Vogl, "Optimization of Surgery Sequencing and Scheduling Decisions under Uncertainty," Health care management science, vol. 10, pp. 13-24, 2007.

[19] C. Granja, K. Dyb, S. R. Bolle, and G. Hartvingsen, "Reduced Elective Surgery Cancellations through Patient Involvement in Pre-Operative Planning in Norway," eTELEMED 2014, The Sixth International Conference on eHealth, Telemedicine, and Social Medicine, 2014, pp. 164-69.

[20] P. Sanjay, A. Dodds, E. Miller, P. J. Arumugam, and A. Woodward, "Cancelled Elective Operations: An Observational Study from a District General Hospital," J Health Organ Manag, vol. 21, pp. 54-58, 2007.

[21] M. B. Ferschl, A. Tung, B. Sweitzer, D. Huo, and D. B. Glick, "Preoperative Clinic Visits Reduce Operating Room Cancellations and Delays," Anesthesiology: The Journal of the American Society of Anesthesiologists, vol. 103, pp. 855-859, 2005.

[22] M. Rai and J. Pandit, "Day of Surgery Cancellations after Nurse-Led Pre-Assessment in an Elective Surgical Centre: The First 2 Years," Anaesthesia, vol. 58, pp. 692-699, 2003.

[23] A. R. Seim, T. Fagerhaug, S. M. Ryen, P. Curran, O. D. Sæther, H. O. Myhre, and W. S. Sandberg, "Causes of Cancellations on the Day of Surgery at Two Major University Hospitals," Surgical Innovation, vol. 16, pp. 173-180, 2009.

[24] C. Rutten, K. Moons, B. Lo, J. Knape, and D. Grobbee, "Limited Effect of Health Council Guideline on Outpatient Preoperative Evaluation Clinics in the Netherlands: An Inventory," Nederlands tijdschrift voor geneeskunde, vol. 145, pp. 25-29, 2001.

[25] R. Busund, "Rapport Fra Prosjekt: Optimal Ressursutnyttelse Av Opperasjonskapasiteten i Unn," Norway 2008 (in Norwegian).

[26] R. R. Cima, M. J. Brown, J. R. Hebl, R. Moore, J. C. Rogers, A. Kollengode, G. J. Amstutz, C. A. Weisbrod, B. J. Narr, and C. Deschamps, "Use of Lean and Six Sigma Methodology to Improve Operating Room Efficiency in a High-Volume Tertiary-Care Academic Medical Center," Journal of the American College of Surgeons, vol. 213, pp. 83-92, 2011.

[27] C. Granja, K. Dyb, S. R. Bolle, and G. Hartvigsen, "On the Development of a Hospital-Patient Web-Based Communication Tool: A Case Study from Norway," MedInfo, 2015, pp. 438-442.

[28] K. Dyb, C. Granja, S. R. Bolle, and G. Hartvigsen, "Digital Users in Pre-Digital Hospital Organisations? An Analysis on the Readiness for Electronic Communication between a Hospital and Surgical Patients," International Journal on Advances in Life Sciences, vol. 8, pp. 39-49, 2016.

[29] K. Dyb, "Patient Reported Reasons for Surgery Cancellations Do Not Necessarily Correspond with Hospitals' Representation of the Same Problem," Health Informatics Vision: From Data via Information to Knowledge, vol. 262, p. 75, 2019.

Detection and Classification Method for a Temporary Change in Walking

Shin Morishima*, Misato Haruta†, Akira Urashima‡, and Tomoji Toriyama§
Faculty of Engineering Toyama Prefectural University
5180 Kurokawa, Imizu-shi, Toyama, Japan 939-0398
Email: *morisima,‡a-urasim,§toriyama@pu-toyama.ac.jp , †t615034@st.pu-toyama.ac.jp

***Abstract*—The global elderly population has grown in recent years and one of the difficulties the elderly face is an increased vulnerability to falls. One of the ways to alleviate this problem is to identify actions that cause falls and to prevent falls using the detection result. Temporary change in walking (i.e., stumble and a stagger) is a typical action. However, existing studies only classify walking or other activities and recognize walking speed, and these studies do not focus on a temporary change in walking. In this paper, we propose a detection method for a change in walking using a change point detection (i.e., anomaly detection for the time series data) and a classification method for the multiple types of change. During the evaluation, four types of anomalous walking videos (i.e. anomaly represents a temporary change.) are used (the total number of videos is 240). As a result, our method can detect anomalous walking in 91.7% of cases and classify four types of detected anomalous walking into three clusters in 89.1% of cases on the basis of each characteristic.**

Keywords–Walking recognition; Classification; Anomaly detection; Human activity recognition.

I. Introduction

The global elderly population has grown in recent years and one of the difficulties the elderly face is an increased vulnerability to falls [1] [2]. Elder people are more likely to be seriously injured by falls. In addition, this issue also leads to an increase in the medical costs, specifically, the cost of falls was approximately $50 billion in the USA in 2015 [1]. Human activity recognition is a research field that is focused on alleviating this problem [3]. Two types of activity recognition can be used to address this problem. The first type is fall detection, which identifies falls soon after they occur. There are already several studies on the topic, which show that this approach prevents injuries due to falls from becoming severe [4] [5]. The second approach is to detect actions that cause falls to prevent falls. Such actions include a stumble and a stagger. These actions are caused by a temporary change in walking. Thus, the recognition of the change is needed. In addition, the classification of change is needed because the change includes actions that are not related to the falls such as a standstill. However, existing studies on walking recognition can only distinguish walking and other activities [6], recognize walking speed [7], and do not focus on a temporary change in walking. Most of the existing studies recognize activity from common features between multiple persons if there is a clear difference between target activities (e.g., walking and sitting). Various methods can be used to identify these activities. However, it is difficult to recognize a change in walking because the difference in each person is larger than the difference of each action. A stumble detection system for powered artificial legs has been proposed in [8]; however, it is difficult to apply it to cases without artificial legs.

In this paper, we propose a detection method for a change in walking using change point detection and anomaly detection for time series data. Furthermore, we propose the classification of multiple types of change using a method that uses clustering of the results of the change point and anomaly detection.

The rest of the paper is organized as follows. Section II presents related works. Section III presents our proposed recognition method. Section IV evaluates our proposed method, and Section V concludes the paper.

II. Related Work

A. Walking Recognition

Walking recognition is the field of human activity recognition. Human activity recognition is defined as the ability to recognize human activities using sensor data [3] [9] [10]. Each recognition method has two steps that are data collection by sensors and the estimation of human activity on the basis of data.

Currently, there are several studies on walking recognition, which focus on the differences in sensors or estimation methods. Khan et al. have detected walking and other five activities (i.e., sitting, standing, washing hands, driving, running) from accelerometer data using change point detection [6]. The abovementioned study uses a genetic algorithm to optimize the parameters of change point detection to improve the accuracy of detection. Thus, it has the high accuracy of 99.4%-99.8% [6]. Trung et al. have classified five types of walking (i.e., walking on flat ground, up/down stairs, and up/down a slope) with a 90.4% accuracy on the basis of accelerometer data using a support vector machine [11]. Davis and Taylor have recognized walking speeds (i.e., normal speed, half of normal speed, and double of normal speed) and classified walking and other eleven activities (e.g., running and skipping) from video-based four joint coordinates data [12]. Haescher et al. have classified walking speed into four classes (i.e., 1, 2.5, 4, and 5 km/h) from capacitive sensor data [7]. The application of this walking recognition includes automated surveillance, monitoring systems to identify people that may be injured or require assistance, and the estimation of the amount of activity [7] [12].

However, contrary to this study, these studies did not focus on a temporary change in walking. The abovementioned studies and our proposed study can be used together to detect walking in several activities using existing methods and to recognize a change in walking using our method.

B. Image-based Human Posture Estimation

In our proposed method, human joint coordinates are used for the input of change point detection. The coordinates can be extracted using the existing method. In this paper, OpenPose [13], which has been proposed by Cao et al. is used to extract the coordinates. OpenPose outputs two-dimensional 25-joint

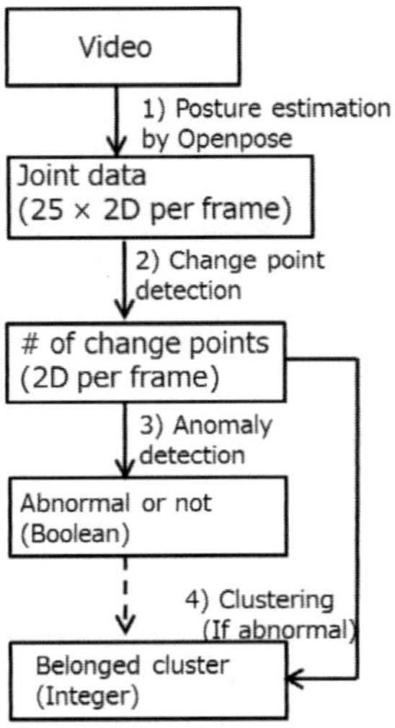

Figure 1. Overview of the proposed method using dataflow.

data per image by the deep learning-based posture estimation. Our proposed method is not constrained by OpenPose. If the coordinates can be acquired, the output of another method can be applied. For example, there are some image-based posture estimation methods such as ArtTrack [14] and DeepCut [15]. Our method can be applied for non-image-based methods such as motion capture or a depth sensor.

III. Recognition Method of a Temporary Walking Change

A. Overview of the Proposed Method

The input of the proposed method is the coordinate data of human joints during walking. The method has three steps. It distinguishes anomaly walk which exhibits behavior that temporarily differs from daily walking and classifies anomaly. Annomaly walking detection can be used to analyze the causes of falls because actions that can cause falls (e.g. stumble and stagger) include the anomalous walking. The input is acquired by videos using OpenPose, as mentioned in Section II. As described in Section IV-A, in this paper, we use a video ($1,920 \times 1,020$ resolution and 30-fps frame rate) of a walking person from the front. A picture of a person is acquired from the front, while the person is walking from the back to the front.

Fig. 1 shows the overview of the proposed method using the dataflow from the video data to classify anomalous walking. The preprocessing of the proposed method is the posture estimation by OpenPose. OpenPose detects a human and estimates posture from the video. OpenPose outputs coordinates of joints, which are two-dimensional 25-point data from one frame of the video.

These joints data are the inputs of the proposed method, and the three steps of the method are shown below.

1) Change point detection determines whether each joint point or relationship between two joints is a change point or not (the meaning of the relationship will be explained in the next section.). The output is the number of change points in each frame because the detection is processed for each joint point and relationship of each frame.
2) Anomaly detection detects anomalus walking using the number of change points. The result is output per the overall walking data (data from one video).
3) Clustering classifies the anomalous walk detected by the second step. It is processed only when input walking data are detected as anomaly.

The processing of each step is performed by unsupervised machine learning. It means that the data labeled by a human are not needed. However, the first and second steps require some preprocessing to determine the parameter of each machine learning algorithm. Preprocessing is the preparation of the standard of the parameter determination. The standard is the result of the change point detection using daily walking data. One or two minutes of walking data are required for each subject. In the evaluation, we use 20 videos of the daily walking data for each subject, and the average length of videos is approximately three seconds. Thus, the total time of the walking data is approximately 1 min. To distinguish daily walking (i.e., input data given by a human and the output data detected by the proposed method), the terms "daily walk" and "normal walk" are used. The details of each step are explained in the next sections.

B. Change Point Detection for the Human Joint Data

Change point detection is processed in each frame data using the target frame and the previous frames consecutive to the target. The data format of the output data of OpenPose is 2-dimensional 25 points data per frame. There are two methods to deal with the data. The data are treated as one 50-dimensional data, or the data are treated as 2-dimensional 25-point data. If the data are processed as 50-dimensional data, the relationships between joints can be considered. However, there is a problem that when some joints considerably change, they affect the entire result. However, if the data are processed as 25-point data, the problem does not occur, however, the relationships cannot be considered. To solve the problem, we propose to use the difference data of all pairs of joints in addition to the 25-point data so that we can deal with the relationships using the sets of two-dimensional data. The total number of data is 325 because the number of pairs is 300. The result of change point detection from these data is output as the number of change points. In this method, we divide the result of joints and differences data to distinguish a change in the movement of joints and the relationships of joints. Thus, the output is one two-dimensional data per a frame.

Fig. 2 shows the change point detection flow as the dataflow. The input data are the joints data of 25 points and the difference data consist of all pairs. We adopt the Multivariate Exponentially Weighted Moving Average (MEWMA) algorithm as the change point detection algorithm because it uses only the target walking data and is not affected by individual differences [16]. The MEWMA algorithm uses the data from the target frame and from previous frames consecutive to the target. If the number of frames is n (i.e., nth frame is the target frame), the MEWMA vector is defined by (1).

$$Z_i = \lambda X_i + (1-\lambda)Z_{i-1} | i = 1, 2, 3, \dots n, Z_0 = 0 \quad (1)$$

X_i is the input vector, which is the coordinate of the joint or the difference of each frame. The change point can be detected

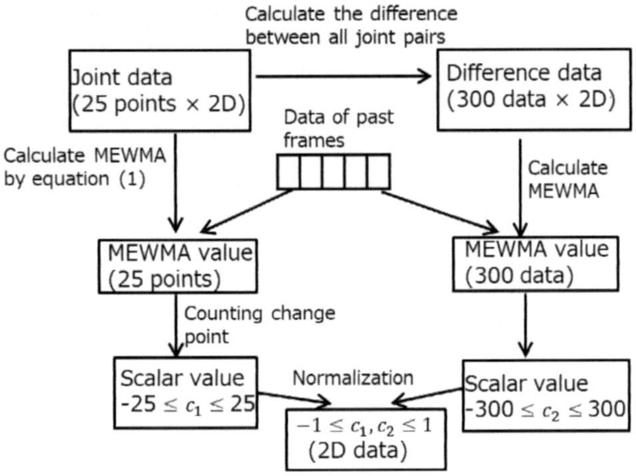

Figure 2. Dataflow of the change point detection.

in equation (2) using the MEWMA vector.

$$T^1 < h_1, T^2 > h_2 | T^2 = Z_n^T \Sigma_n^{-1} Z_n \quad (2)$$

Z_n is the MEWMA vector and Z_n^T its transpose. Σ_n^{-1} is the variance covariance matrix of Z_n. h_1 and h_2 are thresholds ($0 < h_1 < h_2$). h_1 is the case in which a change in the movement during walking becomes small and h_2 is the case when a change in the movement during walking becomes large. The counter of change point is decremented when $T^2 < h_1$, and the counter is incremented when $T^2 > h_2$ to distinguish the cases in which the movement becomes small or large. Using the MEWMA algorithm for all joints and differences, the range of the counter value is $-25 < counter < 25$ (joints) and $-300 < counter < 300$ (differences). Finally, the counting results are normalized by dividing by 25 or 300. The output of the change point detection is the two-dimensional data, and the form is suitable for the anomaly detection mentioned in the next section.

C. Anomaly Detection for the Number of Change Points of Walks

The input data are two-dimensional data per frame obtained from the change point detection. The anomaly is defined as a temporary change in the movement in the daily walk. Thus, the data of daily walking are needed for each person. The data are prepared as a set of the result of the change point detection for the daily walk for a few minutes, and the set is called "normal data".

Anomaly detection is performed in each frame by comparing the normal data and the target frame. Thus, anomaly detection is repeated in each frame by adding the target frame data and removing it after detection. We use a Local Outlier Factor (LOF) as anomaly detection because it can be calculated in each data, and the feature is suitable for repeating [17]. LOF is calculated by comparing local densities. The local density is calculated by reachability distance defined by (3).

$$reachability\text{–}distance_k(p, q) = max(k\text{–}distance(q), d(p, q)) \quad (3)$$

where p and q are the points of two-dimensional data, d(p,q) is the Euclidean distance between p and q, k-distance(q) denotes the Euclidean distance between q and the kth nearest neighbor of q. The local density, which is termed Local Reachability Density (LRD) is defined as (4).

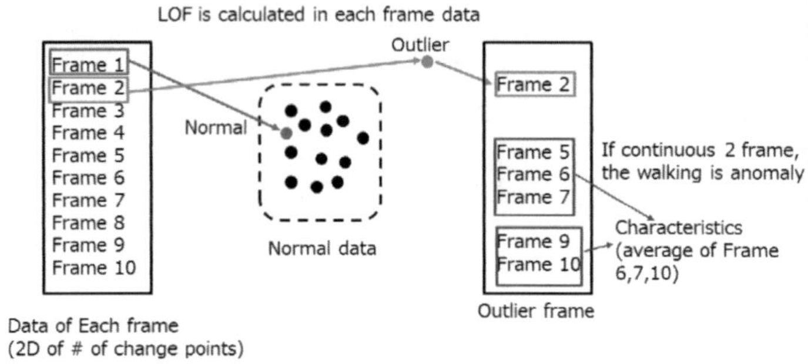

Figure 3. Example of the anomaly detection method.

$$LRD(p) = \left(\frac{\Sigma_{q \in kNN(p)} reachability - distance_k(p, q)}{k}\right)^{-1} \quad (4)$$

where kNN(p) denotes the set of k nearest neighbor of p. LOF is defined using the LRD by (5).

$$LOF(p) = \frac{\Sigma_{q \in kNN(p)} \frac{LRD(q)}{LRD(p)}}{k}. \quad (5)$$

The value of LOF is large when the target is an outlier. LOF shows whether the target frame is an anomaly or not; however, the result of the frame is not used for the direct detection of anomalous walking, because an anomaly frame may appear in the misestimation estimations of OpenPose. Therefore, in this method, the condition of the anomaly walk is the walk that includes two or more continuous anomaly frames. To classify anomaly walks, the characteristics of the walk are defined as the average number of change points of the continuous two or more anomaly frames.

Fig. 3 shows an example of the anomaly detection. In the example, there are ten-frame data. LOF is calculated for each frame and Fig. 3 shows the case of frames 1 and 2. In the field of calculation of LOF, each frame data is deleted after calculation. The frames identified as outliers are collected if there are 2 or more continuous frames and walking is identified as anomalous walking. The characteristics of anomalous walking are the average values of continuous frames excluding the first frame of the continuous frame.

D. Anomaly Walk Classification

The two-dimensional characteristic data of the anomaly walk are acquired from one video. When there are multiple anomalous walking videos, clustering algorithms can be performed for the characteristics and the walk can be classified. In this paper, we use the K-means algorithm, which is a typical clustering algorithm. In the K-means algorithm, the parameter given by a human is only the number of cluster k. In K-means clustering, first, each point, which means the characteristics of walking, is randomly classified into k clusters. Next, the below two steps are repeated until convergence.

1) The center of gravity of each cluster is calculated.

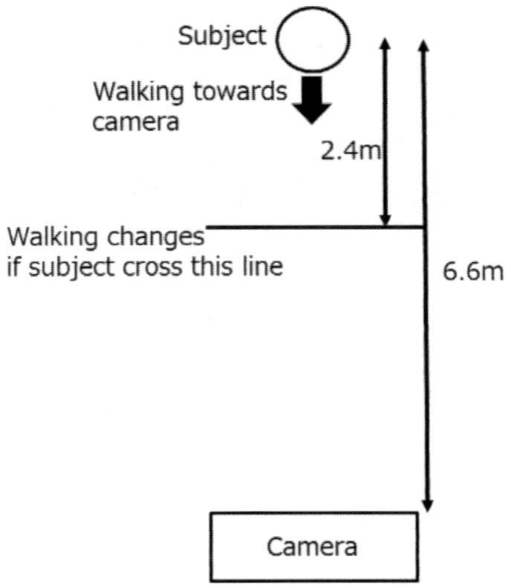

Figure 4. Walking during the evaluation.

2) Each point is reclassified into the cluster which has the nearest center of gravity from the point.

Using processing, the detection of anomaly walks and the classification of the walks is completed from the walking videos. In this method, the longer the video, the higher the probability of the video to include multiple anomalous walking instances, because anomaly detection handles the video as one data unit. The classification method assumes that there is only one anomaly walk per video. Thus, a long video should be divided every several seconds. However, Videos that are too short are also not suitable for the method and the minimum length of the video is two steps because walking has periodicity every two steps.

IV. EVALUATIONS

A. *Evaluation Method and Environment*

We evaluate the accuracy of the detection and classification of anomaly walks in the proposed method. Thus, it is necessary to prepare anomalous walking data for which the correct answer is known. Therefore, we prepare the data of daily walks and change of walking from daily walk to anomaly walk. The subjects are three adults and Fig. 4 shows the details of walking. The walks start 6.6 m away from the camera and change to anomaly walks after 2.4 m of walking. There are the following four anomaly walks.

- Back: Go back one step and start walking again.
- Side: Walk 40 cm from side to side.
- Stop: Stop and start walking again.
- Wide: Walk one large step (1 m).

The resolution of the videos is $1,920 \times 1,020$ and the framerate is 30 fps. Each subject performs 20 times each type of walk (daily and four anomaly walks). Thus, the number of data of each type is 60. Table I shows the average number of frames of each type of walk.

TABLE I. AVERAGE NUMBER OF FRAMES IN EACH WALKING TYPE.

Walking type	Daily walk	Anomaly walk
Daily	91	0
Back	49	83
Side	39	67
Stop	50	58
Wide	39	25

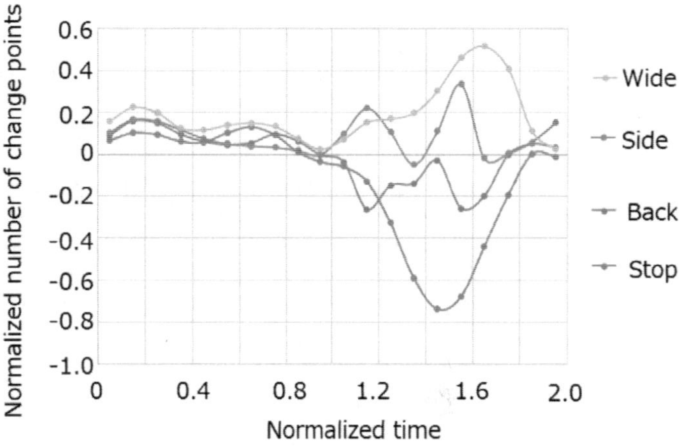

Figure 5. Transitions of the number of change points of the joint data.

B. *Transition of the Number of Change Points in Change Point Detection*

In this section, the transition of the number of change points is shown. It is an interim result of the proposed method; however, it shows an overall trend of each type of anomalous walk. There are five parameters of the MEWMA algorithm. These parameters are detected by a greedy algorithm with a high F value. The order of the parameters of the greedy algorithm is the number of frames, h_1 of joints, h_2 of joints, h_1 of the difference, and h_2 of the difference. The number of frames is 15 (0.5 seconds). The thresholds of the change point are 1/800 (h_1) and 2 times (h_2) the average MEWMA for

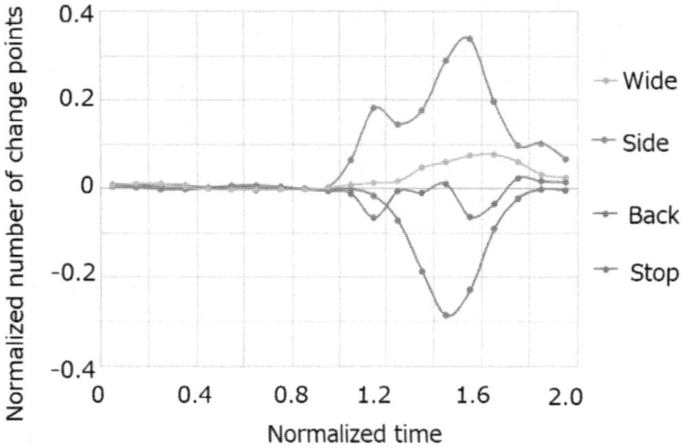

Figure 6. Transitions of the number of change points of the difference data.

TABLE II. RESULT OF ANOMALY DETECTION FOR EACH ANOMALOUS WALKING TYPE.

Walking type	Subject A		Subject B		Subject C		Total	
	TP	FP	TP	FP	TP	FP	TP	FP
Back	14	0	20	6	20	4	54	10
Side	18	0	20	2	20	2	58	4
Stop	20	2	20	1	20	2	60	5
Wide	17	4	18	6	13	2	48	12
Total	69	6	78	15	73	10	220	31

joints and 1/1200 and 11 times for differences data. The data of daily walk is used only for the determination of parameters, and change point detection is performed for the four types of anomaly walks. Fig. 5 and 6 show the transitions of the number of change points. These figures show the average number of change points for 60 walks by the three subjects. The x-axis shows normalized time; 0 to 1 indicates a daily walk, and 1-2 indicates anomalous walks. The boundary between daily and anomaly walks is determined by human confirmation in every frame. The y-axis shows the normalized number of change points. The data are plotted every 0.1 point on the x-axis as the average number of change points in the range. The positive values in the result indicate a change in walking movements to intense movements and the negative values indicate a change in movements to slow movements. Even in daily walk, the values of joints are positive because the foot movement periodically changes. However, the relationships between joints do not change, and the values of difference are almost 0. The figure shows that the tendency of number of the change points changes between daily and anomaly walks and it is different in each type of anomaly walks. In Side and Wide, both of the values are positive; however, a change in the difference in Wide is smaller than that in Side. The change in Wide is only a change in stride, and the change in the positional relationship between the joints of the body is smaller than that in Side. In Stop and Back, the values of Stop are smaller than Back and Stop is characterized by the temporally stopping of the movement of all joints. Thus, the method can detect the difference between each anomaly walk. This means that the method can be used to analyze the trend of each walking type, and the result of change point detection can be used for anomaly detection and classification. However, in actual operation, the types are not given. Thus, analysis should be done after this classification if the method is used for the analysis

C. Anomalous Walking Detection

Table II shows the result of the anomaly detection for each anomalous walking type. The threshold of LOF is 4.5, which is detected by the greedy algorithm after the parameter of MEAWA is detected. The table shows TP (TP is True Positive, which means the detection of anomaly walks as an anomaly) and FP (FP is False Positive, which means the detection of daily walks as an anomaly) of each subject and of all subjects. Each rate of TP and FP is 91.7% and 12.9% in total; TP has high accuracy ($> 90\%$), and FT has medium accuracy ($> 80\%$ *and* $< 90\%$) for standards in [3]. By focusing on individual results of each type (e.g., TP is 100% in Stop case), it is determined that there are differences in the accuracy. This observation shows that there is a possibility of improving the accuracy by optimizing the parameters that we focused on,

TABLE III. RESULT OF THE CLASSIFICATION INTO FOUR CLUSTERS FOR EACH ANOMALOUS WALKING TYPE.

Walking type	Cluster 1	2	3	4
Back	46	0	7	1
Side	0	52	4	2
Stop	50	0	10	0
Wide	0	2	13	33

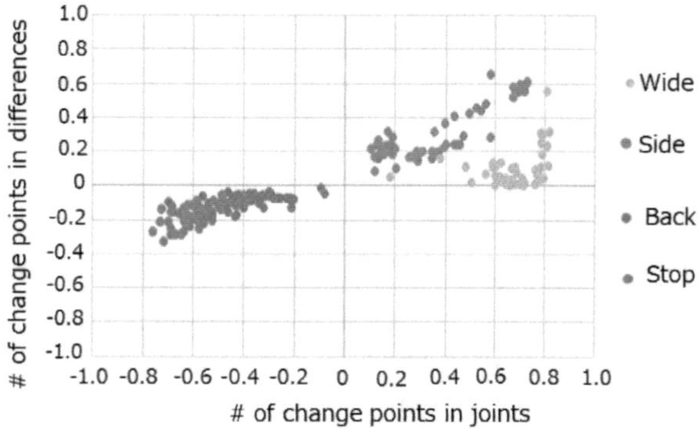

Figure 7. Characteristics of anomaly walks of each walking type.

such as the threshold or window size of each subject and walking type. The optimization is considered effective for the long-term observation of individuals. This will be evaluated in detail in the future.

D. Anomalous Walking Classification

In this section, clustering is performed in the case of TP in anomaly detection. Table III shows the result of clustering into four clusters. The number of clusters is the same as the number of walking types. Each cluster is shown by a number because clustering is unsupervised. In this result, clustering is performed separately for each subject, and we group each cluster of each subject with the highest match rate. Clustering does not classify four types of walks into four clusters. Back and Stop are grouped, and the clusters look divided such as Back and Stop, Side, Wide, and Others. To explain this result, the characteristics of each walking type are plotted in Fig. 7. As shown in the figure, Back and Stop overlapped and created the same cluster. This indicates that the tendencies of movement of Back and Stop are the same, and the difference is only the magnitude of the value. The trend is probably caused by the stopping motion because Back includes the motion of Stop because stopping is needed to switch the front step to the back step. The differences indicate the length of stopping time.

We perform clustering into three clusters owing to the grouping, and Table IV shows the result. In this result, anomaly walks are classified into three clusters (Back and Stop, Side, and Wide), and the accuracy rate is 89.1%. The rate is medium in [3], and it classified walks better than the existing walking recognition methods.

As with anomaly detection, clustering may show individual differences. Fig. 8 shows the plots of the same data of Fig. 7 colored by each subject. For Side and Wide cases, the figure shows that subjects have different tendencies. On the basis of this result, when considering the classification of anomaly walks, it is determined that clustering should not target too

TABLE IV. RESULT OF THE CLASSIFICATION INTO THREE CLUSTERS FOR EACH ANOMALOUS WALKING TYPE.

Walking type	Cluster 1	2	3
Back	46	0	8
Side	0	54	4
Stop	58	0	2
Wide	0	10	38

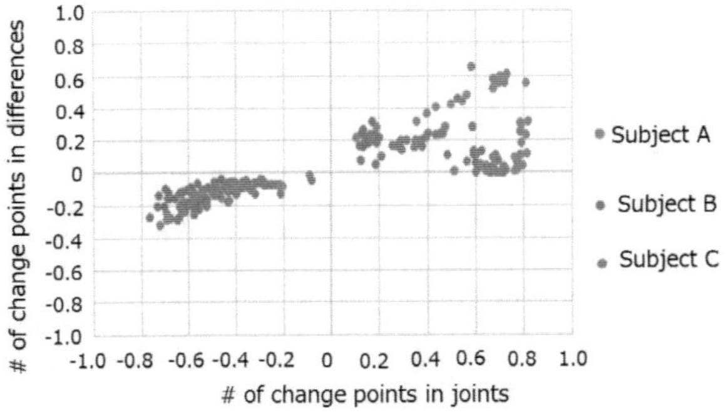

Figure 8. Characteristics of anomaly walks for each subject.

many people at the same time. The result shows the possibility of other applications of the proposed method, if the method can detect or classify the tendency of walking of each subject, it can be used to diagnose diseases, such as paralysis, to measure the effects of rehabilitation.

V. CONCLUSIONS

The detection of a temporary change in walking that causes falls is a way to alleviate the problem. However, existing methods recognize only walking speed or whether a person is walking or not. In this paper, we proposed an anomalous walking detection and classification method by three processing, which are the change point detection, anomaly detection, and clustering. Thus, anomaly walks were detected in 91.7% of cases using 240 videos including change from daily walk to anomaly walk. Furthermore, the detected anomaly walks can be classified into three clusters in 89.1% of cases. The average length of video is 87 frames (30 fps) for the longest motion. The result shows the proposed method can detect and classify a temporary change in walking. This capability can be applied to the analysis of the action that causes falling. In this paper, we evaluated the method by video data, while the required input of the method is the coordinate data of joints. Thus, the method has a wide application range. For example, the method is applicable to wearable technology such as motion sensors, if it can acquire the coordinates of a sufficient number of joints.

ACKNOWLEDGMENT

The authors thank JSPS KAKENHI Grant Number JP19K24350.

REFERENCES

[1] C. S. Florence, G. Bergen, A. Atherly, E. Burns, J. Stevens, and C. Drake, "Medical Costs of Fatal and Nonfatal Falls in Older Adults," Journal of the American Geriatrics Society, vol. 66, no. 4, Mar. 2017, pp. 693–698.

[2] Y. K. Haddad, G. Bergen, and C. Florence, "Estimating the Economic Burden Related to Older Adult Falls by State," Journal of public health management and practice, vol. 25, no. 2, Mar. 2019, pp. E17–E24.

[3] Z. Hussain, M. Sheng, and W. E. Zhang, "Different Approaches for Human Activity Recognition: A Survey," CoRR, vol. abs/1906.05074, Jun. 2019, pp. 1–28.

[4] P. Tsinganos and A. Skodras, "A Smartphone-based Fall Detection System for the Elderly ," in Proc. of the 10th International Symposium on Image and Signal Processing and Analysis, Sep. 2017, pp. 53–58.

[5] S.-H. Cheng, "An intelligent fall detection system using triaxial accelerometer integrated by active RFID," Jul. 2014, pp. 517–522.

[6] N. Khan, S. McClean, S. Zhang, and C. Nugent, "Optimal Parameter Exploration for Online Change-Point Detection in Activity Monitoring Using Genetic Algorithms," Sensors, vol. 16, no. 1784, Nov. 2016, pp. 1–16.

[7] M. Haescher, D. J. C. Matthies, G. Bieber, and B. Urban, "CapWalk: A Capacitive Recognition of Walking-based Activities As a Wearable Assistive Technology," in Proc. of the 8th ACM International Conference on PErvasive Technologies Related to Assistive Environments, Jul. 2015, pp. 35:1–35:8.

[8] F. Zhang, S. E. D′ Andrea, M. J. Nunnery, S. M. Kay, and H. Huang, "Towards Design of a Stumble Detection System for Artificial Legs," IEEE Transactions on Neural Systems and Rehabilitation Engineering, vol. 19, no. 5, Oct. 2011, pp. 567–577.

[9] J. Aggarwal and M. Ryoo, "Human Activity Analysis: A Review," ACM Computing Surveys., vol. 43, no. 3, Apr. 2011, pp. 16:1–16:43.

[10] J. Yang, J. Lee, and J. Choi, "Activity Recognition Based on RFID Object Usage for Smart Mobile Devices," Journal of Computer Science and Technology, vol. 26, no. 2, Mar. 2011, pp. 239–246.

[11] N. T. Trung, Y. Makihara, H. Nagahara, Y. Mukaigawa, and Y. Yagi, "Inertial-sensor-based Walking Action Recognition usingRobust Step Detection and Inter-class Relationships," in Proc. of the 21st International Conference on Pattern Recognition (ICPR 2012), Nov. 2012, pp. 3811–3814.

[12] J. W. Davis and S. R. Taylor, "Analysis and recognition of walking movements," in Proc. of International Conference on Pattern Recognition, Aug. 2002, pp. 315–318.

[13] Z. Cao, T. Simon, S. Wei, and Y. Sheikh, "Realtime Multi-Person 2D Pose Estimation using Part Affinity Fields," CoRR, vol. abs/1611.08050, Nov. 2016.

[14] E. Insafutdinov et al., "ArtTrack: Articulated Multi-Person Tracking in the Wild," in 2017 IEEE Conference on Computer Vision and Pattern Recognition (CVPR), Jul. 2017, pp. 1293–1301.

[15] L. Pishchulin et al., "DeepCut: Joint Subset Partition and Labeling for Multi Person Pose Estimation," in Proc. of the 2016 IEEE Conference on Computer Vision and Pattern Recognition (CVPR), Jun. 2016, pp. 4929–4937.

[16] C. A. Lowry, W. H. Woodall, C. W. Champ, and S. E. Rigdon, "A Multivariate Exponentially Weighted Moving Average Control Chart," Technometrics, vol. 34, Feb. 1992, pp. 46–53.

[17] M. M. Breunig, H.-P. Kriegel, R. T. Ng, and J. Sander, "Lof: Identifying density-based local outliers," in Porc. of the 2000 ACM SIGMOD International Conference on Management of Data, May. 2000, pp. 93–104.

Near-Infrared Mobile Imaging Systems for e-Health

Lighting the Veins

Hüseyin Aşkın Erdem
Computer Engineering Department
Dokuz Eylul University
İzmir, Turkey
e-mail: huseyinaerdem@gmail.com

Işıl Erdem
Civil Engineering Department
Izmir Institute of Technology
İzmir, Turkey
e-mail: isilerdem@iyte.edu.tr

Semih Utku
Computer Engineering Department
Dokuz Eylul University
İzmir, Turkey
e-mail: semih@cs.deu.edu.tr

Abstract—**The evaluation of the methods used in image processing techniques with new perspectives has enabled the development of applications to address a wide range of usage areas. One of these perspectives brought to image processing is spectroscopy. Spectroscopy is a method based on obtaining information about the molecular structure of a substance as a result of its interaction with electromagnetic radiation (light photons). In this study, the interaction of haemoglobin in the blood vein structure with near-infrared light photons is utilized. With this interaction, superficial veins can be detected. The results obtained in this study are used to create a system which presents e-health measurements to the user through virtual environment. The system is created in the form of a mobile e-health application via smart devices or tablets. The data obtained by the near-infrared camera is passed through various image processing filters and stages as the software part of the study. In this way, vein structures are obtained. New images will be taken from the same veins of the user at certain time periods. The images taken at different periods will provide information to the user whether these veins have any problems with blood flow. This information will also be shared with the doctor. This procedure is aimed to create a pre-diagnosis data with the calculated mobile measurements.**

Keywords - image processing; near-infrared light; spectroscopy; e-health; virtual pre-diagnosis environment.

I. INTRODUCTION

As a result of rapidly increasing technological developments, it is revealed that multipurpose systems which can work anywhere any time are extremely necessary. Now, not only desktop computers, but also mobile phones or tablets that we always carry with us as a part of our lives, offer a wide range of applications. Although the most common usage areas of these applications are telephone communication and photo functions, there are also applications covering various fields, such as calculation, notepad, calendar planner, office applications, step counters, and health measurements. Among these applications, health applications are often used for mobile measurement. For example, there are applications which include the measurement of heart rate, the calculation of calorie intake or the daily intake of water, allowing health measurements to be taken.

In terms of wearable devices, specialized devices, such as blood pressure meters or step wristbands, can also be used to make such measurements. Instead of being purchased separately, such devices can be purchased as an integrated system in a single device which is always preferred by the user.

Health measurements taken via electronic devices are usually carried out by means of customized sensors located on the device. These sensors can take measurements from the outside world by means of motion, vibration, temperature, speed, or position detection. There are also special devices operating with infrared light, which are frequently used in hospitals. These devices are connected to the fingers to monitor the heartbeat.

Vascular access, commonly used in blood tests in hospitals, is an invasive process for the patients. It is among the most common applications used in the field of health. In order to perform this process with less pain for the patient, health personnel should open the vascular system, at the first attempt [1]. This depends both on the experience of the health personnel in this process and some of the characteristics of the patient. Characteristics, such as ethnicity (skin color) or adipose tissue may vary from person to person. These characteristics lead to difficulties in detecting the vein to be opened, while increasing the number of trials and giving more pain to the patient. Facilitating vein detection may result in less painful vascular access. To this point, various devices have been developed commercially and are currently being used in hospitals. However, these devices using Near-InfraRed (NIR) light are not suitable for daily use, as they are only intended for a particular purpose (heart rate measurement or vascular access detection) and are very expensive devices.

Among the commercial devices operating with NIR light, which are currently used in hospitals, VeinViewer, which has two different versions: Veinviewer Vision (the smaller one) and Veinviewer Flex (portable), is used to detect veins in vascular access. The device reflects the superficial vein image obtained with the use of NIR light on the patient's skin, so that the anatomical location of the veins remains the same. AccuVein AV300 is an easy-to-use and a hand-held device used for locating and visualizing superficial veins [9][10][22].

Currently used commercial devices capture the image of the vein with the infrared Charge Coupled Device (CCD) camera. The captured image is processed and reflected back (as NIR vein image) to the skin surface onto which the camera focuses [14].

The main method for vein visualization with infrared light used to obtain images in this study is basically similar to the methods used by commercial devices. However, the study differs from these devices in terms of processing the images and determining the blood flow changes in the veins from the processed images. After processing the images, only the Region Of Interest (ROI) in black-and-white colour format can be seen in the NIR image. The vein shapes obtained from the images will be compared with the vein images taken from the same tissue region of the same person in different periods (weeks or months, depending on the disease to be followed). This way, it is aimed to examine whether the blood flow in the veins within the image has changed or not. At this point, the comparison process will be determined by calculating the differences between these veins (more precisely, haemoglobin in the blood) in the next image taken. If any decrease in the lines representing the blood flow within the veins in the images is to be detected, then an obstruction in the vein will be conveyed to the doctor as a preliminary diagnostic data.

NIR devices which are successful in providing visualization in vascular access, often have high costs. Therefore, they are only used for limited applications (vascular access, vascular visualization) for specific targets in private treatment centers [24][25].

Plastic surgeons and dermatologists applying neurotoxins (Botox, Dysport) and fillers can perform more sensitive injections with AccuVein, which is a vein illuminator device. In addition, complications that may occur after the surgery/application (especially bruises, for aesthetic reasons) can be reduced with these devices [23].

Along with advancing imaging technologies, various vein imaging systems involving the use of NIR light have also been developed for different uses. These systems are able to perform identification procedures, from the palm veins or finger veins by performing vein imaging. Specialized ophthalmoscope systems which use NIR light to detect diseases of the eye veins are also available [2].

In this study, which is prepared within the scope of an ongoing PhD thesis, how to design a NIR vein imaging system that can be used as a mobile e-health application is discussed in terms of hardware and software. Also, the features of the cameras and Light Emitting Diodes (LEDs) that can be used to design such a system have been specified. The software steps that should be passed during the processing of the NIR images obtained by using these cameras have been explained in general. In Section 2, literature review about the use of NIR light is given. In Section 3, the basics of NIR imaging are discussed and their hardware characteristics are explained. In Section 4, the software processing stages of the NIR images to be obtained by using the hardware part of the developed system are mentioned. In Section 5, the steps of developing an exemplary NIR imaging system are discussed. Finally, the overall summary of the study is given in Section 6.

In Section 2, the literature review related to the studies involving the usage of NIR light is examined.

II. Literature Review

In the majority of studies involving the use of NIR imaging, vein imaging has been addressed for vascular access or safety. Ai et al. [3] combined images from the vein imaging system with augmented reality environment and aimed to reduce the time of intravenous injection. In this way, they determined that they could increase the ejection success rate and the efficiency by the system they developed. Chaio et al. [4] stated that they could provide better clinical care in vein discovery by using VueTek Scientific Veinsite in their superficial vein visualization. They examined how patient characteristics (ethnicity, age, or adipose tissue) affect the vein visibility by the system they developed. Fernandez and Armada [5] emphasized that the guide system they developed for catheter insertion can be used as a robotic guide system in the future. With their proposed multisensory systems, they identified the surrounding veins and paired them on a three-dimensional reconstructed surface. As a result, a complete representation of the vein region was obtained. Asrar et al. [6] compared three different cameras having different light characteristics (using visible light, infrared light and NIR light) for vein recognition to be used in vascular access. As a result, they stated that the most effective results from these cameras were obtained by NIR camera. Zhong et al. [7] examined smartphone security for authentication and developed the veinDeep system. In this system, vein images that uniquely identify the person were obtained by using an infrared depth sensor. They also underlined that, for authentication security, the disadvantages of fingerprints leaving marks on the surface or the human face being easily registered in the public area are not present in vein patterns. Bazrafkan et al. [8] obtained the vein structure map that they used in the authentication system they developed, by using the vein structure in the middle part of the finger. According to their results, they found that image acquisition and fusion methods provided by the system, which can be easily integrated into handheld devices, are suitable for vein imaging. Şeker and Engin [1] created two separate phantoms, liquid and solid, that will mimic human blood and skin. They used plastic tubes instead of the veins through which the liquid phantom would pass. In this way, they developed a system that can be used in vein network analysis with NIR imaging. The advantages of the system they developed include the utilization of non-ionizing radiation, providing a cost-effective design and providing algorithmic simplicity. Elnasir and Shamsuddin [9] reported that, with the help of the palm vein recognition system, which uses NIR light with a wavelength of 850 nm, penetration up to 3.57 mm depth in the palm area was observed. According to the results of the experiments, they stated that the system they designed gave 99.74% recognition rate and 100% verification rate. Shotri et al. [10] argued that a camera hardware capable of detecting NIR light could be developed by configuring an ordinary webcam to develop a more cost-effective vein imaging system. In addition, the noise on the image due to the use of a webcam with a low imaging capability has been improved by the noise reduction algorithm.

In Section 3, the fundamentals of NIR imaging are discussed and their hardware characteristics are explained.

III. NEAR-INFRARED IMAGING

Light is an electromagnetic radiation including different wavelengths and frequency ranges, which carries energy in the form of photon particles. The diagram formed of these different ranges of the light is known as the electromagnetic spectrum. In this spectrum, "the visible region" that the human eye can detect is the region having wavelength range of about 380-770 nm. The wavelengths of the electromagnetic spectrum covering 380 nm and below are ultraviolet and X-Ray; 770 nm and above are infrared and radio waves [11].

The electromagnetic spectrum region in the wavelengths range of 700-900 nm is defined as the NIR optical window because it is located near the visible region [12][13].

When light energy strikes a matter, the matter exhibits different interactions with respect to the molecules that make it up. As a result of examining these interactions, the method of obtaining information about the molecules that make up the substance is called spectroscopy [11].

The spectroscopy method which provides information from the substance was the basis for imaging of blood veins.

The radiation in the NIR optical window is scattered by direct penetration into the lower layers by tissues, such as skin and adipose layer with low absorption coefficient. On the other hand, blood mostly absorbs these NIR photons due to its haemoglobin molecules. NIR photons are absorbed slightly by oxygenated haemoglobin in arteries and strongly absorbed by deoxygenated haemoglobin [1][12][15].

NIR spectroscopy, in other words, the visualization of blood veins with the use of NIR radiation, requires cameras that have different characteristics than ordinary cameras used in daily life. NIR systems use cameras with infrared-pass filters (which allow these rays to reach the camera lens) instead of the filter that blocks NIR rays in ordinary cameras [14].

The improved work is designed with an external infrared camera integrated into a mobile device (smartphone or tablet). The built-in cameras in all smart devices currently available on the market, designed for daily use, have a filter that allows visible light to pass through, but prevents the passage of NIR light. However, the method of vascular imaging to be used in this study is based on the detection of NIR light. From this point of view, there are two methods for the user to get a NIR image with his/her mobile device. Either in a professional manner, it should be carefully interfered with the mobile device camera (NIR light blocking filter should be removed and replaced with NIR light passing filter), or an external camera should be integrated to a mobile device via USB support. In the study, a USB-powered NIR camera integrated into the mobile device was used. With this NIR camera, the users have the ability to take their own images and send them to the server computer with a mobile application prepared with Android Studio [20] within the scope of the study. The images transferred to the server computer were processed with the MATLAB [21] program and the vein images consisting of black and white colors were recorded. The parts representing the veins in these recorded images were examined. The results (investigating if the pixel values representing the blood vein have changed or not in the next image) obtained by processing these images, which the user will periodically capture and transfer to the server, will be sent to the doctor as preliminary diagnostic data and to the user for information purposes via the server.

From the hardware perspective, in the NIR imaging process, the tissue to be imaged is illuminated by a NIR LED (700-900nm) or laser light. The skin layer passes most of these light photons towards the lower layers, whereas the haemoglobin in the veins absorb it. Therefore, in the NIR image taken by the camera, parts of the reflected photons appear light-colored. The other dark parts represent the veins. [1]

Lingyu and Leedman [15] found that the optimal infrared image was at 800 nm wavelength. They obtained hand dorsum, palm and wrist images using NIR LEDs having this wavelength. Bazrafkan et al. [8] used 10 NIR LEDs with 940 nm wavelength in their studies. Şeker and Engin [1] worked with 875-940 nm NIR LED which they placed in a ring form. Shotri et al. [10] stated that, in order to develop a cost-effective vein imaging system, the best illumination can be provided with an IR sensitive webcam in the middle and concentric LED placement around it. Seymen et al. [16], in order to make the veins more prominent, they suggested making the participants carry 3 kg weighted bags, squeeze elastic ball for 1 minute and put ice on the hands before hand photographs were taken. In this way, they aimed to facilitate vein detection in needle and laser applications.

In Section 4, the software processing parts of the NIR images to be obtained by using the hardware part of the developed system are mentioned.

IV. PROCESSING NEAR-INFRARED IMAGES

Digital image processing is based on the processing and modification of digital images in electronic media by various ways [17].

The processing of an image basically takes 5 steps; (1) The image is transferred to digital media with a hardware, such as camera or scanner that can record in desired format (NIR image, thermal image, etc.), (2) Defects caused by camera or ambient lighting on the raw image transferred to digital media are eliminated by pre-processing (noise removal, contrast enhancement, etc.), (3) In the pre-corrected image, the parts of the target area of interest (defined as ROI) are separated from the rest of the image, (4) Identification of the parts that have formed specific patterns (corner, curve, etc.) on the ROI, (5) Making inferences on the identified patterns by various methods (machine learning, data mining, etc.) [18]. Bazrafkan et al. [8] determined that the vein image was best obtained from the second region in their NIR finger images, by developing a biometric authentication system with a finger vein map. In order to obtain the vein details, first the High Pass Filter in the frequency domain was used and then the median filter was applied. Finally, vein structures were extracted with the Gabor filter. Zhong et al. [7] obtained vein patterns from the NIR image with the adaptive threshold filter and eliminated the background with the depth threshold filter. They created references as key points from the vein patterns they obtained, and compared these reference points with test vein patterns and calculated similarity score. Francis et al. [19] performed the software part of the system developed for the detection of blood veins in the forearm in 9 steps; (1) Convertion of the NIR image to a gray-level image, (2) Noise elimination with median filter, (3) Gaussian filter for smoothing, (4) Adjusting the intensity of images with Contrast Limited Adaptive Histogram Equalization (CLAHE), (5) Detection of vein edges with the Gabor filter, (6) Separation of veins and other tissues by Otsu's thresholding method, (7) Sharpening of vein shapes by morphological procedures, (8) Obtaining the ROI, (9) Enhancing the vein image for real-time application.

In Section 5, the steps of developing an exemplary NIR imaging system are discussed.

V. APPLICATION

The ongoing doctoral dissertation which includes a NIR vein imaging system, consists of two parts as hardware and software, as shown in Figure 1.

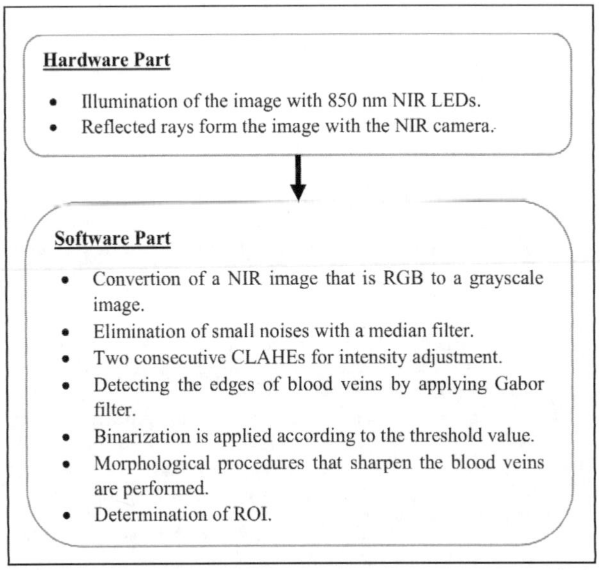

Figure 1. Hardware and software parts of the study, and their contents.

In the hardware part, the tissue to be imaged is illuminated by NIR LEDs having a wavelength of 850 nm. When NIR photons pass through adipose and skin tissue, haemoglobin in the blood absorbs them when they penetrate intravenously. The back-reflected photons are detected by the NIR camera to produce a NIR image of the illuminated portion of the tissue.

In the software part, the steps in Francis et al. [19] work are used. In this section, firstly the Red-Green-Blue (RGB) image obtained in three-dimensional format is rendered two-dimensional (including 0 and 1) as grayscale. Secondly, salt and pepper type small noises are eliminated by applying the median filter. In the third step, in order to adjust intensity, CLAHE algorithm is utilized twice in a row and the contrast of the grayscale image is enhanced. With a re-application of the median filter, the small noise that may occur in this stage is also eliminated. In the fourth step, by applying Gabor filter, the edges of blood veins are detected. In the step five, with the binarization process used, the grayscale image is converted to include only black and white images according to the threshold value, and the blood vein and the surrounding tissue can be extracted from each other. In the sixth step, morphological processes (line-shaped shapes show vein structures better) are applied as pixel insertion or deletion in order to make the blood veins more specific and to show their shapes more sharply. In the last step, the ROI part, which will be studied in more detail on the vein image, is separated from the other sections by considering pixel coordinates.

In this study, the raw NIR image of the forearm with ROI extraction is given in Figure 2. Unlike the working steps of Francis et al.[19], first the ROI was determined and then the image processing part was applied. In Figure 3, the NIR image passing through the filter stages is depicted. The image that has been converted to the black-and-white colour format through the image processing stages is given in Figure 4. The image given in Figure 4 obtained at the stage of vein extraction of the study will be further improved in terms of image processing within the scope of future studies, and the vein structures will be made less discontinuous and more distinctive.

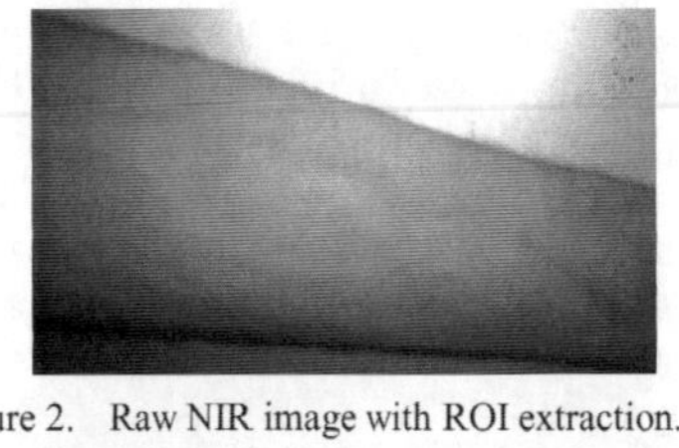

Figure 2. Raw NIR image with ROI extraction.

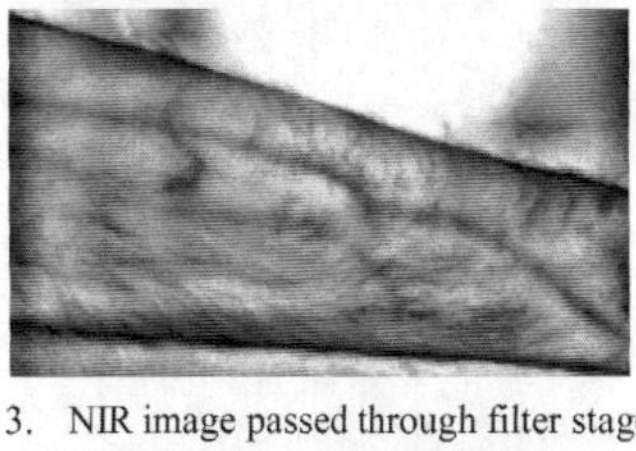

Figure 3. NIR image passed through filter stages.

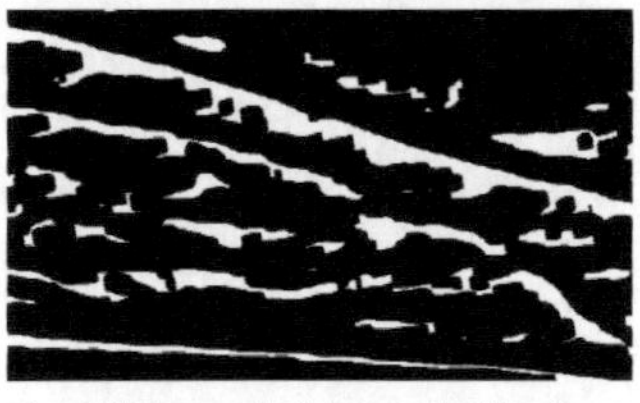

Figure 4. NIR image in black and white format.

In Section 6, an overall summary of the study is given.

VI. Conclusion And Future Work

In commercially or academically developed vein imaging systems, NIR imaging technique is generally used. In terms of user health, the NIR imaging technique is mostly preferred since it does not give harmful beams to the user as in the X-ray or tomography. Moreover, since it is very easy to obtain or create NIR cameras (it is possible to obtain a NIR camera by removing the IR-Cut filter that blocks NIR rays in the ordinary webcam and replacing it with the IR-Pass filter), it is possible to come across numerous studies in the literature with vein imaging systems using such cameras. Most studies on vein imaging are the basis for security systems using vein pattern recognition infrastructure. In this study, which is currently being developed within the scope of a PhD thesis, the hardware and software steps used to develop a vein imaging system to be used for mobile measurements via smart devices are described. In the hardware part, the wavelength range of the NIR technique and the NIR device to be used is specified, and it is explained that the acquired images are generated by the spectroscopy technique. In the software part, in the guidance of a study in the literature which software filters and image processing parts should be applied in order to make the vein images extracted from NIR images more specific is explained and the obtained preliminary examples of the processed images are given.

Within the scope of future studies, the developed system will be used to monitor patients who may have suspected vascular occlusion. Periodically, NIR images can be obtained from patients anytime anywhere by using NIR camera which is integrated with a mobile device (via smartphone or tablet) and personal vein databases will be created in this way. Following the processing of the images and the steps in the software part, the doctor will be notified of any vascular occlusion, and, this way, preliminary information is presented to the doctor and possible delays in treatment are avoided. The blood flow changes detected in the images along with the blood flow rates and numerical values after calculation will be reported both to the doctor and the user via server by e-mail.

References

[1] K. Seker and M. Engin, "Deep tissue near-infrared imaging for vascular network analysis", Journal of Innovative Optical Health Sciences, Journal of Innovative Optical Health Sciences, 10 (3), pp. 1650051.1-12, 2016.

[2] K. Yılmaz, "The design of a new portable ophthalmoscope and utilization of the device for diagnosis and identification", Doctoral Dissertation, Ege University, The Graduate School of Natural and Applied Sciences, 2014.

[3] D. Ai et. al., "Augmented reality based real-time subcutaneous vein imaging system", Biomedical Optics Express, 7 (7), pp. 2565-2585, 2016.

[4] F. B. Chiao et. al., "Vein visualization: Patient characteristic factors and efficacy of a new infrared vein finder technology", British journal of anaesthesia, 110 (6), pp. 966-971, 2013.

[5] R. Fernández and M. Armada, "Multisensory system for the detection and localization of peripheral subcutaneous veins", Sensors, 17 (4), pp. 897,1-16, 2017.

[6] M. Asrar, A. Al-Habaibeh, and M. Houda, "Innovative algorithm to evaluate the capabilities of visual, near infrared, and infrared technologies for the detection of veins for intravenous cannulation", Applied Optics, 55 (34), pp. D67-D75, 2016.

[7] H. Zhong, S. S. Kanhere, and C. T. Chou, "VeinDeep: Smartphone unlock using vein patterns", In Pervasive Computing and Communications (PerCom), IEEE International Conference on, pp. 2-10. IEEE, March 2017.

[8] S. Bazrafkan, T. Nedelcu, C. Costache, and P. Corcoran, "Finger vein biometric: Smartphone footprint prototype with vein map extraction using computational imaging techniques", In Consumer Electronics (ICCE), 2016 IEEE International Conference on, pp. 512-513. IEEE, January 2016.

[9] S. Elnasir and S. M. Shamsuddin, "Palm vein recognition based on 2D-discrete wavelet transform and linear discrimination analysis", Int. J. Advance Soft Compu. Appl, 6 (3), pp. 43-59, 2014.

[10] A. Shrotri, S. C. Rethrekar, M. H. Patil, and S. N. Kore, "IR-webcam imaging and vascular pattern analysis towards hand vein authentication", In Computer and Automation Engineering (ICCAE), 2010 The 2nd International Conference on, 5, pp. 876-880. IEEE, February 2010.

[11] Introduction to spectroscopy. [Online]. Available from: https://edu.rsc.org/download?ac=11384 [retrieved: 03, 2020]

[12] H. Rao, P. Zhang, and C. Sun, "Contrast enhancement for the infrared vein image of leg based on the optical angular spectrum theory", Signal, Image and Video Processing, 11 (3), pp. 423-429, 2017.

[13] S. Crisan, "A novel perspective on hand vein patterns for biometric recognition: Problems, challenges, and implementations", In Biometric Security and Privacy, pp. 21-49. Springer International Publishing, 2017.

[14] S. Sawant, Turn webcam to IR sensitive camera. [Online]. Available from: http://www.instructables.com/id/turn-webcam-to-IR-sensitive-camera/ [retrieved: 03, 2020]

[15] W. Lingyu and G. Leedham, "Near-and far-infrared imaging for vein pattern biometrics", In Video and Signal Based Surveillance, AVSS'06. IEEE International Conference on, pp. 52-52. IEEE, November 2006.

[16] V. Seymen, Ö. F. Boyraz, Ö. Çetin, and M. Z. Yıldız, "Hand-vein pattern detection from infrared images", 2nd International Symposium on Innovative Technologies in Engineering and Science (ISITES), Karabük, pp. 722-729, 2014.

[17] O. Marques, Practical image and video processing using MATLAB: John Wiley & Sons, 2011.

[18] A. McAndrew, An introduction to digital image processing with Matlab: Notes for SCM2511 image processing 1, School of Computer Science and Mathematics, Victoria University of Technology, 2004 [Online]. Available from: http://www.math.hkbu.edu.hk/~zeng/Teaching/math3615/ls.pdf [retrieved: 03, 2020]

[19] M. Francis, A. Jose, and K. K. Avinashe, "A novel technique for forearm blood vein detection and enhancement", Biomedical Research, 28 (7), pp. 2913-2919, 2017.

[20] Android Studio Official Website. [Online]. Available from: https://developer.android.com/studio [retrieved: 03, 2020]

[21] MATLAB Official Website. [Online]. Available from: https://www.mathworks.com/ [retrieved: 03, 2020]

Privacy Preserving Fuzzy Patient Matching Using Homomorphic Encryption

Shiva Ashish Thumparthy
Brainlab
Munich, Germany
email: ashish.thumparthy@brainlab.com

Ilya Sher
Beame.io
Giv'at Shmuel, Israel
email: is@beame.io

Abstract— **Patient record linkage is an important operation that is necessary for identifying similar patients across medical facilities, with the ambition to improve patient outcomes. However, with increased data privacy concerns, these record linkage algorithms must protect sensitive patient demographic information. Previous works have included usage of Bloom filters (susceptible to frequency attacks) and homomorphic encryption (high computational complexity and communication overheads). We propose a record linkage algorithm that utilizes fully homomorphic encryption ciphertext packing for matrices. This ensures that the algorithm remains privacy-preserving and resilient to multiple attack vectors, while allowing multiple patients' records to be compared at once, as opposed to pairwise comparison.**

Keywords- record linkage; homomorphic encryption; ciphertext packing; Bloom filter.

I. INTRODUCTION

Medical record interoperability has been a focus of healthcare systems for purposes of providing better patient outcomes. The objective is to take into consideration patients' comprehensive longitudinal medical histories after aggregating their respective records across medical facilities. However, in order to facilitate the exchange of Electronic Medical Records (EMRs), it is essential to know which records correspond to which real-world patient entities, and at which facilities these records reside. In an age of increasing privacy and confidentiality concerns, the comparison of unencrypted patient demographics is obsolete due to strict guidelines mandating protection of sensitive patient information. Therefore, the aim is to perform statistical patient record linkage in a privacy-preserving setting, while maintaining high linkage quality.

The rest of the work is structured as follows. In Section 2, we describe the background of the patient matching problem. Section 3 describes previous solutions and the basis of our work. Section 4 provides the details of our solution. The dataset used and preliminary results obtained are summarized in Sections 5 and 6, respectively.

II. PROBLEM

In jurisdictions without a nationally issued unique identity number, organizations rely on comparisons of patient demographic attributes to establish linkage using quasi-identifiers such as name, date of birth and recent addresses. Comparison of unencrypted demographics, however, leaks sensitive patient information between medical facilities. In addition, this process leaks critical business information to rival healthcare facilities, which is often thought of as a loss of revenue opportunity. Finally, as is often the case while dealing with data from multiple sources, demographics are subject to data entry errors such as character transpositions. Therefore, an important step in realizing this goal is identifying patients with records across facilities through record linkage algorithms, preferably in a statistical, privacy-preserving fashion that is resilient to data entry inconsistencies, i.e., in a probabilistic manner.

III. SOLUTION

Multiple methods to perform patient matching have been proposed. A common building block to encode patient attributes is the Bloom filter. Bloom filters represent a digest of information without containing any original information in the clear. Originally, Bloom filters contained a digest of each individual attribute, i.e., one attribute per Bloom filter. Subsequently, improvements allowed for the representation of an entire demographic record in one Bloom filter, known as a Record Bloom Filter (RBF). However, Bloom filters were subject to scrutiny due to brute force and frequency attacks, especially in the single-attribute case [1].

Homomorphic encryption of Bloom filters has been proposed in literature to overcome this shortcoming. Homomorphic encryption is a form of encryption that allows computation on encrypted data. The result of the computation, when decrypted, is the same as if the operations had been performed on the unencrypted data. The computing party cannot decipher any data it is performing computation on. Homomorphic encryption has been widely thought of as an important block in solving the patient matching problem, as it is a form of encryption that is considered to be resilient to brute force attacks, even in the quantum-computing realm. However, this form of encryption is highly computationally intensive and is accompanied by large communication overheads.

Randall et al. [2] proposed homomorphic encryption of Bloom filters, which would insure against these attacks at the cost of high computational and communication overheads. This work, in addition to Cruz et al. [3], serves as a basis for our solution to patient matching using homomorphic encryption. Finally, we leverage homomorphic encryption ciphertext packing techniques for matrix multiplication, as proposed by Duong et al. [4] to optimize batch comparison.

IV. METHODOLOGY

We have worked on optimizing homomorphic encryption to the patient matching use case. Our primary contribution is a method of matching lists of Bloom filters at once as opposed to pairwise comparisons of encrypted Bloom filters between two facilities. We achieve this using homomorphic encryption ciphertext packing techniques, which allow for encrypting large volumes of data into a single ciphertext. This allows for computations on entire matrices of data as opposed to individual Bloom filter vectors, thereby reducing the number of ciphertexts required from n (the number of Bloom filter vectors) to 8.

Encrypted records are compared to arrive at an encrypted binary result (match vs non-match) based on an agreed similarity threshold. This ensures that no parties have access to patient demographics, either in plaintext or through brute force attacks on traditional hash-based encoding schemes.

The protocol makes use of two additional third parties called the Decryption Unit (DU) and the Linkage Unit (LU) [2]. The steps to determine record similarity using the Tversky similarity index for two medical facilities is as follows:

1. The DU generates a homomorphic key-pair, and shares the public key with the facilities and the LU.
2. Each facility breaks patient demographics into bigrams, which are then hashed into one RBF per patient. These RBFs are stacked on top of each other to generate a RBF matrix. The RBF matrix is multiplied with the sum of the numerator and denominator of the Tversky index threshold. Two ciphertexts are then generated from the RBF matrix: the first by encryption, the second by inversion and encryption.
3. Two additional vectors are generated at each facility, with the plaintext count of 1s from each row and each column of the RBF matrix. The count vectors are multiplied with the numerator of the Tversky index threshold. They are then encrypted into one ciphertext each using vector packing techniques. All 4 ciphertexts are sent to the LU.
4. The LU randomly picks the row count and encrypted RBF from one facility. It then picks the column count and the inverted-and-encrypted RBF of the other facility. The two RBF ciphertexts are multiplied to generate the multiplied matrix ciphertext. This is sent to the DU, along with the chosen row and column count ciphertexts.
5. The DU decrypts the received ciphertexts. The row count vector is converted into a row matrix (with same dimensions as the multiplied matrix) by making each element a row in the row matrix. Similarly, the column count vector is converted into a column matrix by making each element a column in the column matrix.
6. The DU then subtracts the generated row and column count matrices from the multiplied matrix to generate the plaintext result matrix, which it forwards to both facilities.

V. DATA SOURCES

The patient matching problem in the privacy-preserving setting is hard to realize and implement in the real world. This is mainly due to lack of comprehensive test data and lack of inexpensive data masking techniques. Synthetic data was generated, and errors were introduced at different rates to calculate linkage quality.

VI. RESULTS AND CONCLUSION

The optimized matching algorithm outperforms naïve comparison of encrypted Bloom filters, as shown in Table 1.

TABLE I. TIME TAKEN BY THE PROPOSED APPROCH VS NAÏVE VECTOR HOMOMORPHIC ENCRYPTION OF BLOOM FILTERS

Matrix size	**Time taken (s)**			
	Vector encryption	***Matrix encryption***	***Vector matching***	***Matrix Matching***
4*4	0.0626682	0.0482091	2.07659	0.0584722
8*8	0.125355	0.0412167	8.06762	0.0563123
16*16	0.252727	0.10382	32.1595	0.147115
32*32	0.502159	4.01244	128.199	5.84446

Real-world limitations were taken into consideration while designing a privacy-preserving patient matching solution using homomorphic encryption that significantly lowers computational costs and communication overheads. Future findings will include statistics for larger matrices, using key recryption to reduce the size of keys required.

ACKNOWLEDGMENT

This research was conducted by Brainlab AG, in collaboration with Beame IO towards building patient record registries in Germany.

REFERENCES

[1] P. Christen, R. Schnell, D. Vatsalan, and T. Ranbaduge, "Efficient cryptanalysis of bloom filters for privacy-preserving record linkage", Pacific-Asia Conference on Knowledge Discovery and Data Mining, Springer, Cham, May 2017, pp. 628-640.

[2] S. M. Randall, A. P. Brown, A. M. Ferrante, J. H. Boyd, and J. B. Semmens, "Privacy preserving record linkage using homomorphic encryption", Population Informatics for Big Data, Aug. 2015, 10.

[3] M. S. H. Cruz, T. Amagasa, C. Watanabe, W. Lu, and H. Kitagawa, "Secure similarity joins using fully homomorphic encryption", Proceedings of the 19th International Conference on Information Integration and Web-based Applications & Services, ACM, Dec. 2017, pp 224-233.

[4] D. H. Duong, P. K. Mishra, and M. Yasuda, "Efficient secure matrix multiplication over LWE-based homomorphic encryption", Tatra Mountains mathematical publications, 67(1), Sep. 2016, pp. 69-83.

Reference Design Model for a Smart e-Coach Recommendation System for Lifestyle Support Based on ICT Technologies

[1]Ayan Chatterjee, [1]Martin W. Gerdes, [1]Andreas Prinz, [2]Santiago G. Martinez, [2]Anine Christine Medin
[1]Faculty of Engineering and Science
[2]Faculty of Health and Sports Sciences
University of Agder
Grimstad, Norway
e-mails: {ayan.chatterjee, martin.gerdes, andreas.prinz, santiago.martinez, anine.medin}@uia.no

Abstract—**As acknowledged by the World Health Organization (WHO), the demographic development shows that by 2030, eight out of ten foremost causes of death will be connected to risk conditions of lifestyle diseases, regardless of gender. Chronic illness associated with modifiable lifestyle factors will be accountable for the highest death worldwide. Health behavior change should be given priority to avoid serious damages. An Electronic Coach (e-Coach) system can empower people to achieve a healthy lifestyle with early risk predictions and appropriate tailored lifestyle recommendations. Research in e-Health has the potential to provide methods to improve personal healthcare with Information and Communication Technologies (ICT). An Electronic Health (e-Health) virtual coaching recommendation system can monitor people and convey the appropriate recommendations to improve their lifestyle and prevent against non-communicable or lifestyle diseases. This paper addresses the potential of selected emerging information and communication technologies to make e-Health systems smarter, more collaborative, and more efficient. As a result of the analysis, a reference design is discussed here to develop and validate the performance of a smart e-Coach system utilizing ICT, Internet of Things (IoT), and Artificial Intelligence (AI) to provide individual lifestyle recommendations aiming at a healthier lifestyle to prevent obesity and overweight. The healthcare sector is still looking for collaborative, user-friendly, optimized, cost-effective, reliable systems for health e-Coaching with automatic, meaningful, empirical evidence-based, and personalized lifestyle recommendations. The focus is on this system's implementation and validation, considering obesity and overweight as a case study.**

Keywords- e-Coach; e-Health; AI; IoT; Human-centered design; ICT; Lifestyle diseases; Recommendation; Decision support; Data security and privacy; Ethics; Ontology.

I. Introduction

Globalization, rapid urbanization, poor dietary habits, and a sedentary lifestyle negatively impact people's health across the globe, affecting people from different socioeconomic groups. Key risk factors are excessive use of alcohol, inappropriate nutrition, physical inactivity, surplus consumption of salt, intake of saturated fat, smoking, drinking of excess sugar-containing foods, and beverages. All these factors can contribute extra weight gain, elevated blood glucose level, high blood pressure, higher total cholesterol in the blood, followed by different physiological impacts such as *obesity, cardiovascular disease, hypertension, and diabetes type II,* as illustrated in Figure 1. Non-communicable or lifestyle diseases are the most common cause of death worldwide. It is the leading unconditional probability of dying between ages 30 and 70 years [12]. *'Obesity and overweight'* is one of the major lifestyle diseases thsat lead to other health conditions, such as Cardiovascular Diseases (CVDs), Chronic Obstructive Pulmonary Diseases (COPDs), cancer, diabetes type II, hypertension, and depression. It is a nutritional disorder and represents an excess amount of energy storage as mass in the body.

e-Health monitoring has become increasingly popular providing ICT-based remote and timely care support to its patients and healthcare providers. A *health e-Coach system* is an interactive, secure, monitoring, and caregiving system to produce automatic, meaningful, empirical evidence-based and personalized lifestyle recommendations to attain personal wellness goals. The goals are achieved by maintaining a balance in social life, physical activities (such as sleeping, exercising), eating, and daily stress level. The conferred reference *e-Coach model* will capture personal, physiological, activity (sleep pattern, exercise), nutrition (fruits, vegetables, salads, energy drinks, and alcohol) and contextual data (spatial, temporal and weather) from secure wearable bluetooth enabled devices, manual interactions, feedback, and customized questionnaires over time. The collected time-series data will train machine learning models for behavior analysis, early prediction of wellness trends and risks.

Once the health risk is predicted, the e-Coach system will produce lifestyle recommendations to target individuals to improve their well-being, health, and physical performance by determining and recommending individual lifestyle changes. Emerging e-Coach systems reveal certain limitations in terms of technologies, quality, and performance. To overcome those limitations, an integrated and collaborative e-Coach service is described, addressing the cooperation of secure cloud-based fitness and well-being services with the public health and the care infrastructure.

e-Health research helps us to solve e-Coach associated challenges, such as activity and wellness status determination, an ontology for e-Health information, the selection of an e-Coach strategy, the generation of automated, meaningful, empirical, evidence-based and personalized lifestyle recommendations and their automatic assessment compared to a set goal. This study is deeply focusing on *what to coach* and *how to coach*, including the design, development, testing, and evaluation of observational evidence-based, context-specific, and individual lifestyle recommendations. Here, we have highlighted a reference design model for *e-Coach and recommendation generation for obesity and overweight,* and we are currently working on its implementation and the validation phase.

A high-level reference design is presented in this paper with special consideration of upcoming technologies, such as data ontologies from Semantic Sensor Networks (SSN), IoT, AI, Decision Support Systems (DSS), big data, data mining, and Human-Computer Interaction (HCI). The objective of the considered *health e-Coach* system, as illustrated in Figure 5, can be summarized as follows: a) improve personal healthcare with the help of ICT; b) deliver high quality, evidence-based, secure, cost-effective, timely care to support people for sustaining a healthy lifestyle; c) predict the health risks and generation of risk alerts; d) ensure a rule for personal data security, portability, and remote access; e) normalize the heterogeneous format of personalized data with appropriate e-Health ontology model; f) analyze human psychology, human behavior and; g) follow defined medical guidelines for the generation of lifestyle recommendations.

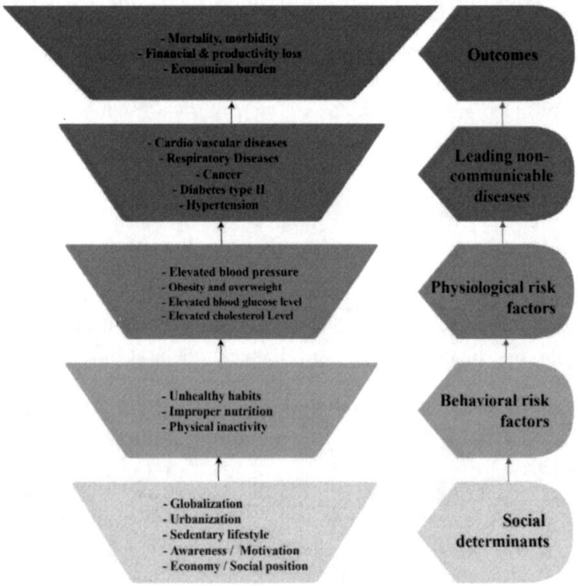

Figure 1. Development of non-communicable or lifestyle diseases.

The main contributions of this paper are as follows:

1) The reference design model of an e-Coach system for guiding obese and overweight people is described. It provides smart and efficient supervision of participants aiming at a healthier lifestyle within a highly secured environment following the medical, security, and human-centric design guidelines.

2) Collection of necessary personal, physiological, activity, nutrition, and context related data following data governance.

3) Derivation of knowledge from raw data and e-Health knowledge representation with ontology and reasoning.

The remainder of this paper is structured as follows. In Section II, we summarize the related work and highlight the differences between our current and existing work. Section III presents health e-Coach design factors. In Section IV, we describe an e-Coach reference model for obesity and overweight. The paper is concluded in Section V.

II. Related Work

Different projects have been conducted by different study groups on *health e-Coach* to generate a personalized recommendation plan for a healthy lifestyle. The idea is to give remote care to participants with a suggestion for a healthy diet and fitness plan to prevent obesity and obesity-associated non-communicable diseases. Behavior and health are strongly linked. Healthy eating and regular physical activity can lead to a healthy lifestyle [1]. Having proper e-Coaching recommendation plan may help people to accomplish health goals to maintain healthy behavior. Research projects have addressed the following two types of recommendations approaching obesity:

(a.) food-based recommendations where a Machine Learning (ML) algorithm with the experience sampling method has been used to develop an interactive support system for dietary recommendations. Sensors are used to collect data related to state (emotion, social activity), context (e.g., location, time, and weather), and collected data are sent to a secured storage over the IoT infrastructure. After that, a decision support system is developed to predict human behavior and generate timely feedback for activity. It uses a classification algorithm (decision tree approach) to warn people against unhealthy food, and a clustering algorithm (Hierarchical Agglomerative) for profiling. It generates feedback (passive feedback or cognitive behavioral therapy or adaptive or active) with adaptive messages (intervention) [4], including proper food suggestions, alerts for allergic or harmful substances, recommendations for relaxation and yoga [3]. The solution for feedback is either Short Message Service (SMS) based, web-based, mobile application based, or combined [4].

(b.) physical activity-based recommendations, where a tracker has been maintained for daily step count, metabolic equivalent of tasks, kilocalories, and distance to reduce sedentary behavior. A healthy diet, enough physical activity to ensure energy balance, not smoking, and limited alcohol are four key factors to prevent obesity significantly. Data has been captured over time and studied (data processing and transformation, such as cleaning, formatting, pre-processing) with ML algorithms to give feedback wheather the target is achieved or not. Based on the result, the personalized and predictive model recommends changing personal behavior, daily routine, or diet plan [5].

Obesity may lead to social isolation or loneliness, or depression [6], that may intensify the chance of premature death. Demisris et al. [7] proposed a theoretical framework for assessment and visualization of older adults' wellness. It is an enhanced screening platform for wellness and a telehealth component captures vital signs and a customized questionnaire. The system also incorporates a gait analysis component. Based on the analysis of multiple health and wellness parameters, the system aims to identify early trends or patterns to help in health risk prediction for overweight people, and recommends a wellness solution with technological tools and techniques. The wellness parameters of the system are physical well-being, social well-being, and spiritual well-being. Gerdes et al. [2] proposed a holistic concept of a coaching system for the *Conceptualization of a Personalized e-Coach for Wellness Promotion*. The paper describes a *Continuous Process of Wellness Management* with a gap to complete the cycle for continuous process. The gap is closed with a *case study of obesity and overweight,* and the reference design model discussed here.

The identified gaps in current health e-Coach studies are summarized as a lack of a.) context-specific empirical, tailored recommendation; b.) medical guidelines to follow to generate lifestyle recommendations; c.) personal, health, and wellness data protection plan (security and accessibility); d.) consent to recruit participants in the study as a part of fair data economy; e.) human-centric designing (co-design, user experience, safety, cost); f.) credibility analysis; g.) complexity analysis; h.) methodology, guidelines, protocols, i.) analysis of efficacy, sustainability, maturity, j.) proper research (legal mobile apps and their data protection policy, ligal body sensors, or actuators to collect important physiological data and storing them at unsafe storage); k.) post-research participant's data security plan.

Our study is focused on mitigating the gaps, as described above, and transform distributed, heterogenous health and wellness sensor data into meaningful information to build a machine learning model for health risk prediction and to generate an effective individualised recommendation in case of obesity, as well as to turn it into a behavioural motivation for an effective human-eCoach-interaction. The concept of the *Reference System for Telehealth and Telecare services* and *System Architecture for Future Telehealth and Telecare Services* [8] has been extended here to design the *health e-Coach* system to guide obese and overweight people.

III. HEALTH E-COACH DESIGN FACTORS

In this section, we first describe the essential factors to be considered while designing, developing, testing and validating the performance of a *health e-Coach* system to achieve the research goal.

A. Perspective

The reference digital *health e-Coach* system will capture personal, physiological (blood pressure, heart rate, blood cholesterol level, blood glucose, height, weight), contextual (time, location, environment, ocial), and behavioral (physical activity, nutrition, sleep) data over time from medically approved secure wearable sensors, customized questionnaires, manual interaction (interviewing), mobile apps, and feedback forms. Data will be stored in a secure storage and will be utilized to train a machine learning model for behavior analysis and the early prediction of wellness trends and risks, as illustrated in Figure 2. Once wellness trends and health risks are predicted, the e-Coach system will deliver corresponding alerts, suggestions, and lifestyle recommendations to its target individuals to improve their well-being, health, and physical performance by determining and recommending optimal individual lifestyle factors (example - healthy diet, regular exercise). The perspectives of the discussed "health e-Coach" system here are to deliver lifestyle recommendations to prevent obesity and overweight, and, thereby, help participants to lead a healthier lifestyle to reduce the risk of chronic illness.

B. Approach

We are going to follow the well-established design-science-research methodology [9] to system design, development, and evaluation of the e-Coach system. The overall tasks can be summarized as follows: (1.) systematic literature review; (2.) design and development (feasibility study, data collection, development of a recommendation engine, e-Coaching through interaction); (3.) trial run; and (4.) model evaluation with performance parameters (goal achievement). The feasibility study includes (a.) what to measure? [independent variables (e.g., demographics) and dependent variables (e.g., BMI)]; (b.) how to measure? [spending time to explore and find the right instruments and methods]; (c.) type of data to be collected [personal, physiological, contextual and behavioral] and determination of data collection process; (d.) policy for the recruitment of the participants; (e.) preparation of consent form; (f.) data security and privacy - General Data Protection Regulation (GDPR); (g.) usage of a standard framework for e-Health lifestyle recommendation and determination of *health e-Coach* efficacy.

C. Data Collection

Personal, physiological, contextual, nutrition, and activity related data are most important in this study to determine the wellness status and we are going to capture a selected set of data in Table I from participants with minimal effort. The frequency of data collection (personal, physiological, contextual, activity, nutrition), the specific target group to be monitored, modelling of an SSN ontology, the appropriate list of sensors and assessment tools, the development of a gateway for data collection, storing of clean, normalized data in the storage (as data volume will be high), and ensuring its security are important. Management of data from continuous observations, management of incomplete data is in focus, as illustrated in Figure 3.

After the collection of necessary data, training, and cross-validation of the machine learning model with appropriate features for health risk prediction are important. The selection of proper machine learning and deep learning algorithms for time-series analysis of collected health and wellnesss data and their comparative study concerning accuracy score, classification report, the confusion matrix are all important and need research focus for generating an effective recommendation plan. Once risk prediction is made, it is necessary to deliver alerts, recommendations, and suggestions to the participants so that they can follow them. This requires interactive human-computer interaction, building trust about the system, integrating offline (human) coaching psychology to the e-Coach, selection of an effective recommendation plan, and improving it further.

TABLE I. OVERVIEW OF TYPE OF DATA TO BE COLLECTED

Personal Data	Age, gender, mobile, email, postcode, income, education, social participation, medical history, food preferences, habit, waist to hip ratio, smartphone app. usage for behavioral tracking, internet connectivity
Physiological Data	Blood pressure, heart rate, blood glucose, total blood cholesterol, weight, height
Contextual Data	Temporal and spatial data, weather data
Activity Data	Raw acceleration, energy expenditure, metabolic rate, physical activity intensity, total steps, activity bouts, sedentary bouts, sleep latency, total sleep time, sleep efficiency
	Duration of activity and non-activity
Nutrition Data	Intake of energy drinks, alcoholic beverages
	Intake of core foods such as vegetables or fruits
	Intake of discretionary foods such as unhealthy "junk foods."

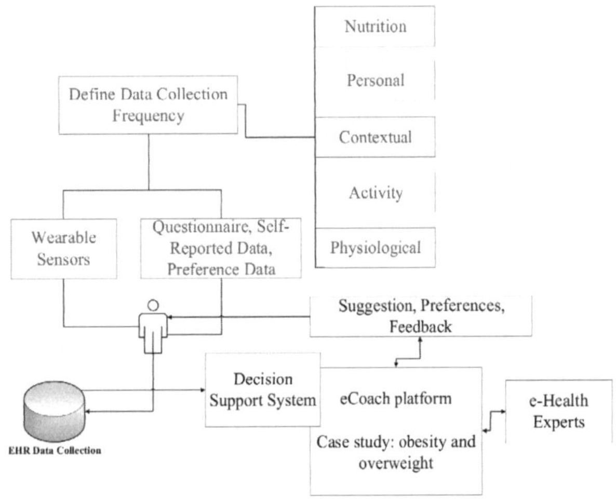

Figure 2. e-Coach overview of obesity and overweight.

D. *Secure Wearable Sensors and Data Security*

Physiological and activity data are generally collected via secure wearable devices. The wearable activity monitors are connected with a smartphone or tab or computer via bluetooth nearfield communication technology. Participants will install a developed mobile application that will transfer the collected activity and physiological data to a private secured storage, in compliance with GDPR or similar. Though different wearable health sensor devices are available on the market, we need to choose only legal devices. Wearable sensor devices should have regulated access control policy, approval by the authorized medical board (example, Food & Drug Administration (FDA), US) to collect, process, and store health related personal data.

In our study, collected personal, physiological, activity, nutrition, and contextual data, as described in Table I, will be processed securely during transmission using encryption, and there will be no possibility of information leakage. Transport Layer Security (TLS), a Virtual Private Network (VPN), authentication (form based, active directory, API key), and authorization (access control) will be used in our model to secure data during transmission and during storage. Data will be stored at a secure storage with Fast Healthcare Interoperability Resources (FHIR) standard and data will be accessed with HTTP based REST API following an API key authentication.

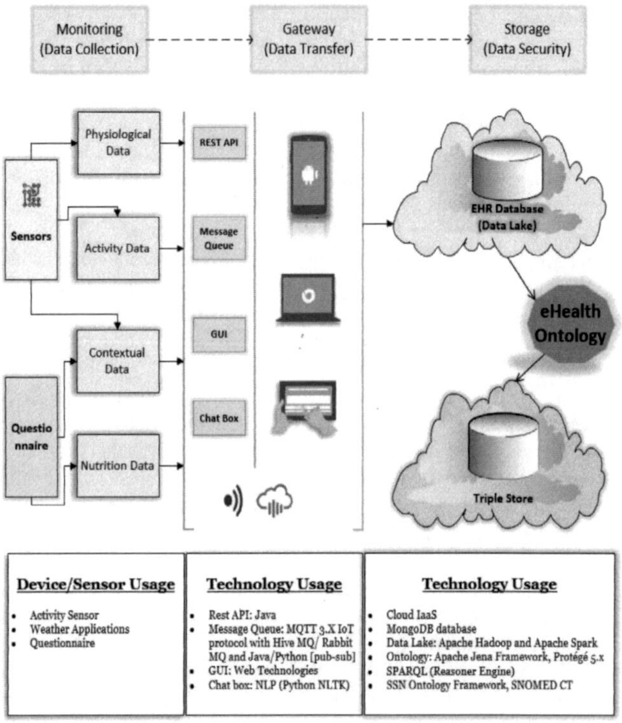

Figure 3. Data collection from obese and overweight participants.

A dedicated platform will be used for collecting, storing, analyzing, and sharing sensitive data in compliance with privacy regulations (GDPR). The API key, access-control list, two-layer authentication with form based authentication (Unique User-id (UUID) or username and password), and active directory security or RSA key are appreciated to protect the system from illegitimate users.

Linking between unique user-id and personal identifiers (such as name/email/phone/postcode) should be stored in a secure environment, and there will be no clue to perform any cross-identification or backtrack legal formulation of any participant. In line with the participant's consent or request, information about the participants must be anonymized by removing personal identifiers or deleted after the successful deployment of the project.

E. *Knowledge Presentation*

Integrating raw, unstructured, high volume data from heterogeneous sources like wearable sensors, questionnaires, mobile applications, interviewing, and creating a compact, machine-understandable, structured, meaningful information from it, is challenging. But, it is important to extract knowledge from data in order to convey tailored lifestyle recommendations.

Semantic web is a concept and a group of technologies that support the development of machine-interpretable data and allows machines to process, explore, and analyze web documents effectively. It has offered an universal framework for describing data with common data formats, specifying data about data with semantic web languages URI, XML/ XML schema, RDF, RDFS, OWL, RIF/ SWRL rules, SPARQL, and assisting better knowledge representation, advanced access, formal analysis of resources, uphold standardization, increase interoperability among heterogeneous sources and networks, data integration, data discovery, and situation awareness. The semantic sensor network incubator group (SSN-XG) has developed an SSN ontology to model sensor devices, systems, and processes observations and to allow its network, sensors, and data to be organized, installed, managed, queried, controlled through high-level specification. It is a domain-independent ontology [10].

OWL (defines what to write) processes web information when RDF (defines how to write with triple-store-resource, property and value) is used for describing. RDFS, which is an extension to RDF, provides data-modeling and structured vocabularies for RDF data. OWL (turtle syntax) with SPARQL API (a query engine for querying and updating RDF models using the SPARQL standards [13]. SPARQL is a query language and a protocol for accessing RDF designed by the W3C RDF data access working group. SPARQL queries RDF graphs.), Protégé (an open-source ontology development tool with OWL support), Apache Jena (a free and open source Java framework for building semantic web, linked data applications and extract knowledge from ontology), SSN Ontology, which gives sensor descriptions and observations, and SNOMED-CT (a systemized nomenclature of medicine since 1965 and designed as an ontology and it is an organized list of a wide variety of clinical terminology defined with unique codes namely ICD) have all been used in our research to model the health e-Coach ontology for the case study: obesity and overweight and the same ontology modelling is beyond the scope of this paper.

F. *Human-Centric Design*

In our research, we will use human-centric approaches to design and develop the desired *health e-Coach* system and make it more usable, utilizing human factors or ergonomics and usability learning and skills [ISO 9241-210 - Ergonomics of Human-Centered System Interaction]. A solid user-centered design that will be used has been discussed here. It follows a cyclic process of *1. plan 2. analyze 3. design and 4. test and evaluate.*

Accomplishing user-centered design is very challenging, and it needs a comprehensive process, beginning with planning, analyzing and closing with testing and improving it further. Such a system can only be accomplished by putting the user experience at the heart of the development process. It is a process that places human needs and limitations at a higher priority compared with other targets during design thinking and implementation. To achieve a successful human-centric design, designers should focus on the following [10]: a.) co-design; b.) define the problem that needs to be solved well; c.) understand user characteristics and needs; d.) perform research on user experience and understand user behavior and impressions; e.) the developed prototypes must be tested and validated by real users to ensure smooth user experience, design consistency, reduction of learning effort, and building trust; f.) a continuous communication between user and the designer/design to gain feedback to ensure further design improvements; g.) the design should be shaped in the user's cognitive knowledge; h.) empathic system experience and according to system development that users love to use daily.

We envision to integrate elements of the human process of coaching in our research, such as feedback/ rating, preference sharing, user-friendly human-computer interaction, intelligent interactive session, problem solving, timely alert (SMS/e-Mail), wellness vision, motivational suggestions, encouragement, assessing human psychology and sentiment based on physiological and contextual data, educating people to make the *human-eCoach-interaction* effective for obesity recommendations. In human centered design, all the human factors, social factors, and technology factors intermingle together under the human activity umbrella. The human centered design process will flow in between three main phases: research, prototype, and implement keeping human at the heart of the design.

G. *Requirement Framework*

The subsequent technical requirement domains have to be considered in the design, development, and evaluation of the discussed *health e-Coach* system and its supporting infrastructure components are as follows (Table II):

TABLE II. REQUIREMENT FRAMEWORK FOR E-COACH SYSTEM

Feasibility	Technical and financial feasibility study to ensure what to measure, how to measure, resource management, deadline, and risk management.
Scalability	The developed solution must support the expected growth in participant numbers, services, and devices.
Data Governance	The developed solution must ensure a fair data economy; it must be trustworthy and allow the implementation of complex, distributed, security policies, and rules following GDPR.
Interoperability	The developed solution must be hosted in a secure domain, but it will enable interactions with other national, international, and third-party domains securely.
Usability	The solution must include a co-design initiative to increase its acceptability and error-free handling.
Reliability, Robustness, Availability	Collection and storing of the needed amount of health and wellness data-keeping participant's burden as low as possible, data security, accurate health risk prediction and real-time, meaningful, evidence-based, context-specific, personalized, risk-free lifestyle recommendations will make the solution reliable.
Training	Proper training (digital literacy) should be provided to increase error-free handling of the system and system acceptability.

H. *Goal*

The goals and focus of the discussed *health e-Coach* system can be summarized as follows – a.) encourage co-design; b.) creation of a compact, intelligible abstraction from massive, raw, unstructured observations for health and wellness data using e-Health ontology; c.) protection against illegitimate access to the system and personal health records; d.) collection of personal, physiological, activity, and nutrition data to build a machine learning model for health risk prediction and to understand if there is any negative change in behavior pattern; e.) model automatic, meaningful, and empirical evidence-based and personalized lifestyle recommendations to achieve health wellness goals. Continuous monitoring of the willing participants and evaluation of time-series monitoring data will produce detail knowledge about the individual relation between nutrition, activity, and Basal Metabolic Rate (BMI). Based on the obtained knowledge, the lifestyle recommendations, as depicted in Figure 4, with regards to nutrition and activity, will be given to the participants to avoid *obesity and overweight*.

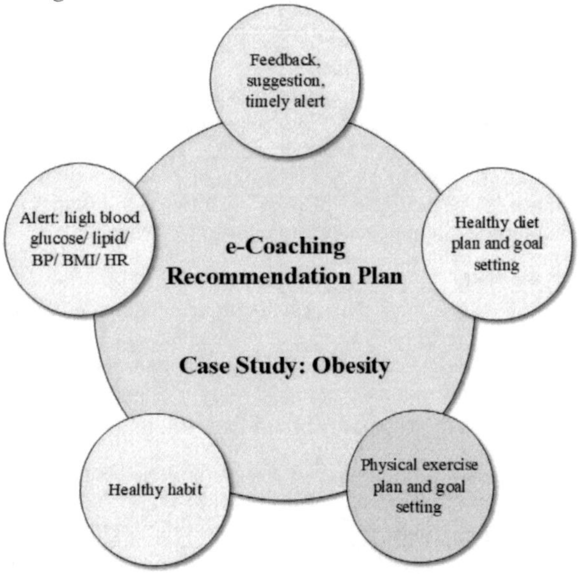

Figure 4. Recommendation plan for obesity and overweight.

IV. PROPOSED MODEL DESCRIPTION AND RESULT

The *holistic obesity and overweight e-Coach model* is based on the co-design initiative, and the participant is located at its core. As illustrated in Figure 5, the complete model is divided into *seven* components – 1.) *formulation* – This includes: systematic literature review to gain complete knowledge about obesity and overweight, its causing factors and associated results, feasibility study (technical and financial), defining *what to coach* and *how to coach*, identification of needed data to be collected, defining frequency of collection, and e-Health ontology design; 2.) *selection of participants* – This includes: advertisement with defined inclusion and exclusion criteria, consent preparation, consent signing, training, participant's account creation, handover of wearable sensor devices; 3.) *infrastructure* – This includes: the needed cloud infrastructure for data collection (network elements, firewall, protocol, active directory authentication, VPN) and secure database storage; 4.) *data collection* – This includes the collection of physiological data, environmental contextual data, personal data, activity data, and nutrition data from participants via secure wearable sensors, manual interaction, interviews, smartphone apps, feedback, and customized questionnaires overtime at defined frequencies; 5.) *decision support system* – This is needed to train a machine learning model for behavior analysis, early prediction of wellness trends and risks; 6.) *e-Coach and recommendation plan* – The e-Coach system will help to model lifestyle recommendations to achieve health wellness goals and will advise about physical activity, nutrition, and other relevant factors for healthier lifestyle coaching. The e-Coach system integrates a proper recommendation plan that follows clinical guidelines; 7.) *human e-Coach interaction* – This menas incorporating elements of the human process of coaching, such as feedback/ rating, preference sharing, user-friendly human-computer interaction, intelligent chat service, problem solving, timely alert (SMS/ e-Mail/ Voice), wellness vision (physical/ social/ emotional/ spiritual), motivational suggestions, encouragement, assessing human psychology and sentiment based on physiological and contextual data, educating people to make the human-e-Coach-interaction effective for lifestyle recommendations.

Health e-Coaching is a continuous process that collects health and wellness data from participants continuously via wearable clinically approved sensors, smartphone apps, questionnaire over time and process them with AI algorithms to predict health risk, behavioral pattern, change in habits, random variation in lifestyle over a week/ month/ day-by-day, and assessment of behavioral data to generate evidence-based, real-time, context-based lifestyle recommendations, suggestions, alerts and followed by, the evaluation of individual goal. The role of ICT experts (researchers/ developers/ technology experts) is to design, develop, test, and evaluate the performance of an e-Coach system in active collaboration with health experts (nutritionists/ nurses/ physicians/ psychologists). The health experts help to design and validate the rules for needed data collection, lifestyle recommendations, health risk prediction, and determination of vital physiological signs.

V. CONCLUSION

We are working on the assumed hypothesis that *'an effective e-Coaching mechanism can prevent obesity and overweight with automatic generation of personalized lifestyle recommendation.'* The data collection and lifestyle recommendation process should neither create any side effects nor any major discomfort to the participants. Hence, researchers need to collect the maximum amount of personal, physiological, activity, nutrition data with a minimum amount of collection, and measurement overhead. The entire measurement process should not be burdensome and, hence, we are going to collect a listed set of data in Table I with the participant's consent. Integrating offline human coaching psychology to an e-Coach, selection of an effective lifestyle recommendation plan, and improving it further to reach the convergence criteria of overweight to normal weight, is a computationally hard problem. This reference design model can be extended to e-Coach people against other lifestyle diseases.

REFERENCES

[1] H. Rutjes, M. C. Willemsen, and W. A. Ijsselsteijn, "Understanding effective coaching on healthy lifestyle by combining theory-and data-driven approaches", InPPT@ PERSUASIVE, pp. 116, 2016.

[2] M. Gerdes, S. Martinez, and D. Tjondronegoro, "Conceptualization of a personalized ecoach for wellness promotion", In Proceedings of the 11th EAI International Conference on Pervasive Computing Technologies for Healthcare, ACM, pp. 365-374, 2017.

[3] A. Fadhil, "Towards Automatic and Personalised Mobile Health Interventions: An Interactive Machine Learning Perspective", arXiv preprint arXiv:1803.01842, 2018.

[4] S. Gerasimos, G. Weiss, B. Boh, L. Lemmens, and A. Roefs, "Machine learning techniques in eating behavior e-coaching", Personal and Ubiquitous Computing, vol. 4, 645-659, 2017.

[5] T. Bdijkhuis et. al, "Personalized physical activity coaching: a machine learning approach", Sensors, vol. 2, 623, 2018.

[6] L. Ellwardt, M. Aartsen, D. Deeg, and N. Steverink, "Does loneliness mediate the relation between social support and cognitive functioning in later life?", Social science & medicine, pp. 116-124, 2013.

[7] D. George, H. J. Thompson, B. Reeder, K. Wilamowska, and O. Zaslavsky, "Using informatics to capture older adults' wellness", International journal of medical informatics, vol. 11, pp. 232-241, 2013.

[8] M. Gerdes, F. Reichert, J. P. Nytun, and R. Fensli, "Reference design for smart collaborative telehealth and telecare services based on iot technologies", International Conference on Computational Science and Computational Intelligence (CSCI), IEEE, pp. 817-820, 2015.

[9] K. Peffers, T. Tuunanen, M. A. Rothenberger, and S. Chatterjee. "A design science research methodology for information systems research", Journal of management information systems, vol. 3, pp. 45-77, 2007.

[10] J. Giacomin, "What Is Human Centred Design?", The Design Journal, pp 606-623, 2014.

[11] N. Lasierra, A. Alesanco, D. O'Sullivan, and J. García. "An autonomic ontology-based approach to manage information in home-based scenarios: From theory to practice", Data & Knowledge Engineering, pp.185-205, 2013.

[12] https://www.who.int/news-room/fact-sheets/detail/obesity-and-overweight, accessed March, 2020.

[13] https://www.w3.org/TR/vocab-ssn/, accessed March, 2020.

[14] M. M Rodgers, V. M. Pai, and R. S. Conroy. "Recent advances in wearable sensors for health monitoring", IEEE Sensors Journal, no. 6, pp. 3119-3126, 2014.

[15] D. Siewiorek, A. Smailagic and A. Dey, "Architecture and applications of virtual coaches", Proceedings of the IEEE, vol. 8, pp. 2472-2488, 2012.

[16] M. Wiesner and D. Pfeifer, "Health recommender systems: concepts, requirements, technical basics and challenges", International journal of environmental research and public health, vol. 3s, pp. 2580-2607, 2014.

[17] A. Chatterjee, M. Gerdes, and S. Martinez, "eHealth Initiatives for The Promotion of Healthy Lifestyle and Allied Implementation Difficulties", IEEE International Conference on Wireless and Mobile Computing, Networking and Communications (WiMob), pp. 1-8, 2019.

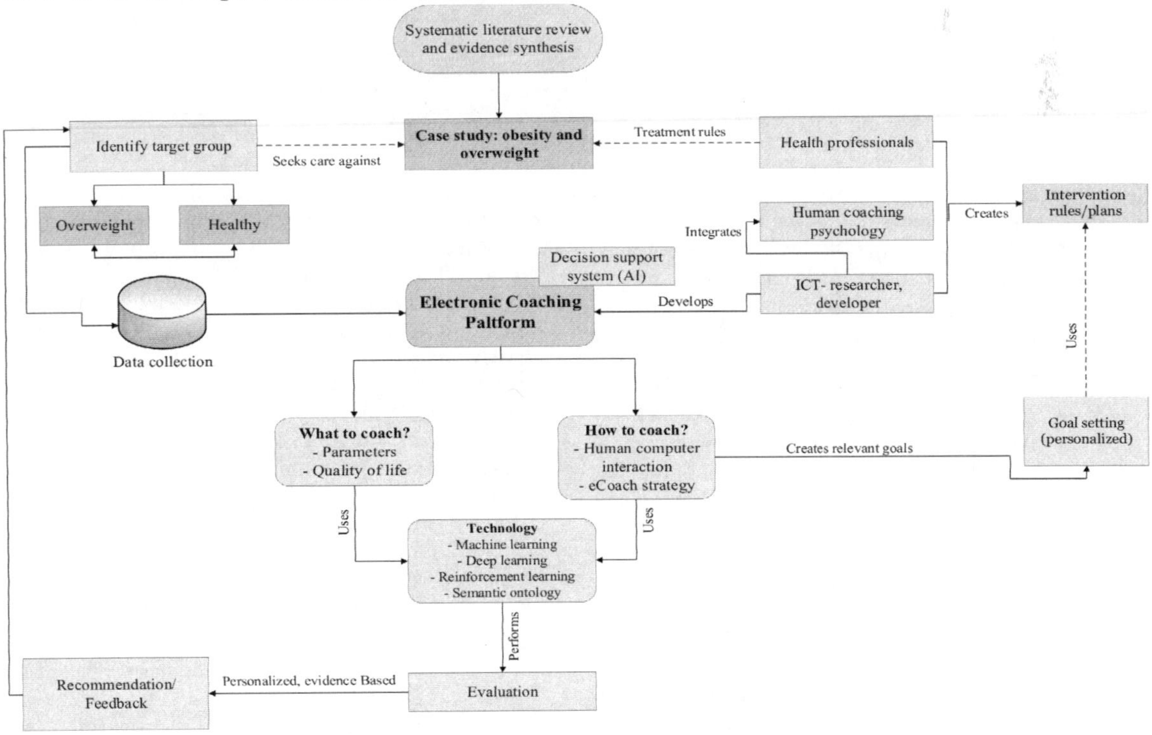

Figure 5. Holistic obesity and overweight e-Coach model.

Combining Patient Pathway Visualisation with Prediction Outcomes for Chemotherapy Treatments

Agastya Silvina and Juliana Bowles

School of Computer Science
University of St Andrews
St Andrews KY16 9SX, UK
email: {as362|jkfb}@st-andrews.ac.uk

Peter Hall

Edinburgh Cancer Research Centre
University of Edinburgh
Crewe Road South, EH4 2XR, UK
email: p.s.hall@ed.ac.uk

Abstract—**The Edinburgh Cancer Centre (ECC) contains National Health Service (NHS) Lothian cancer patient data from multiple resources. However, the lack of proxy between numerous scattered resources hinders the capability to use the information collected in a useful way. ECC data is very varied and includes patient characteristics (e.g., age, weight, height, gender), information on diagnosis (e.g., stage, site, comorbidities) and treatments (e.g., surgery, chemotherapy, radiotherapy). The visualisation of a fraction of ECC data in the form of a patient timeline can aid and enhance the process of observing and identifying the overall condition of the patient, as well as understand how it may compare with cohorts of patients with similar characteristics. We have previously developed machine learning models for predicting treatment outcomes for breast cancer patient data that have undergone chemotherapy. In this paper, we describe, examine, and propose a solution to connect all these aspects and provide a bridge for several resources. This will make it easier for clinicians and other healthcare professionals to support service planning, deliver better quality of care and consequently improve service outcome within NHS Lothian.**

Keywords–Distributed Health Data; Diagnosis; Treatment Timeline; Machine Learning; Oncology.

I. INTRODUCTION

Data is a precious asset in many organisations, but it is often too fragmented and crude to be useful. By integrating data from a variety of sources, we can gain considerably more insight from the information it contains. It is well known that understanding, analysing and building models from complex data has the potential to improve decision making. This is the case for many domains and is particularly relevant in the healthcare domain to improve cancer care.

Data integration is the process of combining/aggregating data from different sources to provide meaningful and valuable information to end-users. There are several ways to perform data integration [1], such as building an enterprise warehouse or creating a proxy server. In the case of the former, it corresponds to creating a centralised database that holds all the business information of an organisation and makes it accessible across the organisation. In the latter case, it corresponds to creating an application which provides the data to the end-users directly from various sources/servers.

Furthermore, the use of healthcare data and data mining/machine learning techniques enables us to develop models which can be used to observe, predict, and analyse the outcome of a specific cancer treatment (for instance, for chemotherapy [2]) by feeding patient characteristics, diagnosis and treatments into the learning algorithm.

The Edinburgh Cancer Centre (ECC) contains NHS Lothian cancer patient data from multiple resources [3]. The Edinburgh Cancer Centre (ECC) is data-rich, however, the use of this resource for improving patient care and cancer treatment has so far been superficial and its potential under-explored. Cancer patient data is currently scattered across disparate systems and repositories. We show several components (some details purposely omitted) of the ECC system in Figure 1.

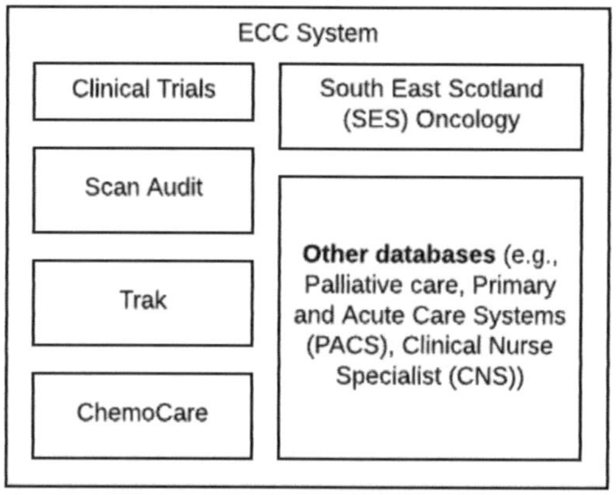

Figure 1. ECC cancer patient data sources

At the moment, it requires significant time and effort to carry out a manual retrospective analysis of data held across multiple systems [3]. In this paper, we describe the work carried out to provide a proxy between the different (sub)systems within ECC. The developed system, called *South East Scotland Oncology Gateway* (*SESO Gateway*), can be seen as one analytical solution for improving the quality and capability of real-time outcomes reporting within South East Scotland using routinely captured and integrated electronic healthcare data. With this application, the users can view the pathway progress at an individual level (e.g., during the clinic and when deciding suitable treatments for a patient), at a cohort basis for analysing a treatment against a set of metrics (e.g., waiting time, following a particular pathway, outcome against disease management protocols) as well as examining the possible outcome of upcoming treatments.

This paper is structured as follows. We present related work in Section II, and describe the data source, analysis and the main features of the system in Sections III and IV. We illustrate the implementation of the system (i.e., front-end, back-end) in Section V. Lastly, we discuss the test results in Section VI and conclude the paper with suggestions for future work in Section VII.

II. Related Work

SESO Gateway continues a line of work by the University of Edinburgh to discover predictors of chemotherapy side effects (e.g., toxicity) in breast cancer patients from a *ChemoCare* data extraction [4]. The data is extracted, transformed, and processed using multivariate logistic regression, as shown in Figure 2.

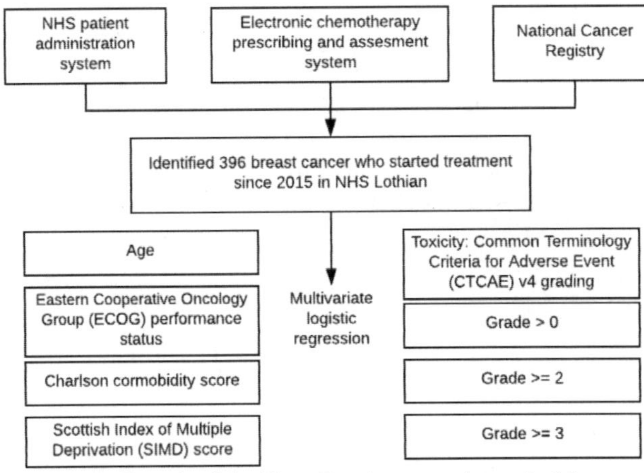

Figure 2. Example of the *ChemoCare* data processing methodology

This study aims to use the data to identify the toxicity rate from commonly used anti-cancer regimens for NHS Lothian patient populations, and find the baseline patient characteristics for predicting excess toxicity. It is possible to profile the pattern of cancer patients, which helps in predicting the side-effects of the chemotherapy regimen. We then add other features to predict and analyse the result of upcoming chemotherapy treatments by using machine learning models that have been developed previously [2].

The use of machine learning models and data visualisation helps users to understand the concept and underlying patterns of patient data. The data visualisation amplifies the benefits of health informatics databases and networks by expanding the capability of clinicians and public health policy-makers to make better decisions [1]. Indeed, multivariate applications like the *SESO Gateways* are becoming more widely developed in the healthcare domain (e.g., IBM Watson [5], Microsoft Research [6], NHS [7]). Integration with existing healthcare systems may help healthcare professionals make decisions, personalise cancer treatments and hence improve the treatment of their patients.

Several existing tools have inspired the development of the *SESO Gateway* (e.g., *LifeLine* [8], NHS Predict [7]). *LifeLine* is a visualisation tool to enhance navigation and analysis of patient records. It provides a general visualisation environment for personal histories. The prototype for this system was developed in 1997 as a research project between IBM Research and the University of Maryland. The basis for modelling the record was a newly operational clinical information system at Kaiser Permanente Colorado.

Predict [7] is an online prognostication tool that helps both patients and clinicians analyse how different treatments for early invasive breast cancer might improve the survival rates of patients after surgery. NHS England developed the tool by training the survival model against women with breast cancer data in England and has been tested on women with breast cancer data from around the world.

III. Use Case

We are developing a dashboard to help oncologists observe, monitor, and analyse the condition of their patients over time. We describe a hypothetical but realistic scenario of a user of the system as follows.

Emma is 38 years old and had been diagnosed with breast cancer. To prevent the spreading of the tumour, she underwent breast surgery. After surgery, chemotherapy treatment was given as a follow-up to her surgery. She comes to the hospital for her chemotherapy appointments. To ensure her recovery is as best as possible, a treatment plan and regimen have been established over several months with chemotherapy treatments in the hospital every three weeks. Emma also has a comorbidity. As any cancer patient on chemotherapy, she might have higher toxicity levels as a result, but it is crucial to guarantee that the scale does not go above level two. Toxicity levels range from 0 (no toxicity) to 5 (very high toxicity).

Emma agrees on using and sharing data between treatment visits via the cancer data gateway and patient portal. Emma determines who in the medical team sees this information: The oncologist/nurse and her GP. Emma is also informed about how to use the web application and pass on relevant information to the clinical team.

Via a user-friendly web application, Emma can provide information on symptoms daily throughout the treatment. Severe reported symptoms can be picked up by the clinical team and acted upon asap. The conditions that are being monitored and provided by patients are *nausea*, *vomiting*, *diarrhoea*, *constipation*, *oral mucositis*, *oesophagitis*, *neurotoxicity*, *hand/foot*, *skin*, *hypersensitivity* and *fatigue*. The information Emma provides is combined with further data on her patient characteristics, cancer information, hospitalisation details and comorbidity.

This combined data will help clinicians adapt treatments better to Emma as an individual patient, keep toxicity levels under control and improve her health outcomes. It uses data from several patients treated over the years with comparable characteristics.

If during the treatment there are signs that toxicity levels are high or that Emma's condition is deteriorating, one of the members of the clinical team (e.g., oncologist, specialist consultant, nurse, GP) will be alerted and contact Emma to intervene.

Overall, Emma can have a more personalised treatment. If a complication arises, the clinical team can act more quickly. Furthermore, Emma's well-being increases as she gets more involved in her treatment plans.

IV. Data Analysis

A. Data Source

As mentioned before, the ECC consists of several scattered databases. Usually, the databases are divided into different categories based on their function: Direct Clinical Care Databases (DCCD) for recording, observing and analysing the direct patient care and Secondary/Derived Databases (SDD) for auditing and reporting purposes.

Table I shows a brief overview of several DCCD and SDD databases. The system features the patient pathway at specific points. Hence, there is some transfer of information within each system (e.g., laboratory, radiology, chemotherapy).

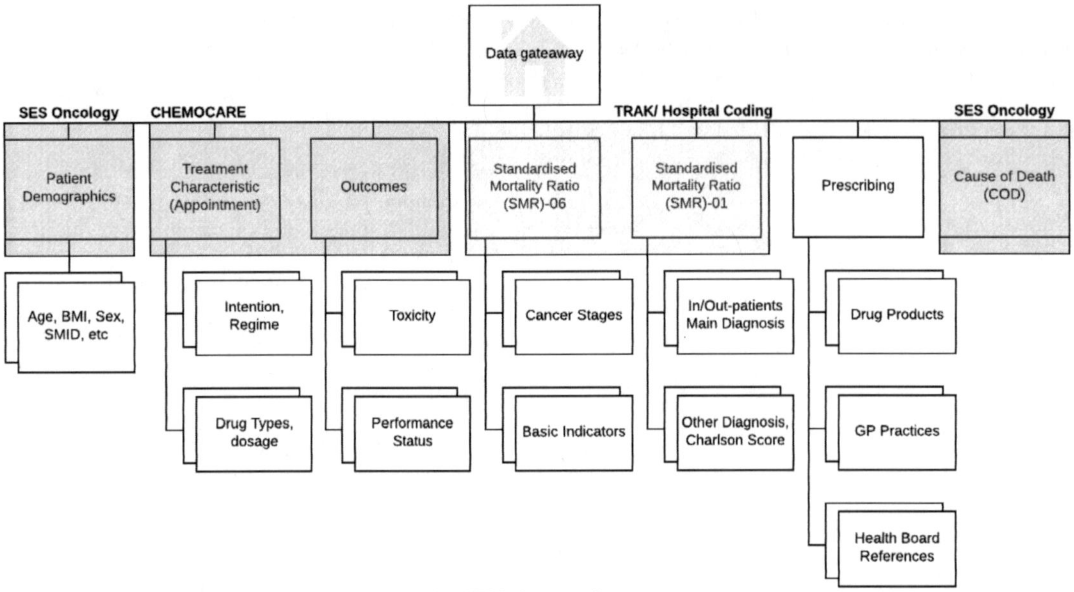

Figure 3. *SESO Gateway* data structure

TABLE I. ECC DATASOURCES

System/Team	*Function*
Trak	Patient Administration System (PAS) (DCCD)
PACS	Radiology Investigation System (DCCD)
ChemoCare	Chemotheraphy ePrescribing System (DCCD)
CNS	Shared patient records (DCCD)
Trak Questionnare	Ensure timely diagnosis and treatment (SDD)
Trak Module	Collect the quality performance indictor dataset (SDD)
Oncology Coding	Collect the whole patient pathway (SDD)
Hospital Coding	Collect the submission to national cancer registry (SDD)

However, because of the lack of proxy, the system has no cohesive view of the patient journey through cancer care. To facilitate this functionality, hospitals usually rely on several teams or manual input.

Figure 3 shows the data structure of the databases connected by the *SESO Gateway*. We developed the *SESO Gateway*, as a proxy server to connect two databases from the DCCD (i.e., *Trak* and *ChemoCare*) and one database from the SDD (i.e., oncology coding) to simplify the observation of the patients' journey through their cancer care for the first version of our application.

B. Patient Information Dashboard

The *SESO Gateway* has two main components: the patient information dashboard and the prognostication tool. The information dashboard facilitates both individual patient analysis (by means of timeline visualisation) and cohort patient analysis.

1) Individual Patient Analysis: The patient timeline summarises the medical records of the patient as a set of lines and icons on an adjustable timeline as shown in Figure 4.

Figure 4 shows an example of patient data spanning about 5 years. The timeline is colour-coded based on the type of event (e.g., follow up, hormone treatment, chemotherapy treatment). The timeline has other features, such as zooming in/out, tab hiding, and scrolling as shown in Figure 5.

With a quick glance, the user can get the idea of the overall patient condition. If the user needs to see the detail of an event, the user can click on the icon. The detail of the event will be shown below the timeline.

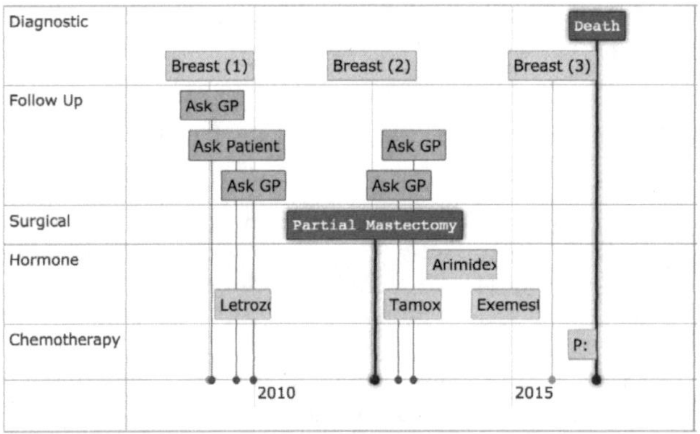

Figure 4. An example of individual patient timeline

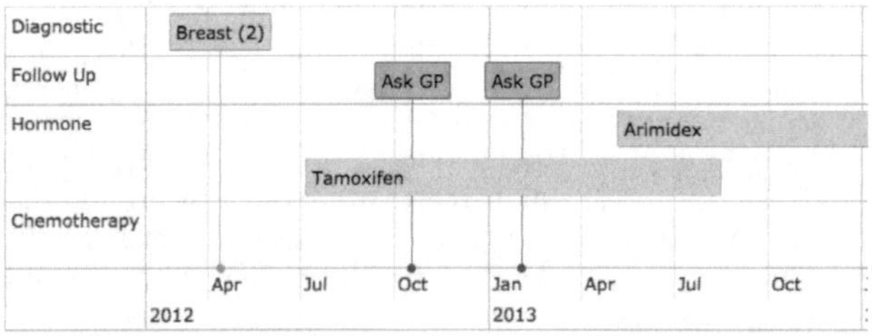

Figure 5. Zoom, hide, and scroll features

2) Cohort Patients Analysis: While several cancer treatments are proven to cure and reduce the (re)occurrence the cancer, the burden of drug-induced toxicity can be substantial. Therefore, there is a need to improve the ability to observe and predict the pattern of several different treatments. To improve the observation of drug effectiveness, we show cohort information in graphs/charts.

Figure 6 shows the wire-frame for the chemotherapy page. It consists of two tab components. The first tab shows the summary obtained from the Oncology DB while the second tab shows the cohort summary from the ChemoCare data extractions. As shown in Figure 7, the graphs show more complex clinical relations (for instance the surgeries that have

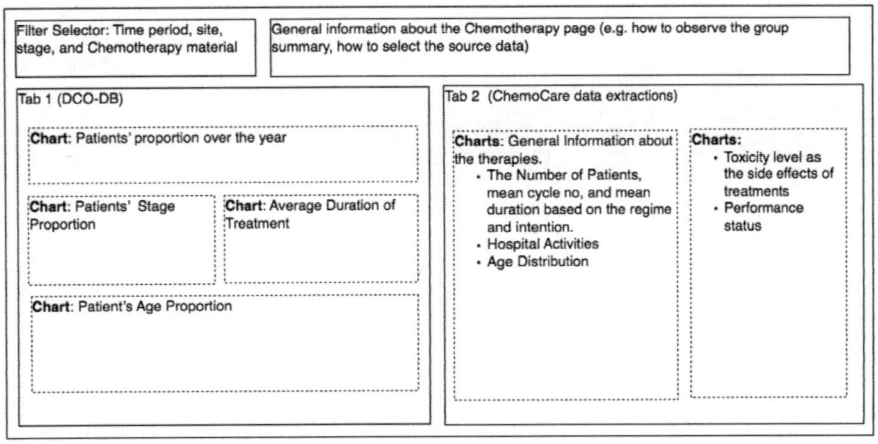

Figure 6. Chemotherapy page wireframe

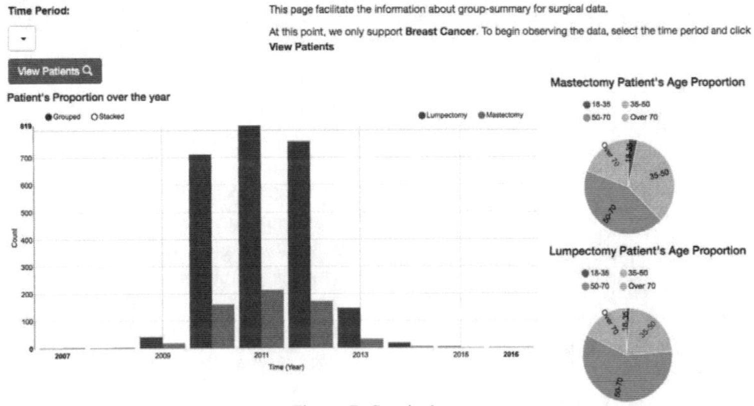

Figure 7. Surgical page

been done). They are useful when there is a point to be made in the shape of the data, or for showing how different things (variables) relate to each other [9]. It helps the user to observe the complex relationships, patterns and trends concisely. The *SESO Gateway* provides the users with graphs and charts for the patients' data within NHS Lothian. These components are shown in several different pages based on the type of treatments (e.g., hormone therapy, chemotherapy, radiotherapy, surgery). The clinical team (e.g., consultants, pharmacists, managers) can use this information for rapid auditing or scope out the data prior to more formal analysis.

C. Machine Learning Models

In our previous paper [2], we developed several models to predict the toxicity outcome of a chemotherapy treatment. We trained the models with the data extracted from *ChemoCare*, which contains the data for 13030 breast cancer treatments (with 933 unique patients). Each patient undergoes two or three different intentions and several regimes/protocols during their cancer care. Each regime might have several cycles (i.e., from 3 cycles to 85 cycles). Table II shows the number of patients in the dataset calculated based on the intention of treatment. Because of the lack of data for the curative intention, for the first version of *SESO Gateway*, we developed models for three different intentions, namely, Adjuvant, Neo-Adjuvant, Palliative.

We used the Random Forest (RF) model for our system. Our RF model performed better than the Hidden Markov Model (HMM) models (i.e., especially in predicting the toxicity outcome during the patients' treatments) [2]. Although the Recurrent Neural Network (RNN) model performed better than the Random Forest model - mainly because of its capability to integrate all the previous treatments for predicting the treatment result - it is prone to over-fitting. This is due to the size of our dataset, as we presently do not have enough data to train the RNN models optimally [10].

TABLE II. TREATMENT INTENTION

Intention	Total patients
Adjuvant	620
Neo-Adjuvant	427
Palliative	483
Curative	17

V. IMPLEMENTATION

The *SESO Gateway* consists of three main components, the user interface (i.e., web-client), the software integration which acts as the proxy between each data source (i.e., Representational State Transfer Application Programming Interface (REST API) services), and the data sources (e.g., the SESCD, *Trak*, and *ChemoCare*). This separation contributes to the modularity of the system, which allows us to easily change each module for any future changes that may happen in the NHS infrastructure.

A. Front-End

The system presentation layer (i.e., Front-End) is a web-client developed using *AngularJS 7* framework [11]. The use of

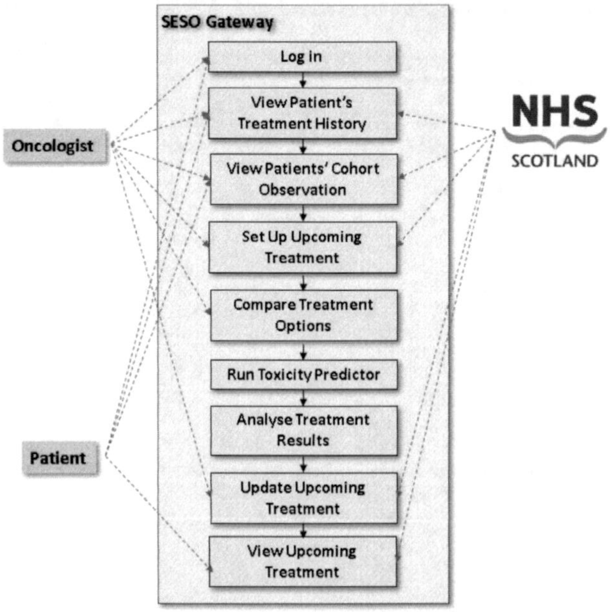

Figure 8. *SESO Gateway* features

the framework helps the application to be more modular (i.e., reusable, Do not Repeat Yourself (DRY), readable, secure).

Our web-client facilitates the user to observe the patients history, cohort data from the NHS data sources. As shown in Figure 8, The web-client has three main features. It shows the event timeline for a particular patient during their treatments in the NHS. It allows the user to observe patterns for various treatments from the cohort observations. Lastly, the web-client has a feature to facilitate the users to analyses the outcome of the upcoming treatments by using the machine learning models that were previously developed.

We implement the *Model-View-View-Model* (MVVM) pattern for the *SESO Gateway* web-client. The MVVM separates the development of the GUI (Graphical User Interface) with the business logic or the data model. Each functionality is separated into different modules. These modules are being integrated into a root module called *Apps*. In *Apps*, we register the built-in *AngularJS* modules (e.g. *ngRoute*, *ngResources*) as well as the custom modules that we create (e.g. *event-timeline*, *group-detail*). It contains the main controller for the *SESO Gateway* client. With the main controller, we generate a Single Page Application (SPA) by only updating specific divisions on the client. We register several *urls* that can be accessed by the users in the Apps.

We separate the modules for the individual timeline, the upcoming treatments analysers, and the cohort/group summaries. With this separation, we have loosely coupled components. Therefore, when we need to change the behaviour of one of the module, we do not need to update any components in the group summaries or the upcoming treatments predictor module.

For the cohort reporting service, we create several different modules/pages based on the type of treatments (e.g. surgical, radiotherapy, hormone therapy, chemotherapy) registered in the *SESO Gateway* for the cohort summary. These modules have a similar structure. For each treatment, we have a module which contains methods to allow the users to apply several filters and then fetch the data from the REST API. The same principle applies to analyse the upcoming treatment feature. Once the data is fetched, we use a third-party library for generating the data visualisation (i.e. *D3* [12]).

B. Back-End

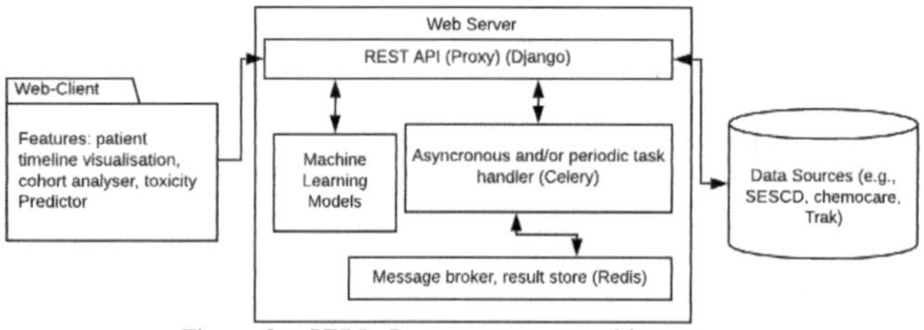

Figure 9. *SESO Gateway* system architecture

The *SESO Gateway* back-end system is a REST API service. The system has an asynchronous and periodic task handler and an internal database as the message broker and asynchronous requests' result storage, as shown in figure 9.

The REST API service of the *SESO Gateway* acts as a proxy between the web-client and the data sources. This uses several database engines to connect a different kind of data sources.

The application consists of several isolated sub-application (i.e., modules) based on the database access and main functionality (e.g., *sescd*, *chemocare*, *toxicity_predictor*). The *sescd* module handles the REST API request for the *SESCD* database, while the *chemocare*, as the name suggests, processes the request for the *ChemoCare* access. Each module has its own REST API for their appropriate function and database access.

To facilitate the functionality to predict the outcome of the upcoming treatment, we serialise the machine learning models which will be loaded the moment the server is up and running. We create a separate API for the user to request an analysis of the upcoming treatment. We develop both *POST* and *GET* request to facilitate this. To analyse the treatment, first the user performs a *POST* request. In the request body, the user specifies the treatment's parameters, such as patient characteristic (e.g., age, weight, height, cancer stage and site), treatment intention (e.g., *adjuvant*), treatment regime (e.g., *FEC 80*) and duration (e.g., 3 months). The request is processed asynchronously by the asynchronous task handler as it might require a longer time to be processed (i.e., in comparison with requesting the patient's medical history). The handler puts the request in the tasks queue and tags it with a unique *task id*. The *task id* is later given to the user/ client as the response of the POST request. Once the *SESO Gateway* finishes predicting the result of the upcoming treatment, the outcome of the prediction is saved in the result storage which can be retrieved by the client via a GET request with the *task id* as the query parameter if the task is complete.

VI. Case Study: Feedback and Evaluation

A. Testing

To evaluate the *SESO Gateway*, we conducted different kinds of testing, organised by the level and purpose of the tests themselves. For the *SESO Gateway*, this included unit-testing, integration-testing and performance-testing.

Our unit tests are performed on the developed module to clarify if the method performs as expected in a set of conditions. There are several different purposes for our unit tests. First, we use them to check the correctness of our queries to retrieve patient data or cohort information from an appropriate data source. We also use it to check if we load the correct machine learning model for our upcoming treatment analyses request. Lastly, we also it for utility methods. Unlike for other forms of testing, we create several mock objects for our unit testing.

For the *SESO Gateway* integration testing, we combine several modules within the application and test those modules as a group. The integration testing takes the modules that have been tested previously in the unit testing. Our integration testing focuses on examining the pathway from the first time the client makes a request to retrieving the correct data from the databases and returning the right response to that request. Similarly, we use the integration testing to simulate the process of the user/client requesting to analyse the upcoming treatment.

Figure 8 shows the potential users of the system. We estimate the number of users between 0 to 150 users accessing the application at the same time. For testing the performance of the *SESO Gateway*, we assume that there could be a time when all users try to access the application simultaneously. Based on this assumption, we need to perform a performance test to observe the robustness of our system.

One issue of *go-live* is performance, such as latency from highly sensitive and slow responses. The performance test allows us to simulate a situation where many users try to access our application at the same time. We use a third party library (i.e., *Gatling-tool* [13]) to perform such a performance test. The tool has the support for HTTP-Protocol and is asynchronous as long as its underlying protocol can be implemented in a non-blocking way. Henceforth, it allows us to do the load-testing of the *SESO Gateway*'s HTTP Server by creating hundred of virtual users.

The *SESO Gateway* is capable of handling 150 simultaneous requests from the user, as requested by the customer, but there is no reason why it cannot process more requests. However, we have not performed any performance test for the system against a million of patient data. The result of performance testing will be highly dependant on the server we access and the environment where we deploy our application.

B. System evaluation

During the development process, we conducted several presentations and demonstrations of the *SESO Gateway* to different users and clinical oncologists. We also conducted a user evaluation of our application. This process allows us to evolve the *SESO Gateway* application to target more concretely the needs and requirements of the end users. Based on users' feedback, we expanded the *SESO Gateway* accordingly, we updated the timeline by adding group functionality as well as better colour-codings after a demonstration to an oncologist. We decided to change the colour of individual timelines and add groups for different kinds of events (e.g., *Diagnosis*, *Surgery*).

One of the oncologists expressed how having an extensive filter for the cohort functions (e.g., filter the patients based on their oncologist which handle the patient) could make the observation faster. We will implement the functionality for generating the list of patients used in the cohort graph in the future because it can further ease the auditing process.

VII. CONCLUSION

Our primary goal is to design a method to accurately visualise patient pathway data from the South-East Scotland Oncology Database. The users can view the pathway progress at an individual level (for example in clinic) as well as a cohort basis to analyse a treatment pathway against a set of metrics (e.g., time between treatment, following a particular pathway, outcome against disease management protocols). With *SESO Gateway*, we have a system which enhances observations for cancer patients in South East Scotland by incorporating a data visualisation tool with additional capabilities to analyse the upcoming treatments. The integration to the new database enables users to effectively see connections and patterns, between patient treatments and their condition, in real-time. Finding these correlations among the data is essential to improve the healthcare systems for ECC.

ACKNOWLEDGEMENT

This research is supported by the NHS Lothian, the DATA Lab, and the EU H2020 project *Serums: Securing Medical Data in Smart Patient-Centric Healthcare Systems* (grant code: 826278). We thank Christina Lilley for her support and contribution to the development of *SESO Gateway*.

REFERENCES

[1] B. Shneiderman, C. Plaisant, and B. W. Hesse, "Improving Healthcare with Interactive Visualization," Computer, vol. 46, no. 5, May 2013, pp. 58–66.

[2] A. Silvina, J. Bowles, and P. Hall, "On Predicting the Outcomes of Chemotherapy Treatments in Breast Cancer," in Artificial Intelligence in Medicine, D. Riaño, S. Wilk, and A. ten Teije, Eds. Cham: Springer International Publishing, 2019, pp. 180–190.

[3] C. Lilley, "Cancer information Programme Plan NHS Lothian," NHS Lothian, 2016.

[4] Z. H. Ng and P. Hall, "Using real-world data to discover predictors of chemotherapy toxicity in breast cancer patients," University of Edinburgh, 2016.

[5] J. L. Malin, "Envisioning watson as a rapid-learning system for oncology," Journal of Oncology Practice, vol. 9, no. 3, 2013, pp. 155–157, pMID: 23942497. [Online]. Available: https://doi.org/10.1200/JOP.2013.001021

[6] "Project Hanover," 2019, URL: https://hanover.azurewebsites.net/ [accessed: 2019-23-09].

[7] G. Wishart, E. Azzato, D. Greenberg, J. Rashbass, O. Kearins, G. Lawrence, C. Caldas, and P. Pharoah, "PREDICT: A new UK prognostic model that predicts survival following surgery for invasive breast cancer," Breast cancer research : BCR, vol. 12, 01 2010, p. R1.

[8] C. Plaisant, R. Mushlin, A. Snyder, J. Li, D. Heller, and B. Shneiderman, "Lifelines: using visualization to enhance navigation and analysis of patient records," The craft of information visualization, 01 2003, pp. 308–312.

[9] S. Powsner and E. Tufte, "Graphical summary of patient status," The Lancet, vol. 344, no. 8919, 1994, pp. 386 – 389, originally published as Volume 2, Issue 8919. [Online]. Available: http://www.sciencedirect.com/science/article/pii/S0140673694914060

[10] M. Z. Alom, T. M. Taha, C. Yakopcic, S. Westberg, P. Sidike, M. S. Nasrin, M. Hasan, B. C. Van Essen, A. A. S. Awwal, and V. K. Asari, "A State-of-the-Art Survey on Deep Learning Theory and Architectures," Electronics, vol. 8, no. 3, 2019. [Online]. Available: https://www.mdpi.com/2079-9292/8/3/292

[11] "AngularJS," 2019, URL: https://angular.io/ [retrieved: 09, 2019].

[12] "D3.js-Data-Driven Documents," 2019, URL: https://d3js.org/ [retrieved: 09, 2019].

[13] "Gatling Open-Source Load Testing," 2019, URL: https://gatling.io/ [retrieved: 09, 2019].

Natural Language Processing, Wearables, and Their Combination in Healthcare: Opportunities, Challenges, and Considerations

Frank Rudzicz[1,2,3,4], Raymond Ng[5], Robert Wu[6], Mohamed Abdalla[2,4], Giuseppe Carenini[5], Kendall Ho[7]

[1] Li Ka Shing Knowledge Institute, St Michael's Hospital, Toronto, Canada,
[2] Department of Computer Science, University of Toronto, Toronto, Canada
[3] Surgical Safety Technologies Inc, Toronto, Canada
[4] Vector Institute for Artificial Intelligence, Toronto, Canada
[5] Department of Computer Science, University of British Columbia, Vancouver, Canada
[6] Toronto General Hospital Research Institute, University Health Network, Toronto, Canada
[7] Department of Emergency Medicine, University of British Columbia, Vancouver, Canada

Email: frank@spoclab.com, rng@cs.ubc.ca, robert.wu@uhn.ca, msa@cs.toronto.edu, carenini@cs.ubc.ca, kendall.ho@ubc.ca

Abstract—**Natural Language Processing (NLP) is a growing field in data analytics that is increasingly applied in healthcare for assessment of patients' electronic records or patients' dialogues to unearth medical issues, emotions, or cognition. Meanwhile, active research is taking place in the use of sensors and wearables in tracking patient phenotypes and physiologic status in diagnosis and management of diseases. This paper's central purpose is to explore how natural language processing can be combined with wearables in healthcare, and how this approach can advance health service delivery and health system transformation. Our recommendations are to stimulate interest and lower barriers in order to implement this novel combination of solutions in healthcare industries, and to do so in consideration of various technical and ethical concerns.**

Keywords - natural language processing; digital health; wearables.

I. INTRODUCTION

Communication between health professionals and patients is a highly important area that affects diagnosis, treatment, and mitigation of acute and chronic diseases. In this era of interprofessional team based care, health professionals need to work together and often asynchronously to ensure smooth continuity of care of patients. Patients themselves can also benefit greatly from optimizing self-management in partnership with health professionals. Underpinning good communication amongst health professionals and patients is the comprehensive capturing and sharing of essential data of the patients in their expressions of symptoms, their physiologic changes, their responses to different types of therapies, and their physical, emotional, and social wellness. The current state of the art in capturing patient data to facilitate accurate tracking of disease progress is based primarily on observational narratives of health professionals, documented episodically in patients' electronic health records, incompletely shared between health professionals and patients, and without any significant patient input. Improvement in these areas conceivably can lead to improvement of disease management through interprofessional collaboration and partnership with patients through patient-centred care.

Modern information and communication technologies can contribute greatly to data capturing and documentation of patient-centric data and its analysis to support diagnosis and management, thereby enriching the therapeutic experiences for health professionals and patients. Data streams that can be tracked include numeric quantities, text, and physiologic signals, which can be valuable for disease management through big data analysis.

This paper will explore opportunities of Natural Language Processing (NLP) and wearables in health as follows. Section 2 gives an overview of NLP opportunities. Section 3 explores NLP in data mining and disease surveillance. Section 4 explores wearables in health. Section 5 posits how NLP and wearables can combine to contribute to health. Section 6 highlights issues and challenges of NLP and wearables in health. Section 7 makes recommendations and conclusions.

II. OPPORTUNITIES OF NLP IN HEALTH

NLP, or computational linguistics, is the study of using computer algorithms and tools to process or analyze natural language data. Key sub-areas of NLP include speech recognition, natural language understanding and natural language generation. In the past two decades, machine learning approaches become more and more popular for NLP, essentially relying more and more on the availability of training data.

NLP is an applied subarea of artificial intelligence that uses Machine Learning (ML) in a variety of tasks, including translation between languages, automatic summarization, speech recognition, information retrieval and search, and general areas of classification. To varying extents, each of these tasks can be expressed within the medical domain, including service delivery, and with each application there are particular considerations that must be taken into account. Proper application of these tools has the potential for significant transformation across the health system; improper application has the risk of significant loss of trust and unintended consequences. This provides the context for the potential impact of natural language processing for big data mining in the electronic health record.

Clinical decision support can assist through enabling better scoring systems, informed through NLP and structured data, such as lab values and vital signs. This can include scores that predict increased risk of an adverse event, including readmission, deterioration in hospital, and 1-year mortality. This, of course, assumes that clinicians will be able to use this information effectively, and that information with regards to deterioration or adverse events can be delivered in an actionable way. This will require better fine-tuning to reduce false alerts, and better design for human-computer interaction.

A computational model that captures patient speech patterns by NLP can be deconstructed into three components: 1) topic modeling to highlight and prioritize topics in patients' communication; 2) sentiment analysis to detect patient changes in affect, such as anxiety or depression, that interconnect with patient state of wellness and need for hospital visits; and 3) discourse coherence analysis to gauge patient coherence in their thinking and detect onset of mental dysfunction or delirium. These three components act synergistically to analyze the totality of the patient's speech pattern and content for optimal. For example, in patients with depression, they ruminate on certain morbid thoughts and topics, use words associated with depression and anxiety, and may have incoherent speech that are difficult for others to understand when the depression is profound or associated with delusions. Topic modeling can pick up the rumination, sentiment analysis can identify words with negative emotional connotations, and discourse coherence can highlight disturbed flow or logic of thinking. Liu *et al.* [1] demonstrated that automated NLP screening could identify 10.3% of 27,002 comments in a microblog were suggestive of suicidality with high precision (0.86), recall (0.78), F-measure (0.86), and accuracy (0.88).

III. NLP IN DATA MINING AND DISEASE SURVEILLANCE

Word embeddings (i.e., 'distributed representations') are dense numeric representations of words which serve as input to a wide array of ML methods. Typically, by learning to do automatic word prediction accurately (through measures of statistical perplexity on training data), these embeddings induce latent dimensions that encode aspects of morphology, syntax, and even semantics. The results therefore can capture meaningful relationships among concepts in the data not afforded by traditional methods.

Publicly-available, pre-trained word embedding models may suffice for certain tasks, but typically training on text within a single domain leads to improved performance [2]. Publicly-available pre-trained clinical word embedding models have been trained on PubMed abstracts and full-text PubMed documents; however, those corpora constitute very different language than clinical notes, limiting transferability.

NLP can uncover information in unstructured data that is not always present (or is inconsistent) in the structured data, including predicted cancer recurrence, whether a patient smokes, and drug treatment patterns [3]. However, traditional text search is typically unsuitable because of spelling differences, synonyms, and general ambiguity. Dubois and Romano [4] trained word embeddings using anonymized notes, treating negative counterparts of words as a single word, since negations are important in evaluation. They also trained embeddings on journal abstracts (MCEMJ) from OHSUMED [5] and evaluated their embeddings on several tasks (e.g., disease and mortality prediction) at the word-, note-, and patient-levels. Embeddings trained on medical notes resulted in greater downstream accuracy than those trained on abstracts.

Unlike other domains, larger data sets (i.e., 'corpora') do not necessarily produce better biomedical models [6]. Wang *et al.* [3] found that word models trained from electronic medical records had the best F1 score (0.90, a geometric mean of recall and precision) on a clinical information extraction task, compared with models trained on larger (but more general) data sets.

Beyond *learning* word embeddings using contextual information during training, it is also important to use context during *inference*, which means modifying outputs in different scenarios. For example, if one knows the author typing a clinical note, the word embedding space may shift. This generalizes to other covariates which may have an effect on the language. Nguyen *et al.* [7], for example, used convolutional neural networks to train an end-to-end deep learning system that learned to extract features from medical records and predicted future risk automatically. Some simple extensions were not taken in that work (e.g., long-term dependencies were simply captured through a max-pooling operation). A more careful consideration of effects that have a temporal component, such as the progression of symptoms or other disease trajectories, ought to enable more accurate predictions.

IV. OPPORTUNITIES FOR WEARABLES IN HEALTH

Commercial devices, such as Apple Watch and Fitbit, have fueled the popularity of qualified self [8] – the measuring of individuals' own physiologic signals so as to gain insights into their own health and wellness. A 2017 survey by the Dublin Quantified self-group [9] found that some of the most common variables being tracked were: steps, sleep, weight, heart rate, diet, and exercise. The reasoning behind the tracking and data usage included "to make change to my life", "I learn something new about myself", "Monitor the effect of a new regime", to motivate myself", and "to make myself accountable". These are powerful reflections of individuals committed and ready to make changes through optimizing self-management.

As a variety of newer and innovative wearables come to the market, the opportunity to apply these devices for continuous tracking of patients for disease diagnosis and management is not only imaginable, but becoming reality. For example, tracking physiologic measures with a wearable necklace was found very promising for predicting hospitalization of patients [10]. Also, continuous monitoring in a hospital ward setting has found a dramatic reduction of unexpected cardiac arrest and length of stay of patients [11]. The Apple watch has recently been certified by Food and Drug Administration (FDA) to be a medical device for diagnosing and tracking atrial fibrillation [12]. The opportunity is ripe to introduce wearables into the health system as an integral part of patient surveillance and monitoring, and incorporate these data into the electronic health record to document patient digital phenotypes [13] and characteristic signatures of their diseases in their well management or exacerbation stages so as to better support and monitor patients' progress. This field of physiological informatics is rapidly rising in recognition and prominence in disease diagnosis, management, and even early prediction [14].

V. COMBINING NLP AND WEARABLES IN HEALTH

The discipline of medicine is built on the foundation of history taking and physical examination of patients in order to carefully deliberate on the constellation of symptoms that the patients experience, and the patterns of these symptoms to suggest the diagnoses and management approaches. With the promise of NLP and wearables, we have now the opportunity to digitize and record these interactions not only narratively in traditional ways, but also digitally in a brand new and innovative approach. This digitization can support diagnosis, management, and data analytics to learn about these individual patients and their personalized patterns of illnesses and their unique responses to therapies. The capturing of this "digital mirror" of the patients is unprecedented, and can truly revolutionize how medicine will be practiced in the future.

The combination of NLP and wearables will tremendously increase the amount of information that we will have about patients in our Emergency departments and admitted to hospital. Wearables will provide continuous real-time data on patient status, including vitals, activity, and sleep. Applying NLP to unstructured data, including clinical notes, and communication between providers, will complement that information. The opportunities could be vast and include significant improvement in the domains of clinical decision support, quality improvement, and research.

VI. CHALLENGES AND CONSIDERATIONS

To develop, implement, and deploy NLP approaches in biomedical contexts, it is necessary to define a principled framework for systematic identification and response to ethical dilemmas that may arise. As biomedical NLP lies at the intersection of NLP and biomedicine, it inherits ethical principles and conceptual frameworks developed for both parent-fields, such as those laid out in the International Code of Medical Ethics (adopted by the World Medical Organization) and ACM's Code of Ethics. These, and others, serve as a good starting point to investigate and further develop ethical questions that may arise in biomedical NLP applications. However, critically, these existing frameworks do not cover emerging ethical concerns, which fall beyond the scope of many established frameworks. We take the approach of Hirst [15] in discussing possible areas of concern when it comes biomedical NLP by exploring three areas: 1) concerns regarding how research is carried out, 2) concerns regarding motivations for the research, and 3) concerns about the unintended consequences of our research.

First, regarding how research is performed, we inherit the general ethical guidelines of research ethics. While formulations may differ in detail, the foundations are largely agreed upon. Here, we highlight possible issues facing biomedical NLP researchers which affect how research should be performed. One such issue is the problem of demographic skew. Because of the data available for training, ML algorithms often underperform for minority demographics [16]-[18] and this can have life-altering consequences for our patients. Additionally, upholding the privacy of patients is of vital importance and greatly affects how we do research. However, in addition to basic precautionary steps, such as encrypting hard-drives and password protection, we must take additional steps to protect the data we work with. The literature presents many ways that the privacy of patients can be compromised, including adversarial attacks [19], re-identification from released models [20], and re-identification of hidden variables from de-identified datasets [21]. We must be aware of the vulnerabilities of the techniques used and take the appropriate measures to protect our patients.

Second, regarding the motivation for research, we must, as our parent fields do, proceed with the intention of improving patients' lives and alleviating suffering. This was at the core of Hippocratic oath, which was modernized at the Declaration of Geneva, and also adapted for the general researcher by many including Joseph Rotblat who proposed a "Hippocratic Oath for Scientists" [22]. As scientists, we must place the well-being of patients first and take actions in line with the aforementioned principles. Concretely, in a biomedical application of NLP, this could translate to being aware of the issue of over- or under-exposure [23]. Overexposure, the increased mainstream attention to topics in health, helps to create biases that lead to discrimination regarding project choices, and approaches. To those of us applying biomedical research to clinical settings, we must ensure not to over-complicate our work. The most novel approaches in the literature often lack the rigorous testing required by healthcare systems. When the results of various systems are comparable, we should opt to use the simpler and more studied of those systems. To those pushing the state of the research, we must not 'oversell' our results and the capabilities of our systems. Doing so not only hurts trust in our work when they fail to match the oversold promises, but also distract from the more useful discussions possible around our work regarding limitations and possible improvements.

Third, we must be aware of the unintended consequences of our research and work to mitigate the negative effects of our work. Researchers should be aware that ML methods serve to further propagate the biases implicitly and explicitly present in our society [24] unless methods to remove and prevent such bias are taken [25]. Accurately determining predisposition and risk for a disease given publicly available text, such as social media, would enable medical professionals to improve quality of life but would also allow insurance companies the opportunity to deny coverage thereby *reducing* quality of life.

Using patient speech patterns to build models that inform disease management raises several important questions. AI systems for NLP currently generate imprecise analyses of patients' explicit meaning and implicit affect and cognition. Who will be responsible for erroneous therapeutic decisions? How will these percolate into therapeutic decisions? Would NLP allow health professionals to be more empathic to these patients and sensitive to their physical and mental management needs? If information collected regarding management may reveal secondary or incidental findings (e.g., emotional symptoms, signs of domestic abuse) that may be shared with clinicians outside of the initial or immediate treatment team, how will patient consent, privacy and confidentiality be handled? In our multicultural communities, how will NLP from patients of different linguistic histories or first languages affect efficacy, and how will sources of bias in the data be mitigated against?

Meanwhile, the applications of wearables in healthcare also have their challenges. For health system to adopt these technologies, they need to be medical grade requiring institutional approval, such as FDA. Even though consumers' grade devices can still lead to positive effects, such as step tracking to promote exercises, the variability of the different step trackers in the market place [26] make these devices difficult to fully be deployed in a health context. As more wearable devices come to the market that are medically certified, their application will likely be more pervasive.

Even though there are many observational studies and case series about the use of wearables in healthcare that are promising, convincing evidence, such as randomized controlled trials, have so far not demonstrated unequivocal health benefits for patient management [27]. While many of these studies focus on outcomes in health, some authors argue that the patient experience and health system benefits should also be taken into account [28].

VII. CONCLUSION

This paper has identified a rapidly evolving domain of research and areas of exploration to this field. Beyond research, we also need to identify approaches towards change management and knowledge translation at this early stage of NLP and wearables in healthcare, so that when evidence of their benefits are clearly demonstrated, adoption of these practices will not be prolonged to delay reaping of the patients and health system advantages in using these technologies. At the same breath, we should not enthusiastically adopt these technologies now without judicious research and analysis to truly tease out their benefits, so as to ensure evidence-informed policy translation in incorporating these technologies into health system provision of care.

A governance, education, and ethics body, aimed at researchers, should collate and disperse general information regarding issues and concerns of AI in health, including the application of NLP and wearables. This includes best practices and policies regarding analysis of data. Such an organization would serve as i) a central hub for knowledge, hosting common issues, their solutions, and ii) an active educational force, holding workshops, and developing curricula.

The potential impact of NLP in healthcare is massive and largely untapped. However, with this potential good comes a lot of potential harm, and researchers should devote energy to ensuring that minimal harm befalls their patients and society at large.

REFERENCES

[1] X. Liu et al. "Proactive Suicide Prevention Online (PSPO): Machine Identification and Crisis Management for Chinese Social Media Users With Suicidal Thoughts and Behaviors," *J Med Internet Res*, 21(5):e11705, 2019. doi: 10.2196/11705

[2] S. Pyysalo, F. Ginter, H. Moen, T. Salakoski, and S. Ananiadou, "Distributional semantics resources for biomedical text processing," In LBM 2013, pp. 39–44, 2013.

[3] Y. Wang et al., A comparison of word embeddings for the biomedical natural language processing. pp.1–21, 2018. URL http://arxiv.org/abs/1802.00400 [retrieved: January 2020].

[4] S. Dubois and N. Romano, Learning effective embeddings from medical notes. Technical report, 2017. URL https://web.stanford.edu/class/cs224n/reports/2744372.pdf [retrieved: January 2020].

[5] Xmdv Tool. Ohsumed DataSet, URL http://davis.wpi.edu/xmdv/datasets/ohsumed.html [retrieved: January 2020].

[6] B. Chiu, G. Crichton, A. Korhonen, and S. Pyysalo, "How to train good word embeddings for biomedical NLP," Proceedings of the 15th Workshop on Biomedical Natural Language Processing, pp. 166–174, 2016.

[7] P. Nguyen, T. Tran, N. Wickramasinghe, and S. Venkatesh, "Deepr: A convolutional net for medical records," *IEEE Journal of Biomedical and Health Informatics*, 21(1), pp.22–30, 2017.

[8] Quantified Self. Quantified Self: Self knowledge through numbers. URL http://quantifiedself.com/ [retrieved: January 2020].

[9] Quantified Self. Dispatch from QS Dublin: Results from a QS Community Survey. URL http://quantifiedself.com/2017/02/quantified-self-health-wellness-now-future-dispatch-qs-dublin/ [retrieved: January 2020].

[10] R. M. Khandwalla et al., "Predicting heart failure events with home monitoring: use of a novel, wearable necklace to measure stroke volume, cardiac output and thoracic impedance," *J Am Coll Cardiol* 2016;67:1296. doi:10.1016/S0735-1097(16)31297-9

[11] H. Brown, J. Terrence, P. Vasquez, D. W. Bates, and E. Zimlichma, "Continuous monitoring in an inpatient medical-surgical unit: a controlled clinical trial," *Am J Med* 2014;127:226–32. doi:10.1016/j.amjmed.2013.12.004

[12] Healthcare IT News. Apple unveils Watch Series 4 with FDA-approved ECG. URL https://www.healthcareitnews.com/news/apple-unveils-watch-series-4-fda-approved-ecg [retrieved: January 2020].

[13] S. Gehrmann et al., Comparing deep learning and concept extraction based methods for patient phenotyping from clinical narratives. *PLoS ONE*, 13(2): pp.1–19, 2018.

[14] J. Bai, L. Shen, H. Sun, and B. Shen, "Physiological Informatics: Collection and Analyses of Data from Wearable Sensors and Smartphone for Healthcare," *Adv Exp Med Biol*, 1028:17-37, 2017. doi: 10.1007/978-981-10-6041-0_2

[15] H. Hirst, "Keynote: NLP ethics in a dangerous time," In Proceedings of the First ACL Workshop on Ethics in Natural Language Processing, 2017.

[16] J. Anna, D. Hovy, and A. Søgaard, "Challenges of studying and processing dialects in social media," In Proceedings of the Workshop on Noisy User-generated Text, pp.9-18, 2015.

[17] D. Hovy, "Demographic factors improve classification performance," In Proceedings of the 53rd Annual Meeting of the Association for Computational Linguistics and the 7th International Joint Conference on Natural Language Processing. 1, pp. 752-762, 2015.

[18] R. Tatman, "Gender and dialect bias in YouTube's automatic captions," In Proceedings of the First ACL Workshop on Ethics in Natural Language Processing, pp. 53-59, 2017.

[19] J. Ebrahimi, A. Rao, D. Lowd and D. Dou, "HotFlip: white-box adversarial examples for text classification," In Proceedings of the 56th Annual Meeting of the Association for Computational Linguistic. 2, pp. 31-36, 2018.

[20] M. Fredrikson, S. Jha, and T. Ristenpart, "Model inversion attacks that exploit confidence information and basic countermeasures," In Proceedings of the 22nd ACM SIGSAC Conference on Computer and Communications Security, pp. 1322-1333, 2015.

[21] A. Narayanan and V. Shmatikov, "Robust de-anonymization of large sparse datasets," In Security and Privacy, IEEE Symposium, pp. 111-125, 2008.

[22] J. Rotblat, "A Hippocratic Oath for scientists," *Science*, 286, pp.1475-1475, 1999.

[23] D. Hovy and S.L. Spruit, "The social impact of natural language processing," In Proceedings of the 54th Annual Meeting of the Association for Computational Linguistics, 2, pp.591-598, 2016.

[24] S. Kiritchenko and S. Mohammad, "Examining gender and race bias in two hundred sentiment analysis systems," In Proceedings of the Seventh Joint Conference on Lexical and Computational Semantics, pp. 43-53, 2018.

[25] T. Bolukbasi, K.W. Chang, J. Y. Zou, V. Saligrama, and A.T. Kalai, "Man is to computer programmer as woman is to homemaker? debiasing word embeddings," In 30th Conference on Neural Information Processing Systems, 2016.

[26] P. Alinia et al., "How accurate is your activity tracker? A Comparative Study of Step Counts in Low-Intensity Physical Activities," *JMIR mHealth uHealth* 2017;5:e106. doi:10.2196/mhealth.6321

[27] A. Vegesna, M. Tran, M. Angelaccio, and S. Arcona, "Remote patient monitoring via non-invasive digital technologies: A Systematic Review," *Telemed e-Health* 2017;23: pp.3–17. doi:10.1089/tmj.2016.0051.

[28] P. Zanaboni et al., "Methods to evaluate the effects of internet-based digital health interventions for citizens: Systematic review of reviews," *J Med Internet Res* 2018;20. doi:10.2196/10202.

Relationship Between Breath Regulation and Stroke Volume with Exercise Intensity: a Pilot Study

Wei-Chen Lai
Institute of Biomedical Engineering
College of Electrical and Computer Engineering
National Chiao Tung University
Hsinchu, Taiwan R.O.C.
e-mail: jasmine9130@gmail.com

Po-Hsun Huang
Institute of Computer Science and Engineering,
College of Computer Science
National Chiao Tung University
Hsinchu, Taiwan R.O.C.
e-mail: sdk58927@gmail.com

Tzu-Chien Hsiao
Department of Computer Science, Institute of Biomedical Engineering,
Biomedical Electronics Translational Research Center
College of Computer Science
National Chiao Tung University
Hsinchu, Taiwan R.O.C.
e-mail: labview@cs.nctu.edu.tw

***Abstract*—Exercise is a physical activity that can enhance or maintain physical ability and health. For oxygen extraction, the ventilation will be affected by the breathing response. Cardiac Output (CO) amounts to blood by the heart pumping through the circulatory system. This blood ensures that oxygen is delivered to active tissue. During inspiration or expiration, Stroke Volume (SV) also changes. As we can see, the cardiopulmonary reactions during exercise face dynamic changes. In this work, we used a non-invasive method to measure SV and Breathing Frequency (BF) in order to investigate cardiopulmonary interactions observed during incremental exercising. The result in this experiment showed that SV would decrease to a stable value in the last stage of exercise. CO and BF would increase. SV and BF have opposite changes during incremental exercise. To the best of our knowledge, this study is the first one to use non-invasive methods to observe the changes in hemodynamic parameters during incremental exercising. The benefit of a non-invasive approach is that it is more suitable for home activities.**

Keywords- Breath Frequency; Stroke Volume; Incremental Test

I. INTRODUCTION

Physical activity helps people maintain health and physical ability. During exercise, the first observed response is breathing and heart rate. People regulate their breathing frequency and the depth of breathing unconsciously or consciously. Breathing is influenced by the rhythm of exercising. For example, athletes have their own rhythm of breathing to maintain during exercising [1]. Athletes' ability to integrate the breathing action effectively without disturbing performance is crucial to getting a better performance [2]. However, breathing patterns change dynamically during exercising.

For oxygen extraction, the ventilation is influenced by the breathing response [3]. CO amounts to blood by the heart-pumping thorough the circulatory system. The blood delivers oxygen to active tissue. The heart and lungs are in the intrathoracic space, and the gas exchanges the volume and pressure when humans exercise [3]. During expiration, the pressure increases in the chest, helping to drain off blood in the ventricle and increasing SV temporarily. On the other hand, during inspiration, the negative intrathoracic pressure makes the blood drain difficultly and decrease blood volume in the ventricle [3]. To transport blood to limbs and return to the heart, the skeletal muscle support venous return during exercising [4]. Therefore, the cardiovascular and respiratory systems are integrated during exercising. Furthermore, the responses interact in circulation during dynamic exercising.

Regulation of breathing can provide support during cycling when the BF cooperates with the frequency of pedaling. The BF increases at perceived resistance at a high level. Humans also maintain exercising continuously. According to the previous study, breathing patterns and CO are helpful for understanding mechanisms during exercising. Breathing can affect HR and SV. By inspiration and expiration, the gas pressure changes the SV volume. However, to the best of our knowledge, few papers discuss SV and BF during exercising. In this study, we aim to observe CO, SV, and HR during exercising. We also observe the breathing frequency and SV situation during exercising.

The rest of the paper is structured as follows. Section II presents the materials and the method used. Section III presents the results, and Section IV the discussion. We conclude in Section V.

II. MATERIALS AND METHOD

A. Ethical approval and Participants

This study was reviewed and approved by the Research Ethics Committee for Human Subject Protection, National Chiao Tung University (NCTU-REC-107-092). All participants received detailed information about the study objectives and experiment process before doing the experiment. All of the participants had to write an informed consent.

Two healthy participants were included in this experiment. Participants did not have any history of respiratory and cardiovascular diseases. Before the experiment, each participant had their blood pressure measured in the sitting position to make sure they could be included this experiment.

B. *Equipment*

The Respiratory Inductive Plethysmography (RIP) (Ambu Sleepmate Ripmate Inductance Belt Thorax, Ambu Inc., USA) is a non-invasive senor to measure Thoracic Wall Movement (TWM) and Abdominal Wall Movement (AWM) to observe the response in breathing. Two elastic belts are placed on the chest and abdomen. It is suitable for people to monitor breathing unobtrusively during exercise [5]. It can record Thoracoabdominal Movement (TAM) to observe the breathing feature, including breathing frequency.

The Impedance Cardiography (ICG) (AESCULON, Osypka, Germany) is a non-invasive way to measure hemodynamic parameters, such as stroke volume (SV) by bioimpedance [6]. ICG can provide continuous measures in blood flow. It places sensors on neck and chest. The sensor detects the electrical change and impedance change when blood passes through the vessel.

As stated above, the participants were evaluated by ICG and RIP in this experiment. We observed the situation during the experiment.

C. *Experimental protocol*

All participants had their blood pressure, SV and breathing signal measured during resting. After three minutes of resting, all subjects were asked to ride an upright-seated cycle ergometer. In cycling, the pedal should be maintained continuously ot 60 beats per minute (bpm). Stroke volume and breathing were recorded by ICG and RIP. The workload increased by 25 W every 3 minutes until they could not manage it any more. After finishing riding, participants were asked to cool down and rest.

D. *The instantaneous breathing frequency Calculation*

During exercising, the BF was in a dynamic situation. First, the TWM signal used Empirical Mode Decomposition (EMD) to decompose into many Intrinsic Mode Functions (IMFs) [7]. The algorithm of EMD is shown in Figure 1. However, the problem with EMD is that mode mixing is involved in IMFs. The Complete Ensemble Empirical Mode Decomposition (CEEMD) added a pair of white noise to the signal to solve the mode-mixing problem [8]. The algorithm of CEEMD is shown in Figure 2. After decomposing the signal, a significant test is used in this procedure to ensure a dominant component. Therefore, Normalize Direct Quadrature (NDQ) is used to decompose the dominant component [9]. NDQ is shown in Figure 3. This procedure of analysis was developed by a software platform (LabVIEW version 2018, National Instruments Corp, Austin, USA).

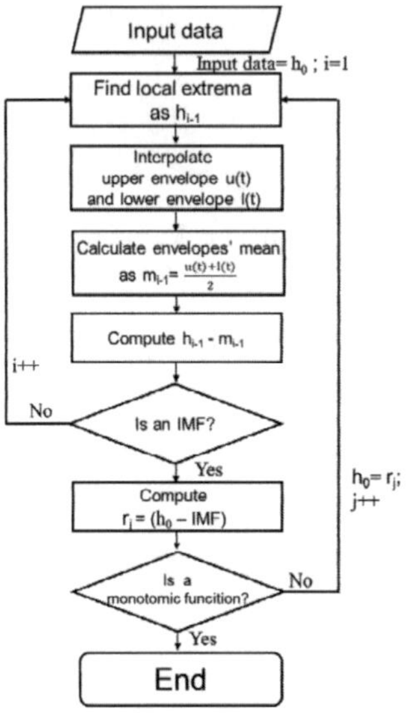

Figure 1. The algorithm of EMD

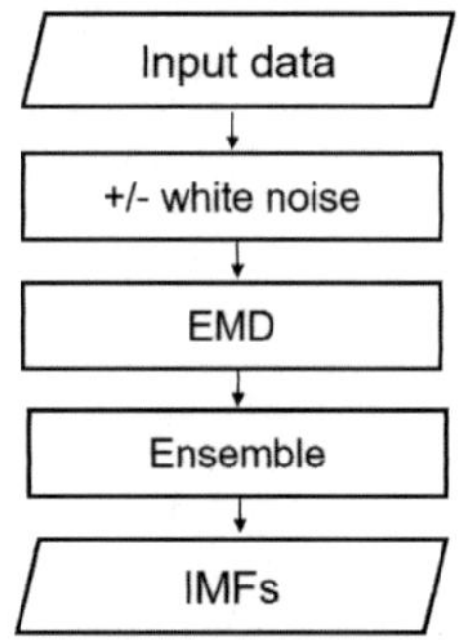

Figure 2. The algorithm of CEEMD

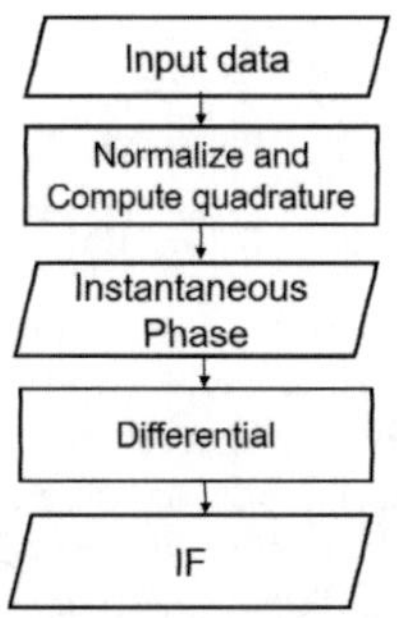

Figure 3. The algorithm of NDQ

III. Result

A. *Participants description*

In this experiment, participants have different riding times. Participants are one man and one woman. The participants description is shown in Table I.

TABLE I. PARTICIPATE DESCRIPTION

Sex	Male	Woman
Age	23	30
Height	168 cm	153 cm
Weight	66 kg	58 kg
Riding time	21 minutes	18 minutes

B. *The result of the experiment*

The result of the experiment is shown in Table II. The change during exercising is shown in Figures 4 to 7. In Figure 4 and Figure 5, CO increases during exercising. By contrast, SV decreases until the last stage. SV decreases in stable volume. HR increases over the exercising time. In Figure 6 and Figure 7, BF and SV had the opposite trend during exercising. In Figure 8 and Figure 9, HR and BF had the same trend during exercising.

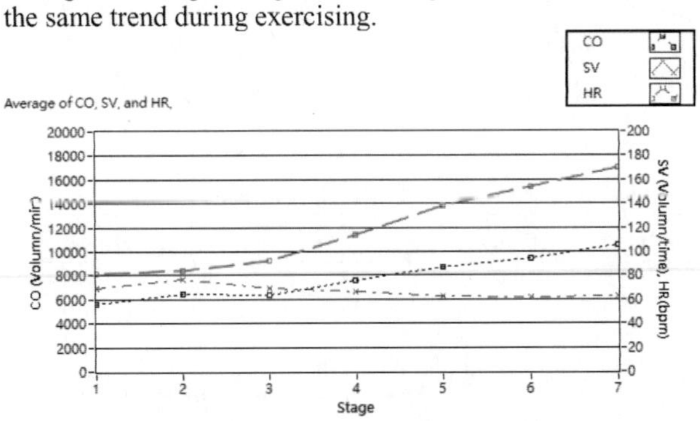

Figure 4. The CO, SV, and HR change in man

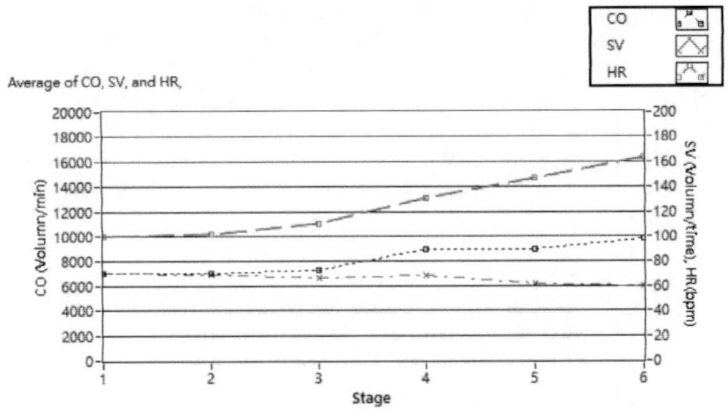

Figure 5. The CO, SV, and HR change in woman

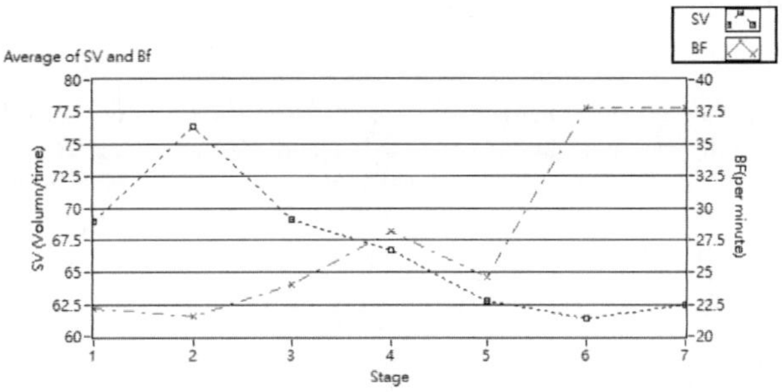

Figure 6. SV and BF change in the experiment

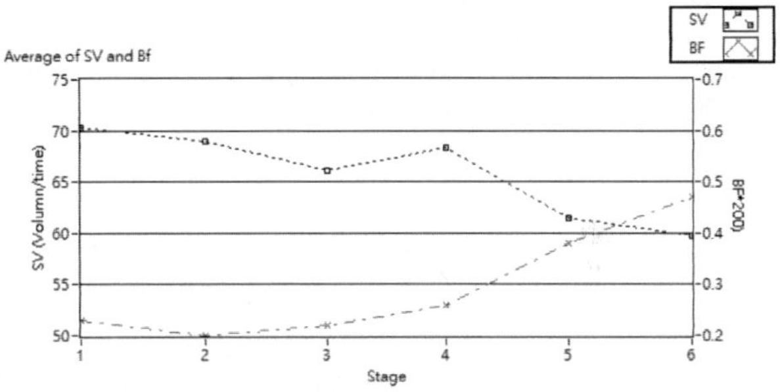

Figure 7. SV and BF change in woman

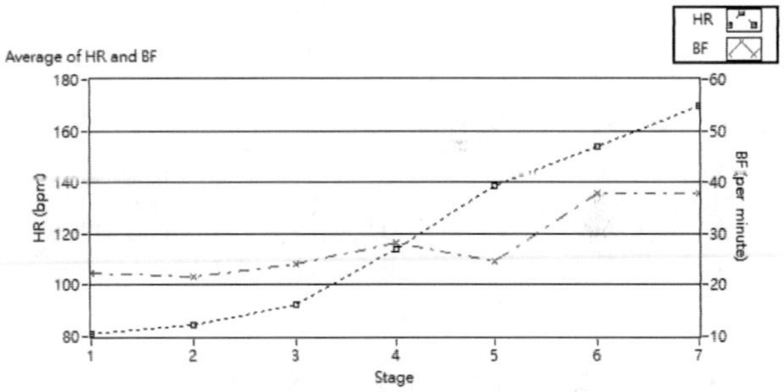

Figure 8. HR and BF change in man

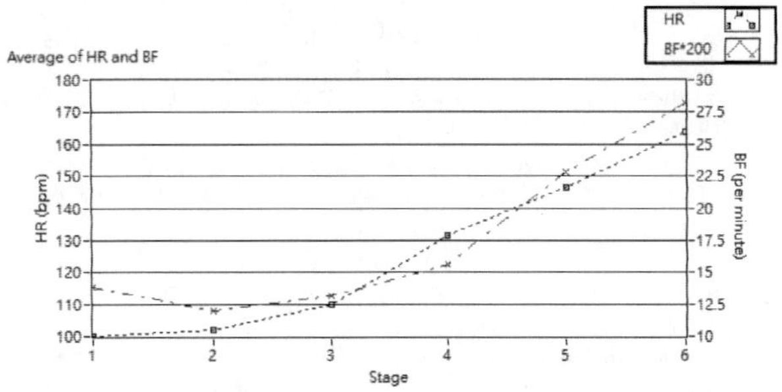

Figure 9. HR and BF change in woman

IV. Discussion

In this experiment, we observe the change in BF, SV, CO, and HR during incremental exercise. SV decreases during exercise. CO, HR, and BF had an increasing trend. The reduction of SV was associated with HR in this study. BF affects hemondynamics [10]. HR reacted to temperature and breathing. In our experiment, we excluded the temperature situation. We show that BF and HR are influenced during exercising. SV is associated with BF and HR.

In this study, we had some limitations in this experiment. Ventilation is also an important parameter to figure the mechanism in blood flow [12]. Chantler et al. use the elastance of the vessel and derive the index of SV to figure out the mechanism of the cardiovascular system during exercise [13]. There is still a big issue in stroke volume and breath during exercise. In future work, we will increase the number of subjects and provide more data to enhance the result. Moreover, we will use body mass index to separate and analyze different phases during exercising to explore the change.

V. Conclusion

In this study, we present a result of the breathing frequency and stroke volume situation during exercising. The result is an opposite change between breathing frequency and SV. HR interacts with breathing and indirectly influences stroke volume. In health care, we use the non-invasive way to measure BF and SV. It can help people to exercising at home. It can monitor the response during exercise. We show that HR and CO are increasing during exercising.

Acknowledgment

This work was fully supported by Taiwan Ministry of Science and Technology under grant numbers MOST 108-2221-E-009-087.

TABLE II THE BREATHING FREQUENCY AND STROKE VOLUME DURING EXERCISING

Sex	Male		Female	
Parameter	Stroke volume (ml)	Breathing frequency (Hz)	Stroke volume (ml)	Breathing frequency(Hz)
Stage 1	69.00 ± 80.90	0.37 ± 0.23	70.31 ± 100.13	0.23 ± 0.16
Stage 2	76.35 ± 84.25	0.36 ± 0.09	68.89 ± 102.11	0.20 ± 0.12
Stage 3	69.07 ± 95.52	0.4 ± 0.21	66.13 ± 109.89	0.22 ± 0.11
Stage 4	66.71 ± 113.91	0.47 ± 0.28	68.51 ± 137.28	0.26 ± 0.10
Stage 5	62.79 ± 141.47	0.41 ± 0.25	61.38 ± 146.46	0.38 ± 0.12
Stage 6	61.34 ± 154.12	0.63 ± 0.39	59.71 ± 166.99	0.47 ± 0.18
Stage 7	62.50 ± 169.86	0.63 ± 0.29		

References

[1] F. Garlando, J. Konl, E. A. Koller, and P. Pietsch "Effect of coupling the breathing-and cycling rhythms on oxygen uptake during bicycle ergometry." European journal of Applied Physiology and Occupational Physiology, Vol. 54, pp. 497-501,1985.

[2] M. A. Daley, D. M. Bramble, and D. R. Carrier "Impact loading and locomotor-respiratory coordination significantly influence breathing dynamics in running humans." PloS one, Vol. 8, pp. e70752, 2013

[3] C. M. Peters and A. W.Sheel "Pulmonary Physiology and Response to Exercise." Exercise and Sports Pulmonology, Springer, Cham, pp. 3-17, 2019

[4] R. Willeput , C. Rondeux, and A. De Troyer " Breathing affects venous return form legs in human." Journal of Applied Physiology, Vol 57, pp. 971-976, 1984.

[5] C. F. Clarenbach, O. Senn, T. Brack, M. Kohler, and K. E. Bloch "Monitoring of Ventilation During Exercise by a Portable Respiratory Inductive Plethysmograph" Chest, Vol. 128, pp. 1282-1290, 2005.

[6] D. P. Bernstein. "Impedance cardiography: Pulsatile blood flow and the biophysical and electrodynamic basis for the stroke volume equations" Journal of Electrical Bioimpedance, Vol. 1, pp. 2-17, 2019.

[7] N. E. Huang et al., "The empirical mode decomposition and the Hilbert spectrum for nonlinear and non-stationary time series analysis," Proceedings of the Royal Society of London. Series A: Mathematical, Physical and Engineering Sciences, Vol. 454, pp. 903-995, 1998.

[8] Y. C. Chen and T. C. Hsiao "Towards estimation of respiratory muscle effort with respiratory inductance plethysmography signals and complementary ensemble empirical mode decomposition", Medical and biological engineering & computing, Vol.56, pp.1293-1303, 2017.

[9] N. E. Huang et al., "On instantaneous frequency," Advances in Adaptive Data Analysis, Vol. 1, pp. 177–229, 2009.

[10] M. A. Russo et al., "The physiological effects of slow breathing in the healthy human." Breathe, Vol.13, pp. 298-309, 2017.

[11] R. G. Fritzsche, D. M. Santarelli, and D. O'Rourke "Stroke volume decline during prolonged exercise is influenced by the increase in heart rate" ournal of Applied Physiology, Vol.86, pp. 799-805, 1999.

[12] A. W. Sheel and L.M. Romer "Ventilation and Respiratory Mechanics." Comprehensive Physiology, Vol. 2, pp. 1093-1042, 2011.

[13] P. D. Chantler et al., " Use of the Frank-Starling mechansiam during exercise is linked to exercise-induced changes in arterial load" American Journal of Physiology-Heart and Circulatory Physiology, Vol. 302, pp. H349-H355

Electrocardiography Signal Decomposition Using a Novel Modulated Ensemble Empirical Mode Decomposition Method

Chun-Hsiang Huang
Institute of Biomedical Engineering, College of Electrical and Computer Engineering
National Chiao Tung University
Hsinchu, Taiwan R.O.C.
e-mail: s09930720@gmail.com

Po-Hsun Huang
Institute of Computer Science and Engineering, College of Computer Science
National Chiao Tung University
Hsinchu, Taiwan R.O.C.
e-mail: sdk58927@gmail.com

Tzu-Chien Hsiao
Department of Computer Science, Institute of Biomedical
National Chiao Tung University
Hsinchu, Taiwan R.O.C.
e-mail: labview@cs.nctu.edu.tw

Abstract—**Electrocardiography (ECG) is an important test in the diagnosis of heart disease. The analysis of T waves in the ECG is an essential clinical tool, however, it is often difficult to extract the T waves from the ECG. Empirical Mode Decomposition (EMD) can decompose nonlinear and nonstationary signals, but this method suffers from the problem of mode mixing. Ensemble Empirical Mode Decomposition (EEMD) can solve the mode mixing problem but generates a new problem, namely, that of the reconstruction error. Moreover, noise may remain in the decomposed signals and pollute the waveforms. Therefore, we propose a new method based on EEMD to solve these problems and to decompose the T wave from ECG. The results show that the T waves' waveforms can be successfully decomposed in the fourth Intrinsic Mode Functions (IMFs) in all 3 cases studied, namely, no noise, power line noise, and Gaussian white noise.**

Keywords- Electrocardiography; Empirical Mode Decomposition; Ensemble Empirical Mode Decomposition; Reconstruction error.

I. INTRODUCTION

Electrocardiography (ECG) is an important biomedical signal which can help to diagnose heart diseases [1]. ECG usually consists of several waveforms. These waveforms are labeled as P wave, QRS-complex, and T wave. Figure 1 shows the ECG and the waveforms. The P wave represents the depolarization of the atria. The QRS-complex is caused by the depolarization of the right and left ventricles. Because the ventricle muscles are bigger than the atria, the QRS-complex's amplitude is usually larger than the P wave. The T wave is generated by the ventricular repolarization, so the T wave is after the QRS-complex. The times or segments between these waveforms are also important features of ECG. The PR-interval is the time cost of the impulse from the sinus node to the atrioventricular node. The PR-interval can help to evaluate the function of the atrioventricular node. The ST-segment represents the period of ventricles depolarization and usually isoelectric. The depression or elevation of ST-segment may represent a cardiac abnormality. The QT-interval is from the beginning of the QRS complex to the end of the T wave and varies with the heart rate. The corrected QT-interval is the QT-interval divided by the square root of the RR-interval. Prolonged Corrected QT-interval is a risk factor for ventricular tachyarrhythmia and sudden death. Therefore, feature extraction and detection of ECG are critical.

For ECG feature extraction and detection, the R wave is important for its timing. Most common feature extraction methods mark QRS-complex firstly and then search the P wave forwardly and the T wave backwardly [2][3]. Unlike the R wave, T wave's amplitude and shape are also as important as its timing. But, the noise like power line noise and Gaussian white noise will affect ECG [4]. The T wave will be influenced, too. Getting ECG features without noise is very important [5]. Many diseases are diagnosed by T wave's feature. For example, T-wave inversion, biphasic T-wave, T-wave alternans, etc. However, most judgments need a long time signal analysis and, moreover, these judgments always depend on doctors and professionals. Therefore, extracting the T wave is necessary and helpful for judging the performance of the heart system [6].

Mahmoodabadi et al. use a multiresolution wavelet to decompose ECG and extract features of ECG [7]. However, the wavelet is not adaptively [8]. The selection of the mother wavelet will limit the performance of the wavelet analysis. Pal and Mitra use Empirical Mode Decomposition (EMD) [9] to detect the QRS-complex of ECG [10]. EMD is a decomposition method that can decompose nonlinear and nonstationary signals into Intrinsic Mode Functions (IMFs) adaptively. However, it has several problems like boundary effect and mode mixing problem. The boundary effect is caused by that extrema value is hard to define at boundary, and will cause IMFs are distorted at the boundary. The mode mixing problem usually occurs when patterns appear intermittently and will cause two issues. The first issue is the

IMFs of EMD will have different scale patterns mixed in. The second issue is the same scale pattern may be separated in different IMFs. So, the ECG, which is one of the signals with intermittent pattern, will hardly to decompose well by EMD. Ensemble Empirical Mode Decomposition (EEMD) is a method which solves the aforementioned problem of the EMD [11]. Unfortunately, EEMD adds noise into signal and will remain noise in it [12]. Because the noise remains in, EEMD is not a good choice to decompose ECG.

In this study, we propose a new method based on EMD and EEMD. The new method will not add noise to the original signal. The Gaussian white noise will only assist to get the reference points which are treated as the extrema points. The modified part will solve the mode mixing problem and avoid adding external noise into the ECG signal. Furthermore, we use this new method to decompose the T wave from ECG. This new method is temporarily named as modulated ensemble empirical mode decomposition.

In Section 2, we will introduce the proposed method and the testing data. In Section 3, we will decompose ECG and simulated ECG by the proposed method. Otherwise, we will add some external noise to test whether the new method is influenced by noises. In Section 4, we will discuss the result and the main IMF of the T wave. In Section 5, the conclusion and further work will be proposed.

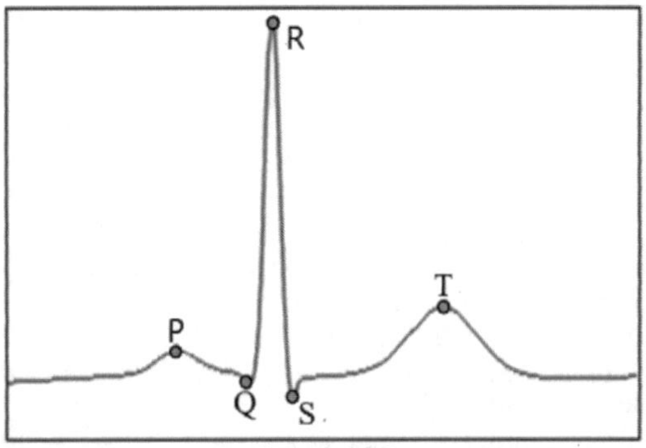

Figure 1. ECG

II. MATERIAL AND METHODS

A. Data

In this study, we verify our method by simulated data and PhysioNet QT database ECG. Simulated data is generated by LabVIEW "Simulate ECG". The database ECG will take one-minute data and test (A) original signal, (B) signal with 60 Hz power line and (C) signal with Gaussian white noise. The sampling rate is 250 Hz for all signals.

B. Methods

Our method is modified from EMD and EEMD. The detailed processes of EMD and EEMD are shown in Figure 2 and Figure 3.

$SD(.)$ means standard deviation. SD_{input} is the setting of stop criteria. $M(t)$ is the mean envelope. $r_i(t)$ is IMF or residue. Compared to EMD, EEMD adds white noise and averages each IMF to reduce the influence of noise adding.

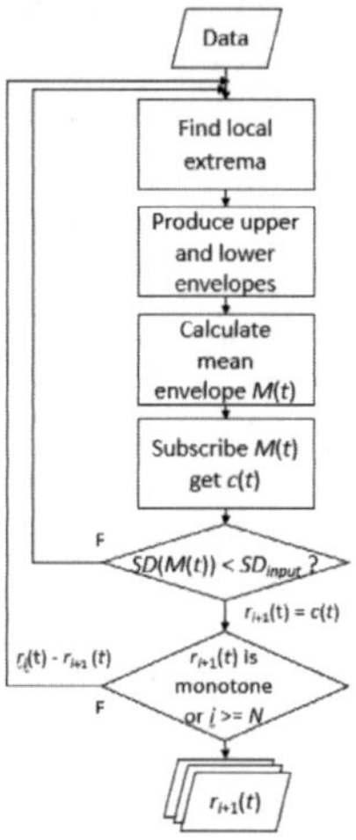

Figure 2. EMD algorithm procedure

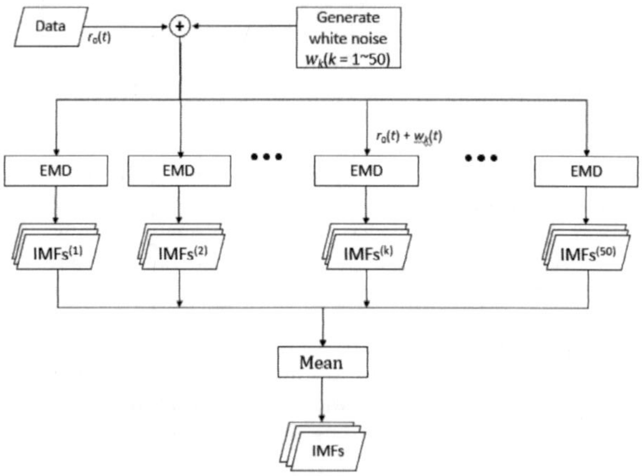

Figure 3. EEMD algorithm procedure

Adding local extrema by adding controlled noise is the main idea of the noise-assisted variations of EMD [13]. Adding noise is a method to solve the mode mixing problem. However, the main idea is gaining local extrema to assist in getting local means. Therefore, we find other points and treat these points as local extrema. Using these upper and lower envelopes' reference points can also find the local mean of different time scales. To find the upper and lower envelopes' reference points, we use Gaussian white noise to assist. Different from other methods, our method only picks up the timing of the extrema after adding Gaussian white noise. Then, we use these timings to get the reference points of upper and lower envelopes. Figure 4 illustrates the proposed method.

III. RESULT

A. Simulated ECG

Figure 5(a) is the simulated ECG which is generated by LabVIEW "Simulate ECG". Figure 5(b) shows IMF_1 to

IMF_4 which are decomposed by the proposed method. The QRS-complex timing can be found easily in IMF_1. The T wave might be decomposed in IMF_4.

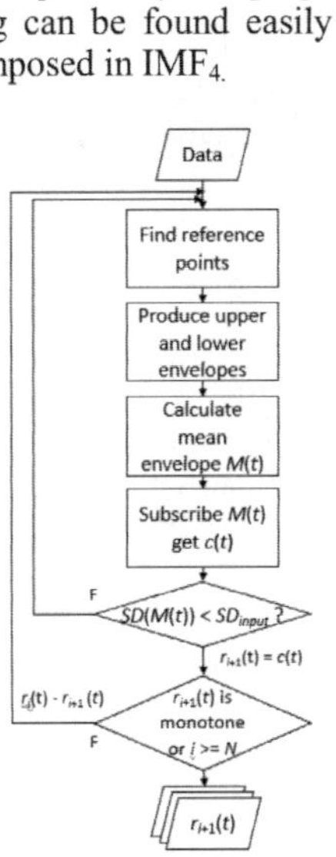

Figure 4. Modulated EMD algorithm procedure

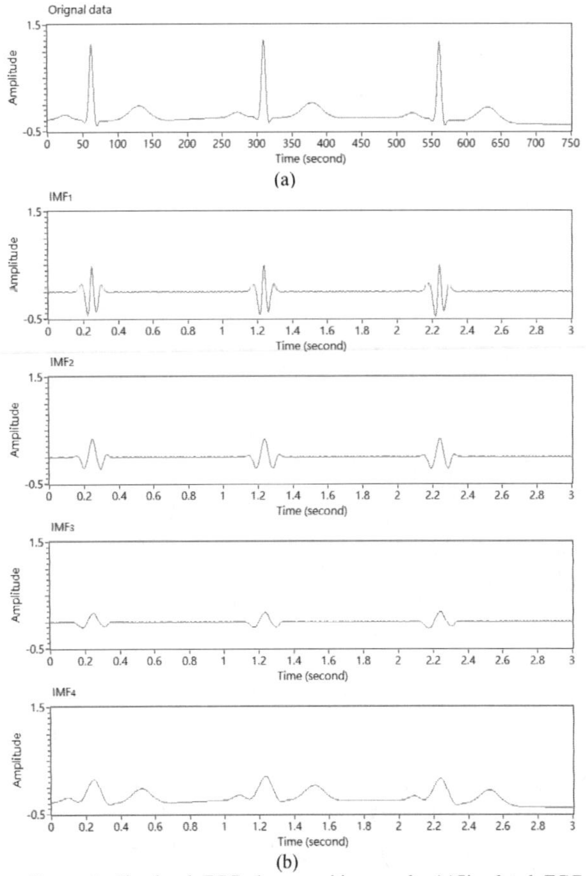

Figure 5. Simulated ECG decomposition result: (a)Simulated ECG which is generated by LabVIEW "Simulate ECG" and (b) decomposition results.

B. QT Database ECG

Because the main purpose is extracting T waves, comparing T wave and IMFs is necessary. T wave timing is obtained by PhysioNet QT database annotation. Figure 6(a) shows the ECG signal sel30 and the T wave is marked by a thick line.

Figure 6(b) shows the results of decomposition and the same periods of T wave are marked by a thick line. IMF_1 might make QRS-Timing obviously. IMF_4 can clearly show the T wave.

C. QT Database ECG with Power Line Noise

The power line is a common noise of the ECG signal and might influence the T wave's waveform. To test whether decomposition can remove the power line noise, the power line is added into signals. The adding power line's amplitude is 1/10 of the ECG signal and the frequency is 60 Hz.

The ECG signal sel30 with power line noise is shown in Figure 7(a) and the T wave period is also marked. Figure 7(b) shows that power line noise is almost dissembled in IMF_1. Therefore, IMF_4 can still find clear T waves. The QRS-complex timing can be checked in IMF_2.

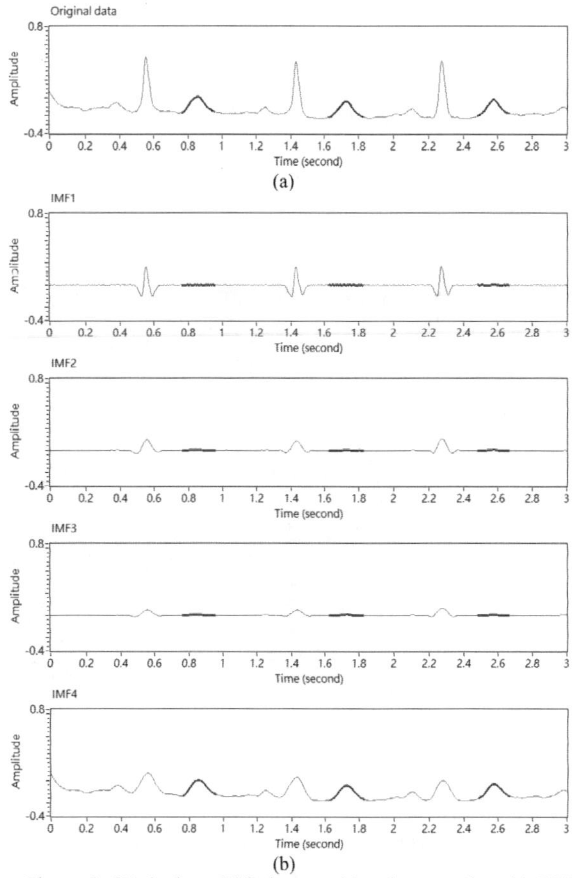

Figure 6. QT database ECG decomposition demonstration: (a) ECG signal sel30 and (b) decomposition results. The thick line marks the T wave period.

D. QT Database ECG with Gaussian White Noise

Gaussian white noise is also a common noise in the ECG signal. To test the decomposition method, Gaussian white noise is added to the ECG signal. The added Gaussian white noise's standard deviation is 1/10 of ECG's standard deviation. Figure 8(a) shows the ECG signal sel30 with Gaussian white noise. Figure 8(b) shows the decomposition results. The T wave can still be seen in IMF_4 clearly. The R-wave timing can be easily marked from IMF_2.

IV. DISCUSSION

Comparing Figure 6, Figure 7, and Figure 8, adding white noise will increase the difficulty of the QRS-complex timing extracting. Moreover, if adding noise in ECG, the IMF number of judgment QRS-complex timing will be changed. Otherwise, the T wave seems to be decomposed in IMF_4. No matter if adding power line noise, Gaussian white noise, or none, the T wave's waveform remains good. Noise seems to be decomposed to other IMFs except for IMF_4. The T wave is always decomposed in IMF_4. It has an advantage of automatically finding IMF with T wave. However, it still needs some judgment indicator to confirm the result.

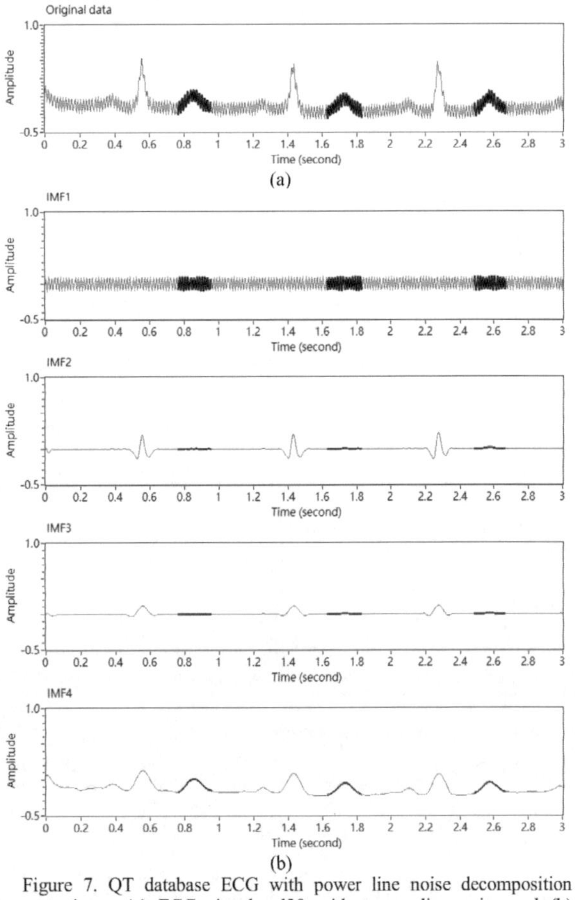

Figure 7. QT database ECG with power line noise decomposition demonstration : (a) ECG signal sel30 with power line noise and (b) decomposition results. The thick line marks the T wave period.

To judge whether decomposition results can extract the T wave, calculating the Correlation Coefficient (r) and the Root Mean Square Error (RMSE) between decomposition results in the T wave period is helpful. These two judgment indicators can check whether the T wave's phase and amplitude are retained.

Figure 9 shows the result of 105 records in the QT database. T waves are decomposed in IMF_4. Most of IMF_4's correlation coefficients are more than 0.9 and most of RMSE are less than 0.1. The result shows that T waves remain good in IMF_4.

Figure 10 and Figure 11 show the results of 105 records with power line noise and 105 records with Gaussian white noise. The results are as good as the results of the clear ECG signals.

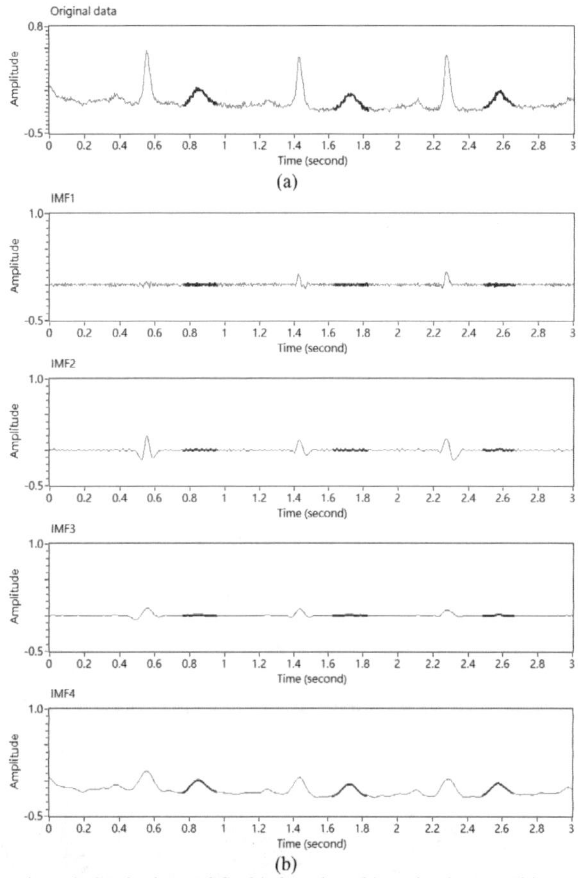

Figure 8. QT database ECG with Gaussian white noise decomposition demonstration : (a) ECG signal sel30 with Gaussian white noise and (b) decomposition results. The thick line marks the T wave period.

Table 1 shows the result of all records with treatments. The IMF_4's correlation coefficients are all over 0.95 and RMSE are all less than 0.04. The proposed method seems to extract the T wave in the IMF_4 and retains the waveform and phase well. Moreover, with the method we proposed, adding slight noise will not influence the T wave extracting. The results show that our method is useful for decomposing T waves. This might be helpful for several heart diseases diagnosing like T-wave inversion, biphasic T-wave, T-wave alternans, etc.

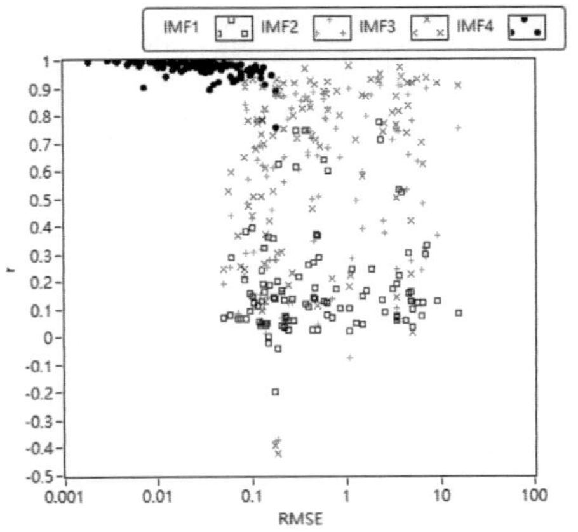

Figure 9. Correlation coefficient and RMSE between ECG signal decomposition results in T wave periods

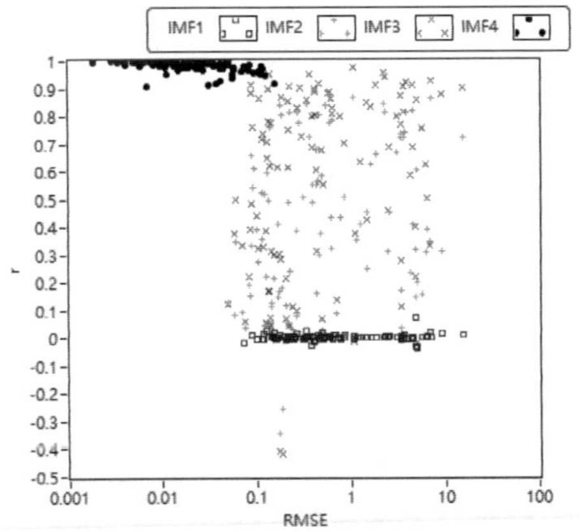

Figure 10. Correlation coefficient and RMSE between decomposition results of ECG signal with power line noise in T wave periods.

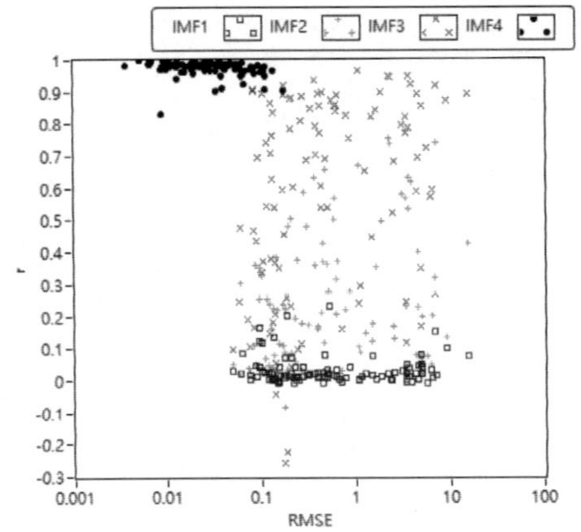

Figure 11. Correlation coefficient and RMSE between decomposition results of ECG signal with Gaussian white noise in T wave periods.

V. CONCLUSION

The results of ECG signal decomposition show that the proposed method extracts T wave well and is helpful for detecting QRS-complex timing. Furthermore, the new decomposition method has less influence on power line noise and Gaussian white noise. The proposed method might help ECG feature extraction and detection.

Although the new method can help to decompose ECG signals, how to automatically mark ECG's features is the next important research.

TABLE I. CORRELATION COEFFICIENT AND RMSE

Judgment indicator	**Signal treatments**		
	Original signal	***Add power line noise***	***Add Gaussian white noise***
r	0.97±0.03	0.98±0.02	0.98±0.02
RMSE	0.04±0.04	0.03±0.03	0.04±0.03

correlation coefficient and RMSE between IMF_4 and T wave in T wave period

VI. ACKNOWLEDGMENT

This work was fully supported by Taiwan Ministry of Science and Technology under grant numbers MOST 108-2221-E-009-087.

REFERENCES

[1] T. M. Nazmy, H. EL-Messiry, and B. Al-Bokhity, "Adaptive Neuro-Fuzzy inference system for classification of ECG signals", *The 7th International Conference on Informatics and Systems (INFOS)*, pp. 1-6, 2010

[2] A. Espiritu-Santo-Rincon and C. Carbajal-Fernandez, "ECG feature extraction via waveform segmentation", *IEEE International Conference on Electrical Engineering Computing Science and Automatic Control (CCE)*, pp. 250-255, September 2010.

[3] T. Khan, N. Sultana, R. Reza, and R. Mostafa, "ECG feature extraction in temporal domain and detection of various heart conditions", *2015 InternationalConference on Electrical Engineering and Information Communication Technology (ICEEICT)*, pp. 1-6, May 2015.

[4] S. Jain, V. Bajaj, and A. Kumar, "Riemann Liouville fractional integral based empirical mode decomposition for ECG denoising", *IEEE J. Biomed. Health Inform.*, vol. 22, no. 4, pp. 1133-1139, 2018.

[5] V. Seena and J. Yomas, "A review on feature extraction and denoising of ECG signal using wavelet transform", *Devices Circuits and Systems*, pp. 1-6, March 2014.

[6] D. Zavantis, E. Mastora, and G. Manis, "Robust automatic detection of P wave and T wave in electrocardiogram", 2017 *Computing in Cardiology (CinC)*, pp. 1-4, 2017.

[7] S. Z. Mahmoodabadi, A. Ahmadian, M. D. Abolhasani, M. Eslami, and J.H. Bidgoli, "ECG feature extraction based on multiresolution wavelet transform", *Proc. 27th Annual International Conference of the IEEE Engineering in Medicine and Biology Society*, pp. 3902-3905, January 2005.

[8] T. Liu, Z. Luo, J. Huang, and S. Yan, "A comparative study of four kinds of adaptive decomposition algorithms and their

applications", *Sensors*, vol. 18, no. 7, pp. 2120-2170, July 2018.

[9] N. E. Huang, et al., "The empirical mode decomposition and the Hilbert spectrum for nonlinear and non-stationary time series analysis," *Proc. R. Soc. Lond. A*, vol. 454, pp. 903-995, 1998.

[10] S. Pal and M. Mitra, "Empirical mode decomposition based ECG enhancement and QRS detection", *Comput. Biol. Med.*, vol. 42, no. 1, pp. 83-92, January 2012.

[11] Z. Wu and N. E. Huang, "Ensemble empirical mode decomposition: A noise-assisted data analysis method", *Advances in Adaptive Data Analysis*, vol. 1, no. 1, pp. 1-41, 2009.

[12] M. E. Torres, M. A. Colominas, G. Schlotthauer, and P. Flandrin, "A complete ensemble empirical mode decomposition with adaptive noise", *Proc. IEEE Int. Conf. Acoust. Speech Signal Process.*, pp. 4144-4147, May 2011.

[13] M. A. Colominas, G. Schlotthauer, and M. E. Torres, "Improved complete ensemble EMD: A suitable tool for biomedical signal processing", *Biomed. Signal Process. Control*, vol. 14, pp. 19-29, November 2014

Towards Estimations of Continuous Cardiac Output with Impedance Cardiography: a Pilot Study

Chao-Ting Liu
Institute of Biomedical Engineering,
National Chiao Tung University
Hsinchu, Taiwan R.O.C.
e-mail: dfr1507@gmail.com

Po-Hsun Huang
Institute of Computer Science and Engineering,
National Chiao Tung University
Hsinchu, Taiwan R.O.C.
e-mail: sdk58927@gmail.com

Tzu-Chien Hsiao
Department of Computer Science, Institute of Biomedical Engineering,
Biomedical Electronics Translational Research Center
National Chiao Tung University
Hsinchu, Taiwan R.O.C.
e-mail: labview@cs.nctu.edu.tw

Abstract—**Impedance Cardiography (ICG) is a non-invasive method to estimate the cardiac output, which is very helpful for the prognosis of cardiovascular disease and the diagnosis of disease. By applying a stable high-frequency and low-intensity current to the human, we can measure the impedance variation of the human body, and, therefore, get the ICG signal to judge the possible characteristic points and estimate the cardiac output. The measurement of ICG in long-term care is limited due to the cost involved and the need for professional evaluation. For this reason, in this paper, we develop an impedance signal measurement system suitable for long-term care. We use a modified circuit system and LabVIEW software to detect the impedance variation of the human body and the characteristic points of ICG. At present, our preliminary results have been able to detect possible points B and C, for the detection of point X. We need to compare and confirm these results using Electrocardiography (ECG). In the future, we will use the feature points detected by this system to estimate the cardiac output and develop it into a portable instrument for long-term care.**

Keywords-Impedance Cardiography (ICG); Cardiac Output (CO); Stroke Volume (SV).

I. INTRODUCTION

In recent years, cardiovascular diseases have seriously affected people's health and have led to an increase in mortality [1]. Therefore, the prediction and management of cardiovascular diseases become very important. In the past, it has been proposed that cardiac output can be estimated by thermal dilution [2], however, this is not an option for long-term monitoring because it is an invasive method.

Impedance Cardiography (ICG) is a non-invasive, continuous, easy and accurate method to evaluate left ventricular stroke volume and cardiac output [3], and it is also suitable for long-term monitoring of cardiac activity. In the past, other non-invasive evaluation methods have been proposed to measure cardiac output, such as Doppler ultrasound [4]. Although the measurement results are accurate, Doppler ultrasound is not suitable for long-term care because of the high cost and the need for professional analysis.

The principle of ICG instrument is to apply a constant low current (1mA-5mA) and high frequency (50kHz-100kHz) signal to the outer electrode by means of four electrodes placed, then Ohm's law is used to measure the voltage signal of the external electrode to estimate the chest impedance of human body [5]-[7]. The cardiac output can be evaluated by the characteristic points and the method proposed by Kubicek [3][8].

Cardiac output is the volume of blood entering the aorta from the left ventricle every minute, and stroke volume is the volume of blood ejected from the left ventricle in each systole. Cardiac output is the product of cardiac stroke and heart rate [9], as (1).

$$CO = SV \times HR \quad (1)$$

Cardiac output is important information to evaluate cardiac health, and ICG is a simple way to evaluate cardiac output.

The typical waveforms of impedance variation (delta Z), ICG and Electrocardiography (ECG) are shown in Figure 1 [10].

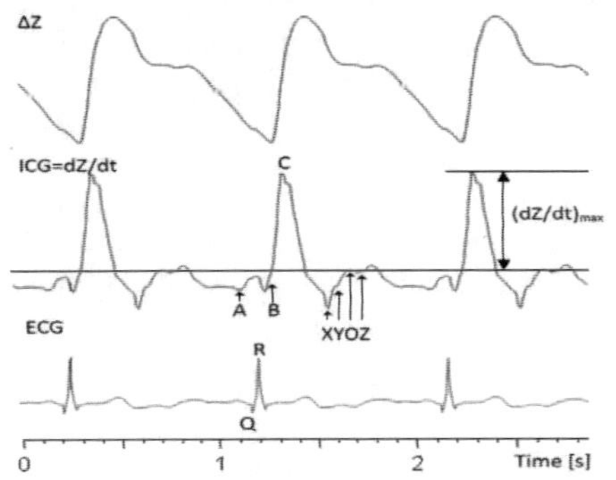

Figure 1. Impedance variation, ICG typical signal and ECG signal

A typical ICG can distinguish the following characteristic points [11]:

TABLE I. CHARACTERISTIC POINT

Characteristic point	description
A	This point is related to atrial contraction.
B	Corresponding to the opening of the aorta, this point occurs at the zero crossing point before ((dZ/dt)max), which is an important information to judge the pre-ejection period (PEP) and Left Ventricular Ejection Time (LVET).
C	((dZ/dt)max): This point is the maximum amplitude of this signal, which can reflect the maximum speed of impedance change, which is related to the maximum ejection speed of the heart.
X	Corresponding to aortic closure, it is the most negative point of ((dZ/dt)max) signal, occurring after point C.
Y	Corresponding to pulmonary artery closure.
O	It is related to mitral valve opening and volume change in diastolic period.
Z	This point is related to the third heart sound after the O point.

The time relationship between the characteristic points of ICG and ECG signal is as follows [11]:

TABLE II. CHARACTERISTIC PERIOD

Characteristic period	Description
Q-B	Q point in the ECG and B point in the ICG (the opening of the aortic), which is the pre-ejection period (PEP).
Q-C	Time interval between Q point in ECG and ((dZ/dt)max)in ICG, which can be used to calculate the heather index for cardiac contract.
B-X	The time interval between the opening of the aorta and aortic closure. This segment is the Left Ventricular Ejection Time (LVET), which is the ICG signal feature.

After obtaining these characteristic points and time interval relations, we can estimate SV by Kubicek's method [3][8][12] and then evaluate the Systolic Interval and activity of the heart.

$$SV = \frac{\rho L^2}{Z_0^2} \times LVET \times \frac{dZ}{dt}(\max) \tag{2}$$

where ρ is the blood resistivity of human body in Ω-cm, L is the distance between two sensing electrodes in cm, Zo is the basal impedance in Ω, (dZ/dt)max is the maximum value of impedance change in Ω/s, and $LVET$ is the Left Ventricular Ejection Time in s.

We have made a preliminary introduction to our research. In Section 2, we will show our system design and experimental process. In Section 3, the results of the experiment are shown. In Section 4, we discuss the problems we are facing at present. In Section 5, we summarize our current research results and contributions.

II. MATERIAL AND METHOD

A. System Description

ICG circuit system is mainly divided into three blocks (1) transmitter (2) receiver (3) software Layer. Figure 2 is the framework diagram of the ICG circuit system; Figure 3 is the ECG framework and position of electrode corresponding to the body.

(1) Transmitter

First of all, we use a 12V lead-acid battery as the DC power supply of the oscillator which can produce stable frequency, and then connect the Voltage Controlled Current Source (VCCS) circuit to make the sinusoidal signal output stable current [7]. In order to ensure safety, we design an additional voltage limiting circuit as the protection circuit, and then apply the stable current signal to the outer electrode of the four electrodes.

(2) Receiver

Then, we receive the voltage signal of the human body from the inner electrode, using ad633 as an instrument amplifier, and use 60kHZ and 90khz band pass filter of 8 orders as filtering signal. Finally, we use NI USB-6210 to do ADC and input the signal into the computer.

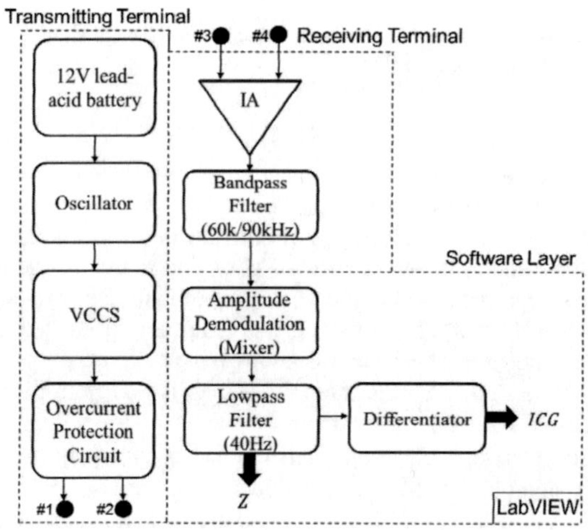

Figure 2. Framework diagram of ICG circuit system

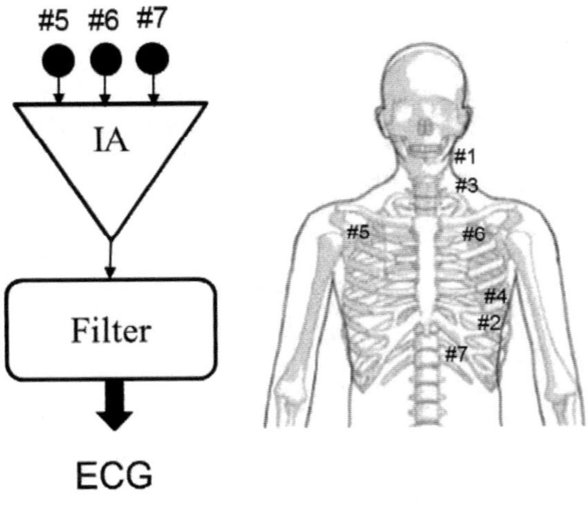

Figure 3. ECG and position of electrode corresponding to body

(3) Software Layer

We use LabVIEW as software layer and analysis, the received signal is demodulated by mixer. Then we use 8 order's 40 Hz low pass filter to get the modulation signal (delta Z). Then, we use the obtained modulation signal (delta Z) do differential and multiplied by a negative sign to obtain the ICG signal [13].

The following is the process of mixer:

- Demodulated by mixer

$$X_{Carrier} = V_C \cos(2\pi f_C t) \tag{3}$$

$$X_{AM} = V_{AM}(1 + m\cos(\pi f_m t))V_C \cos(2\pi f_C t) \tag{4}$$

(3) multiplying (4) → (5)

$$X_{Out} = \frac{V_{AM}V_C^2}{2} + \frac{V_{AM}V_C^2}{2}\cos(2\pi f_m t) + \frac{V_{AM}V_C^2}{2}(1 + m\cos(2\pi f_m t)\cos(2(2\pi f_C t)) \tag{5}$$

(5) through a 40Hz low pass filter → modulation signal where $X_{CARRIER}$ is the carrier signal, V_C is voltage amplitude of carrier signal, f_C is the frequency of carrier signal, X_{AM} is modulated signal, V_{AM} is voltage amplitude of the modulated signal, f_m is the frequency of modulation signal, t is time.

B. *Subject data collection and experiment process*

This research's data was extracted from one subject; no history of cardiovascular disease was in this subject.

Before the experiment, the electrodes should be placed first as shown in Figure 3. Four electrode positions required for ICG: Lead 1 is placed on the left upper neck. Lead 2 is placed 3 cm below lead 1. Lead 3 is placed at the point where the xiphoid is cut to the left of the chest. Lead 4 is placed 3 cm below lead 3 [6]. The positions of the ECG electrode (Lead 5-7) are as shown in Figure 3.

The experimental procedure is: first, let the subject rest for 3 minutes to measure the baseline, and then do 30 seconds of rhythmic breathing to acquire ECG and ICG signal.

III. RESULT

Figures 4 and 5 show the signals when we measure only the ICG of the subjects. Figures 6 and 7 show the signals when we measured the ECG and ICG of the subjects at the same time.

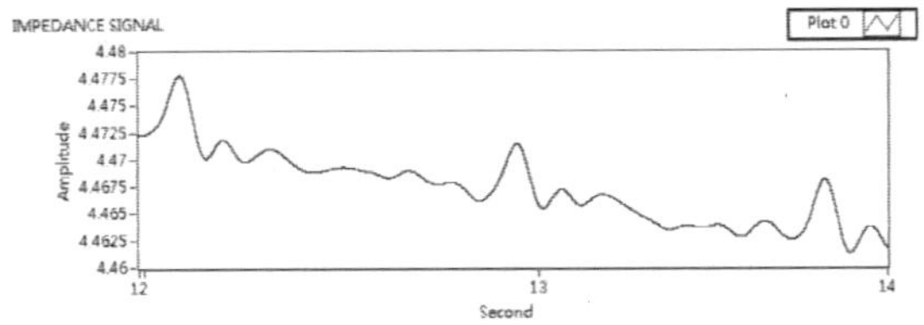

Figure 4. Impedance variation (delta Z)

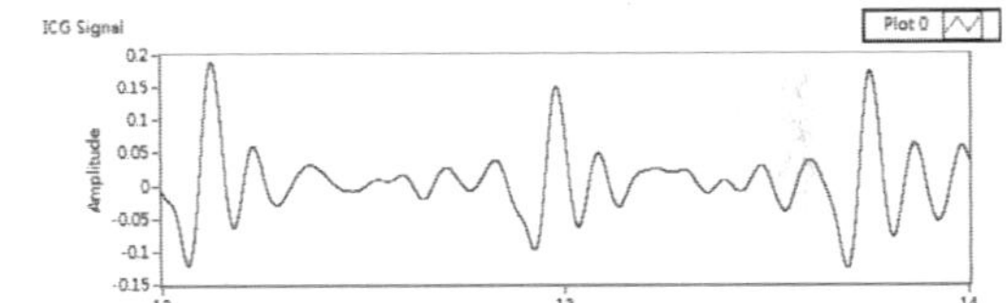

Figure 5. Impedance Cardiography (ICG)

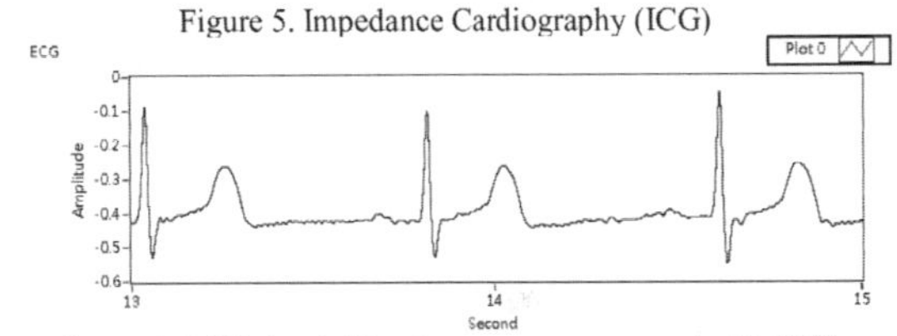

Figure 6. ECG signal (Simultaneous measurement with ICG)

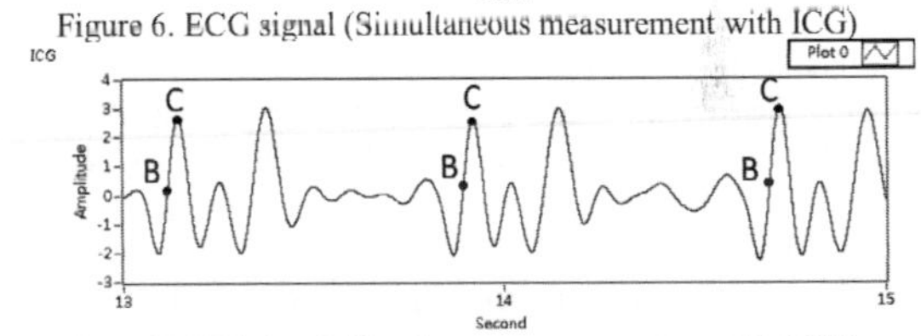

Figure 7. ICG signal (Simultaneous measurement with ECG)

The change of human body impedance was shown in Figure 4. The maximum impedance can be observed in each cardiac cycle. From the three cardiac cycles in Figure 5, we can observe the ICG that obtained by letting delta Z do differential and multiplied by a negative sign.

Figure 6 is the ECG signal measured with ICG at the same time. Figure 7 is the ICG signal measured with ECG at the same time. We may determine the possible point B and point C, and the interval between point C of ICG is similar to the RR interval of ECG. But point x cannot be accurately determined. We need to find out the possible x point by comparing ICG with ECG.

IV. DISCUSSION

By comparing the signal characteristics of Figure 4 and Figure 5 with those of previous papers, we can detect possible B and C points, but we cannot accurately determine x points. Therefore, we hope to get possible x points by comparing with ECG signals.

According to [5] [11], we can confirm point B of ICG from point R of ECG. By comparing Figures 6 and 7, we can see that point B (when the aorta is opened) does occur after point R of ECG, and point C occurs at the highest point after point B, so point B and point C may be correct. Then, we can compare the x-point of ICG by the end of the T-wave of ECG [11] [14]. From Figures 6 and 7, we may find that at the end of the T-wave of ECG, the X point corresponding to this time is almost the most negative value after zero crossing, so this is the possible x point.

Although it is possible to detect the possible feature points of ICG through the simultaneous measurement of ECG and ICG, there are still several deficiencies in this study. First of all, there is the problem of signal coupling in the measurement. It can be seen from the ICG signal that the influence of ECG signal-coupling may affect the judgment of feature points. Second, the number of subjects is too small, there is only one subject at present, it is necessary to measure the data of multiple subjects for comparison, which can increase the reliability of signal and feature point detection

Finally, the problem of signal distortion will affect the judgment of feature points; the circuit design can be modified to increase the stability of the instrument.

V. CONCLUSION

This study shows that the impedance variation and ICG may be detected by the modified circuit, despite the fact that the possible characteristic points of ICG can be detected at present. However, there are still several deficiencies that have not been resolved, so it is impossible to accurately estimate cardiac output. In the future, we will solve these deficiencies, so as to estimate the accurate cardiac output, and then develop into portable instruments, which will help long-term care for the tracking and diagnosis of diseases.

ACKNOWLEDGMENT

This work was fully supported by Taiwan Ministry of Science and Technology under grant numbers MOST 108-2221-E-009-087.

REFERENCES

[1] G. A. Roth et al.,"Global, regional, and national burden of cardiovascular diseases for 10 causes, 1990 to 2015," Journal of the American College of Cardiology, 70(1), pp. 1-25, 2017.

[2] M. A. Branthwaite and R. D. Bradley, "Measurement of cardiac output by thermal dilution in man," Journal of Applied Physiology, 24(3), pp.434-438, 1968.

[3] W. G. Kubicek, "Development and Evaluation of an Impedance Cardiac Output System," Aerosp Med, 37, pp. 1208-1212, 1966.

[4] J. B. Mark et al., "Continuous noninvasive monitoring of cardiac output with esophageal Doppler ultrasound during cardiac surgery," Anesthesia and analgesia, 65(10), pp. 1013-1020, 1986.

[5] X. Hu et al., "Adaptive Filtering and Characteristics Extraction for Impedance Cardiography," Journal of Fiber bioengineering and Informatics, 7(1), pp. 81-90, 2014.

[6] L. Y. Shyu, C. Y. Chiang, C. P. Liu, and W. C. Hu, "Portable Impedance Cardiography System for Real-Time Noninvasive Cardiac Output Measurement," Journal of medical and biological Engineering, 20(4), pp. 193-202, 2000.

[7] H. Yazdanian, A. Mahnam, M. Edrisi, and M. A. Esfahani, "Design and Implementation of a Portable Impedance Cardiography System for Noninvasive Stroke Volume Monitoring," Journal of medical signals and sensors, 6(1), 47-56, 2016.

[8] W. G. Kubicek et al., "The Minnesota impedance cardiograph-theory and applications," Bio-Medical Engineering, 9(9), pp. 410-416, 1974.

[9] D. G. Newman and R. Callister, "The non-invasive assessment of stroke volume and cardiac output by impedance cardiography: a review," Aviation, space, and environmental medicine, 70(8), pp. 780-789, 1999.

[10] J. Malmivuo and R. Plonsey, "Principles and applications of bioelectric and biomagnetic fields," Oxford University Press, 1995.

[11] G. Cybulski, "Ambulatory impedance cardiography - The Systems and their applications," Springer, Berlin, Heidelberg, pp. 39-56, 2011.

[12] D. P. Bernstein and H. J. M. Lemmens, "Stroke volume equation for impedance cardiography," Medical and Biological Engineering and Computing, 43(4), pp. 443-450, 2005.

[13] W. G. Kubicek, "On the Source of Peak First Time Derivative (dZ/dt) During Impedance Cardiography," Annals of Biomedical Engineering, 17: pp. 459-462, 1989.

[14] B. E. Hurwitz et al., "Signal fidelity requirements for deriving impedance cardiographie measures of cardiac function over a broad heart rate range," Biological psychology, 36(1-2), pp. 3-21, 1993.

Towards a National Clinical Decision Support Framework for Norway: Expert Assessment and Proposed Architecture

Luis Marco-Ruiz, Kristian Malm-Nicolaisen, Alexandra Makhlysheva, Rune Pedersen
Norwegian Centre for E-health Research
University Hospital North Norway
Tromsø, Norway
emails: [luis.marco.ruiz, alexandra.makhlysheva, kristian.nicolaisen, rune.pedersen]@ehealthresearch.no

Abstract— **Computerized Clinical Decision Support (CCDS) is an important component for making available latest medical evidence at the point of care. With the aim to gain knowledge on the best approaches for implementing a nation-wide CCDS, we performed a survey by sending an extensive questionnaire about CCDS architectures, use of standards, and use of terminologies to international experts and CCDS vendors. The responses to the questionnaire were analyzed and mapped to the requirements of the Norwegian health IT context. With this correspondence between responses and requirements, we designed a national architecture with the components needed for providing CCDS at a national scale in Norway. The architecture leverages different components aiming to allow supporting several standards and terminologies, performing both national and local governance, reusing CCDS functionality, and adapting CCDS modules to local contexts.**

Keywords- E-health Services; Clinical Decision Support; Clinical Guidelines; Computerized Clinical Decision Support; Interoperability; Software Architecture; Electronic Health Records.

I. INTRODUCTION

The Learning Healthcare System concept has motivated efforts in different areas of the health Information Technology (IT) realm worldwide [1]–[3]. Examples are the general adoption of Electronic Health Records (EHRs), the development of health data analytics platforms, the development of data reuse networks, and the definition of large-scale clinical decision support frameworks [3]–[5]. In this context, Norway has allocated significant funds to build momentum for advancing medical informatics [6]–[9]. EHRs have been adopted to enable data reuse including highly structured formats, such as openEHR [10]; projects for data reuse have been funded setting the basis for what later has become a national primary care research network [7]; and several initiatives have provided knowledge on best approaches for Computerized Clinical Decision Support (CCDS) interventions [11]–[14]. CCDS systems are software systems designed to provide useful information at the point of the clinical workflow when it is needed. Previous projects about CCDS have provided very valuable knowledge specifying requirements and success factors during EHR adoption. However, when it comes to building a large-scale national infrastructure to govern and manage CCDS systems, knowledge about their architecture and organization is still needed. On the one hand, it is needed to better understand how to optimally leverage the plethora of technologies, clinical information standards, and biomedical terminologies; and, on the other hand, it is needed to determine how to organize the development and management of CCDS algorithms. The Norwegian context, is complex as a result of a mixture of legacy and recently introduced Health Information Systems (HIS) that operate using different clinical information standards and terminologies. For example, the main vendor of hospital information systems relies on a free text EHR, and has been evolving it into an openEHR-based EHR [10]. The Central Norway Regional Health Authority, recently procured Epic for its introduction in the incoming years. Regarding terminologies, Norway uses the International Classification of Diseases (ICD-10), the International Classification of Primary Care (ICPC-2), and other well established terminologies [15]. In addition, Norway has recently become a member of SNOMED International [16]. Currently, the aim is to pilot parts of SNOMED-CT for clinical use in the Central Norway Health Region. This scenario poses requirements for CCDS systems that need to be able to work in a multi-standard environment with various Clinical Information Models (CIMs) and terminologies.

In order to design strategies for this scenario, the Norwegian Directorate of E-health (NDE) requested a survey from the Norwegian Centre for E-health Research in order to get information about CCDS, clinical information standards, and biomedical ontologies. Thus, in 2017, we performed a study enquiring CCDS experts and vendors about different topics that directly affect CCDS infrastructures. A summary of the results was published in two reports [17] [18]. In this paper, we provide a more detailed and deep analysis of the responses, and we map them to the Norwegian scenario drawing an architecture that leverages, on the one hand, the advise from experts and, on the other hand, the requirements of the Norwegian scenario. The remaining of the paper is organized as follows. Section 2 presents the methods for performing the survey and analyzing the results. Section 3 presents the results of the analysis of the questionnaires by inductively analyzing them and collating the information to form main categories of recommendations, observations, and future perspectives. It also provides the architecture designed using the information gathered for covering the Norwegian requirements regarding the CCDS domain. Finally, Section 4 discusses the architecture proposed and relates it to previous studies.

II. METHODS

We performed a survey as part of two studies performed for the Norwegian Directorate of E-health during 2017 [17] [18]. The survey was sent to 25 experts and vendors on CCDS. Among the representatives invited to participate, there were researchers, consultants, and vendors. Thirteen respondents were vendors and 12 were experts in CCDS or biomedical ontologies. The invited participants were known companies, researchers, and consultants of the CCDS arena. It was a requirement for the invited participants to be involved in CCDS projects that had been deployed in an operational environment, and not only in research academic projects.

Of the 25 invited participants, 11 responded agreeing to participate. Of these, finally 9 provided the completed questionnaire requested. We sent a word document with all the questions to be answered that the respondents could edit freely without a word limit. The complete questionnaires are available at the previously published reports [17] [18].

Eight participants provided extensive information about the topics and the architecture of their CCDS interventions. Respondents provided schemas and detailed descriptions in their CCDS interventions. In addition, all participants shared impressions and their future vision regarding the CCDS arena. Among the participants that agreed to participate, we received 8 completed questionnaires. Four questionnaires corresponded to researchers involved in small CDS companies and research, and 4 corresponded to major vendors. One of the respondents was an expert on biomedical ontologies who participated as an active developer of them, but that did not have experience with CCDS. This respondent completed only the part of the questionnaire associated with ontologies and not CCDS. We collated the results by marking the topics that appeared in each of the responses. We proceeded inductively grouping the topics identified in main categories and subcategories. This categorization was used to report the results that follow.

III. RESULTS

A. Overview

We identified the following main categories and subcategories: a) *software architecture and information standards*, with the subcategories architecture, and *clinical information standards*; b) *biomedical terminologies*, with subcategories *role of terminologies in CCDS* and *ontology-based terminologies*; c) *Organization, governance, and shared development*, with subcategories *authoring*, *governance frameworks*, and *local adaption/customization*; and, d) *Knowledge base (KB)*. The best design factors identified in the survey were used to develop the proposed architecture. The following describes, first, each of the categories identified in the survey and, second, the architecture developed from the survey results and adapted to the Norwegian context.

B. Software Architecture and Information Standards

1) Architecture

Respondents considered the following aspects if the architecture as critical: speed, concurrent support for many clients, high availability and error tolerance, support for interoperability standards, and support for both EHR and population queries. With this regards, all the respondents agreed that, in general, Service Oriented Architectures (SOAs) were the optimal choice for fulfilling the mentioned requirements in large and distributed CCDS environments. That is because SOAs encapsulate CCDS functionality making it available to various clients through a Web service, thus enabling concurrent support, easier fault tolerance, interoperability, and support for queries over disparate systems.

More specifically, the use of RESTful stateless Web service architectures is also seen as beneficial for simplifying the architecture. Both synchronous and asynchronous ways of operating are needed. One respondent wrote that, in general, SOA is better but it is important to understand the requirements because the optimal architecture may be a mixture of some approaches (SOA, stand-alone, process oriented, etc.).

Another advantage of the encapsulation provided by SOAs concerns the inference engine and logic specification mechanisms. When it comes to the decision algorithm, respondents considered that SOA allows encapsulating any algorithm implemented in any technology inside the Web service. This algorithm is then exposed through a standard API based on information standards. This way, SOA alleviates the need of a separate formalism (e.g., Arden syntax) that is later translated into the language used by the inference engine. For example, openCDS operates directly with JBoss Drools instead of using another specification formalism for abstracting the logic representation [19].

One respondent considered that, although for reusing CCDS functionality SOA was the optimal solution, embedding the CCDS functionality within the EHR had advantages too. The reason is that embedded CCDS provides better performance due to the possibility of performing pre-processing of data structures that makes significant data are rapidly available when a CCDS triggers.

2) Clinical Information Standards

Standardization of CCDS was considered essential by all respondents but with subtle differences. Most respondents considered standardization of the CCDS data schema, a.k.a. Virtual Medical Record (VMR), and the SOA payload as paramount for all CCDS systems. Standardization was seen as a way to communicate the payload of Web service messages in and out of the CCDS Web service in a normalized way that all clients can understand. This approach makes the CCDS client (often an EHR) responsible for committing to that standard data format.

Recalling the previous section, one of the respondents indicated the advantage for embedding CCDS into the EHR, but acknowledged the need for standardization in a national-level CCDS. Another respondent leveraged both views by relying on openEHR for both the VMR and the EHR. This approach can directly reference the same standard information schema that the EHR relies on. In order to implement such design, both the EHR and the CCDS system need to work on the same set of openEHR archetypes or rely on effective abstraction mechanisms. However, this is not always easy to achieve having CCDS and EHR operating at different levels of granularity. It is likely that several standards will coexist and that transformation mechanisms will need to be provided as discussed below.

The role of HL7 Fast Healthcare Interoperability Resources (FHIR) is considered very relevant as a standard for sharing patient data extracts with CCDS Web services. For CCDS interoperability, FHIR is the preferred standard by respondents since they claimed it to be the one with the highest acceptance rate across vendors. In addition to the FHIR standard, there is the library for authentication and integration, SMART [20]. SMART on FHIR was seen as a positive but not critical addition on top of FHIR. One of the respondents considered it to be particularly useful for authentication; while other pointed out the value regarding its applications deployment framework. For example, one respondent remarked the possibility of using SMART for showing context-specific data with graphical user interfaces (with relevant data, literature, etc.). Regarding the implementation of the VMR, one respondent mentioned HL7 vMR as a very comprehensible standard; however, the same respondent also pointed out that it had a low adoption rate. Noteworthy, the open source initiative openCDS has developed conversion mechanisms from HL7 vMR to FHIR [19]. Two respondents used proprietary formats, and one used openEHR archetypes.

A standard mentioned for embedding CCDS functionality in the EHR was CDS Hooks [21]. Respondents considered CDS Hooks a useful and disruptive standard for embedding CCDS requests in the appropriate part of the clinical workflow. Thus, it may allow some context awareness if it is used in the appropriate way. One respondent pointed out that CDS Hooks still needed to be extended and constrained. Actually, while writing, it is a Standard for Trial Use (STU) release in its version 1.0 [21]. Several respondents pointed out that for new CCDS developments CDS Hooks should be seriously considered since it allowed to perform workflow aware CDS actions (triggering at a precise point of the clinical workflow) and accelerate CCDS implementations.

Another relevant topic is the management of CCDS systems in environments where several information standards coexist. This is the Norwegian case where openEHR is used for structuring EHRs in secondary healthcare and FHIR is recommended for EHR extracts exchange [22]. In this case, the respondents pointed out the complexity of the scenario and the need for implementing transformation methods among different standards. To that end, several solutions were proposed. One respondent mentioned that, for this kind of complex scenarios where iso-semantic models were present, a common agnostic representation would be needed. Another respondent indicated that when several information models are present, the messages will need to be written using a normalized model of choice, which is preferred by the national centralized system. Client systems not supporting such a model natively will need to transform to and from that model in order to consume the national CCDS.

C. Biomedical Terminologies and Ontologies

1)Role of terminologies in CCDS

Respondents considered that terminologies play a critical role. Respondents agreed that the CCDS main function is to avoid ambiguity and allow identifying the same concept in multiple ways, thus providing a standardized way to define CDSS criteria at the points of care, population management queries, and predictive analytics. For coherence, the management of terminologies should be centralized if possible. Terminologies are essential for expressing the semantics of patient data and the recommendations captured within terminology concepts, therefore the inferences over terminology concepts are needed within any CCDS. Terminology concepts have been used for content binding providing the subset of possible values to fill a specific slot in a CIM, but also as a semantic binding for specifying the meaning of a specific element of the CIM. The use of CIMs in combination with terminologies is not a clear issue and, as we pointed out in previous studies, it is still dependent on the needs for automatic interpretation of clinical data [23] [24].

With regards to terminologies, the situation is similar to the one of CIMs. Proprietary terminologies will need to be mapped to the reference value sets used by the national CCDS. The value sets used by guidelines need to be governed and maintained.

2) Ontology-based terminologies

We asked our respondents about the use of ontology-based terminologies. We use the word ontology in the computer science sense, i.e., classifications with some description logics or model-theoretic semantics underpinning. The only ontology used by the respondents was SNOMED-CT.

Most respondents considered that ontology-based terminologies such as SNOMED-CT could be useful for the maintenance of complex terminologies, the maintenance of CCDS rules, and the definition of mappings across terms from different terminologies. Regarding mappings among terminologies, respondents also pointed out that mappings and transformations should be made with caution. One respondent warned that, for example, “topical steroids and systemic steroids have different contradictions"; thus, "mapping all drugs that contain steroids may lead to inappropriate recommendations from CCDS systems". The same respondent made two observations about the definitions of value sets using SNOMED-CT. The respondent pointed out that SNOMED-CT has many inconsistencies in its logical structure; and, for this reason, a better way to proceed is to manually curate the concepts into explicit value sets than using logic definitions over SNOMED-CT for defining subsets.

We asked our respondents about the use of other ontology-based terminologies, e.g., for allowing EHRs to support genomic medicine, but all of them agreed on the fact that this aspect is considered as not crucial at the moment. SNOMED-CT has sparse support for molecular biology and more specific ontologies would be needed for that. One respondent actually said that simpler CCDS functionality that does not require complex semantic analysis should be first implemented.

Curating and pre-processing of ontologies into the internal CCDS format is common. Also, implementing support for third party terminologies is needed due to the amount of proprietary code systems. Other vendor considered a better solution to be in an inside-system embedded for performance without a specific Terminology Server (TS).

The use of the logic underpinning of SNOMED-CT was rather sparse. As the expert in ontologies indicated, currently the formal semantics of SNOMED-CT are used for the maintenance of the terminology itself, e.g., when defining new concepts. The only use from its underlying logic model was subsumption. Subsumption ("is-a" relationships) reasoning is considered useful for facilitating the setup and maintenance of rules in CCDS, but respondents observed that in order to truly use this capability, a supportive infrastructure is needed. SNOMED queries would require the SNOMED OWL representation and a classifier.

D. Organization, Governance, and Shared Development

1)Authoring

For implementing a national CCDS infrastructure respondents indicated that there should be a sufficient collaboration among clinical centers. Those centers should discuss about the national clinical practice guidelines to base the computerized CCDS on. According to them, authoring tools are needed so that different stakeholders can collaborate in a distributed manner having discussions and defining clinical decision algorithms. Respondents considered that a national portal with narrative and semi-structured guidelines can be helpful in the CCDS development. In addition, two respondents indicated that measuring and monitoring the impact of CCDS interventions would be needed to clarify their effect and decide on their long-term maintainability. Another respondent recommended the locally deployed CCDS environment for performance.

For Computer Interpretable Guidelines (CIGs) one respondent recommended to start by defining the goals that the CCDS intervention pursues and then, once the goal is clear, to identify the steps towards improving that goal. Other vendor with CIGs implementation experience pointed out that for each clinical guideline, a medical specialist is appointed. That specialist is often a national or regional leading figure that already has active participation in guidelines development. That specialist is the one responsible for the acceptance and follow-up of the deployment.

2) Governance frameworks

Respondents agreed that the architecture for the governance framework should enable access to terminology services, access to evidence-based guidance, and access to editorial tools for the development and maintenance of CCDS content. Respondents pointed out that CCDSs need to be shared and contrasted among organizations. To that end, organizations should gradually incorporate more CCDS modules performing pilot interventions and running studies to evaluate them. Thus, gradual adoption of CCDS was considered as an important factor.

A mixed model for governance was recommended. On the one hand, a centralized governance body for guidelines development and governance should be settled. On the other hand, smaller local governance bodies should exist in the institutions that could actually make use of the CCDS services. The editorial teams should have cross-membership between the central and local governance bodies. The centralized workgroup would coordinate subgroups and delegate work to the other CDS editorial groups.

3) Local adaption/customization

One respondent recommended for a maximum standardization without too much localization to work defining guidelines incrementally from narrative to structured format. The same respondent proposed a layered organization of rules. In such organization, the most internal levels of the CDSS represent goals, while more external levels represent recommendations. The latter are adapted as local workflows (indicating when and who to show recommendations to).

Respondents also agreed on clinical guidelines to be built via consensus, and once their content is agreed, they should be pushed to the EHR with the consent from clinical users. Consequently, respondents recommended to start with non-controversial content and develop reusable CCDS modules from parts that are not dependent on the local context. These modules will become the building blocks of more complex CCDS to be adapted in the local context.

E. Knowledge Base and Inference Engine

In this section, we summarize the responses with regards to the KB and the inference engine. By KB we mean the set of rules that conform a CCDS algorithm to provide a recommendation (in the case of rule-based systems); or the statistical model that performs an estimation/prediction (in the case of statistical or machine learning models).

Regarding the specification of medical decision algorithms, rule-based and logic-based methods (i.e., rules-based and ontology-based ones) are seen as the most intuitive and efficient ones. Respondents considered logic-based methods as intuitive and simpler for knowledge management. Graph structure formalisms such as GLIF were considered difficult to implement and integrate. In addition, latest approaches such as FHIR Plan Definition are considered more flexible and easier to translate to different inference engines. Several respondents considered that the formalism of specification could be the one provided by the technology to perform the inferences (e.g., JBoss Drools). To that end, it should be hosted in a Web service, and it should be made accessible through a standard API.

Respondents considered statistical methods to be important but more effort intensive for certain scenarios. Some systems report to use rules but trigger the invocation of a statistical model when necessary. Nevertheless, all respondent pointed out that the use of machine learning and statistical models are becoming increasingly relevant and those should be considered when appropriate.

F. Proposed National Architecture for Norway

Following the opinions and comments from the respondents, we draw a general architecture covering the requirements to be fulfilled in the Norwegian scenario for building a nation-scale CCDS framework. The schema in Figure 1 depicts the architecture proposed for a national centralized CCDS service.

At the top of the figure, it is shown a national governance committee that uses a common online authoring tool for designing CCDS modules based on best practices. CCDS modules are minimal stand-alone algorithms that serve a CCDS purpose in a context and that may be combined forming more complex CCDS flows. The national governance and editorial committee depicted use semantic interoperability resources provided by external repositories. These resources are of two main types: a) externally curated and approved terminology value sets; and, b) CIMs that may be FHIR profiles approved for national use, or openEHR archetypes published by the Norwegian Clinical Knowledge Manager. The governance and editorial committee use these artifacts imported through the authoring tool as the data schemas referenced by the decision logic that they define. This coupling of algorithms, CIMs, and terminologies, should be done through a proper graphical user interface that shows only the relevant information. Fine-grained technical details should be managed by the backend automatically.

The Norwegian context currently has organizations that base their developments on openEHR , FHIR, or both. The authoring tool allows for building CCDS algorithms that reference one standard or the other, thus creating a CCDS library where some algorithms use openEHR as a VMR and the others use FHIR as a VMR. For example, algorithms aiming for performing CDS in hospitals that belong to the regions working with an openEHR-based EHR should be written taking nationally published archetypes as a basis. Conversely, regions or services using FHIR may ask the editorial committee to prioritize the design of the algorithm using FHIR as an information schema for the VMR. Since CCDS modules require lots of professionals to be developed, their cost is very high. This poses a need for performing transformations among openEHR and FHIR in order to allow clients to use algorithms regardless of the standard they are based on. In Figure 1, the cloud between the FHIR and openEHR boxes represents transformation software. This way, a system operating in FHIR should be able to invoke a CCDS algorithm that was designed using an openEHR VMR. The client will invoke the FHIR endpoint (interaction n) and, internally, the received FHIR payload will be transformed into openEHR compliant extracts (interaction i) to execute the algorithm. Once a response is provided, data will be transformed back into FHIR compliant data (interaction i) and returned in the SOA payload to the client (interaction n). For openEHR clients with the need to invoke a FHIR-based CCDS algorithm the situation is the inverse.

When an algorithm is not logic- or rule- based, the statistical models component is invoked in a similar way to the described above (see interaction lines j and k). Statistical models should be imported in Predictive Model Markup Language and approved by the governance and editorial board. In some cases, as described by one respondent, rule or logic-based algorithms delegate some part of the computation to a statistical model. In this case, the presented architecture allows the models that contain logic-based models to do so (see interaction lines *i* and *m*).

The Web service layer represents the interface offered online to the clients of the CCDS framework. Two main endpoints should be offered, one based on FHIR compliant payload, and the other based on openEHR-compliant payload. This way, different iso-semantic models used by clients can be utilized. For example, Hospital C represents a hospital in Central Norway health region where openEHR is not adopted, thus interoperation is based on FHIR. The national framework should allow such hospital to consume a CCDS module developed by other regions that rely on openEHR. This is possible by using the transformation mechanisms previously described.

Finally, the human-like figures at the extremes of the bottom of the figure represent local governance and editorial committees that are responsible for approving and, if needed, adapting certain CCDS module to their organization local context. As recommended by the respondents, these committees should coordinate their actions with the national governance and editorial committee in order to properly escalate the CCDS interventions and developments (see interaction lines *a* and *b*). When adaption to the local context has been performed, the new modified modules should be made available for contrasting them with other deployments. This is shown in the layered architecture represented in the two logic CCDS modules. In that approach, CCDS modules are layered, containing on the top layer a "core" set of the main goals to fulfill. Secondly, there is an implementation of the parts of the CCDS actions that are common to all contexts and that are uncontroversial. Finally, these logic components are specialized each into a more superficial layer that is exposed to the clients. These superficial layers are the ones adapted to the local context of each organization when needed. This adaption is performed with exhaustive control from the committees in order to minimize deviations from the original algorithm and guarantee future scalability and governance. It is important to understand that, although the figure shows only one Web service, there may be several instances of the CCDS Web service with different collections of the CCDS modules available. Currently, this is achievable by using containers technologies such as Docker and Kubernetes.

IV. CONCLUSION

The proposed architecture attempts to agree with many of the principles for already published best practices. For example, Kawamoto et al. [25] recommended the SOA architecture for large-scale CCDS. We believe that the recent developments in openCDS will help in setting a reference for that [25]. Actually, many of the components required to build a national CCDS infrastructure could be borrowed from openCDS and openEHR GDL [26]. Since both are open-source projects, many of their components could be merged to build the multi-standard framework presented.

Regarding the internal structure of CCDS modules, the layered architecture of CCDS modules is based on the concepts proposed by Boxwala et al. [27], but in the most internal layer, instead of narratives, we prioritize the inclusion of the main goals to achieve in the clinical setting as recommended by one respondent and proposed elsewhere [28]. The transformation mechanisms between FHIR and openEHR can be based on previous research [29].

Despite machine-learning has recently received lots of attention, our respondents considered that for CCDS there are some "low hanging fruits" to be focused on before building complex artificial intelligence frameworks at national scale. In addition, as pointed by Fox [28], logic formalisms have demonstrated to be as good as Bayesian methods for specifying medical knowledge. This does not imply that we should not perform research in machine learning. But it means that we still need to be able to deploy large-scale CCDS frameworks where the most pressing challenges are related to governance, adaption to local contexts, and different information and knowledge representation formats. We believe that once these requirements are clear and a proper edition and governance framework is in place, most machine-learning algorithms will fit in the framework. These algorithms will complement logic-based CCDS modules when required, thus leveraging the best from both logic and statistical methods.

ACKNOWLEDGMENT

This study was funded by the Norwegian Centre for E-health Research. We would like to thank all the respondents for the time devoted to explain their valuable opinions, experiences, and points of view.

REFERENCES

[1] C. Friedman and M. Rigby, "Conceptualising and creating a global learning health system," *Int. J. Med. Inf.*, vol. 82, no. 4, pp. e63-71, Apr. 2013, doi: 10.1016/j.ijmedinf.2012.05.010.

[2] Institute of Medicine (US) Roundtable on Evidence-Based Medicine, *The Learning Healthcare System: Workshop Summary*. Washington (DC): National Academies Press (US), 2007.

[3] A. Budrionis and J. G. Bellika, "The Learning Healthcare System: Where are we now? A systematic review," *J. Biomed. Inform.*, vol. 64, pp. 87–92, December 2016, doi: 10.1016/j.jbi.2016.09.018.

[4] "PCORnet," *PCORnet*. [Online]. Available: http://www.pcornet.org. [Accessed: 19-Mar-2020].

[5] B. Haarbrandt *et al.*, "HiGHmed – An Open Platform Approach to Enhance Care and Research across Institutional Boundaries," *Methods Inf. Med.*, vol. 57, no. S 1, pp. e66–e81, Jul. 2018, doi: 10.3414/ME18-02-0002.

[6] A. Budrionis, L. Marco-Ruiz, K. Y. Yigzaw, and J. G. Bellika, "Building a Learning Healthcare System in North Norway," in *Proceedings from The 14th Scandinavian Conference on Health Informatics 2016, Gothenburg, Sweden, April 6-7*, 2016, pp. 1–5.

[7] "PraksisNett," *Universitetet i Bergen*. [Online]. Available: https://www.uib.no/nb/praksisnett. [Accessed: 24-Nov-2019].

[8] K. Y. Yigzaw, A. Michalas, and J. G. Bellika, "Secure and scalable deduplication of horizontally partitioned health data for privacy-preserving distributed statistical computation," *BMC Med. Inform. Decis. Mak.*, vol. 17, no. 1, p. 1, March 2017, doi: 10.1186/s12911-016-0389-x.

[9] S. Van de Velde *et al.*, "A systematic review of trials evaluating success factors of interventions with computerised clinical decision support," *Implement. Sci.*, vol. 13, no. 1, p. 114, Aug. 2018, doi: 10.1186/s13012-018-0790-1.

[10] B. Christensen and G. Ellingsen, "Evaluating Model-Driven Development for large-scale EHRs through the openEHR approach," *Int. J. Med. Inf.*, vol. 89, pp. 43–54, May 2016, doi: 10.1016/j.ijmedinf.2016.02.004.

[11] S. Van de Velde *et al.*, "The GUIDES checklist: development of a tool to improve the successful use of guideline-based computerised clinical decision support," *Implement. Sci.*, vol. 13, no. 1, p. 86, Jun. 2018, doi: 10.1186/s13012-018-0772-3.

[12] L. Marco Ruiz, J. A. Maldonado, R. Karlsen, and J. G. Bellika, "Multidisciplinary Modelling of Symptoms and Signs with Archetypes and SNOMEDCT for Clinical Decision Support," in *Stud Health Technol Inform.*, Madrid, 2015.

[13] L. Marco-Ruiz, J. A. Maldonado, V. Traver, R. Karlsen, and J. G. Bellika, "Meta-architecture for the interoperability and knowledge management of archetype-based clinical decision support systems," in *2014 IEEE-EMBS International Conference on Biomedical and Health Informatics (BHI)*, 2014, pp. 517–521, doi: 10.1109/BHI.2014.6864416.

[14] L. Marco-Ruiz *et al.*, "Combining Multivariate Statistics and the Think-Aloud Protocol to Assess Human-Computer Interaction Barriers in Symptom Checkers," *J. Biomed. Inform.*, Sep. 2017, doi: 10.1016/j.jbi.2017.09.002.

[15] R. Pedersen, C. Granja, and L. M. Ruiz, "The Value of Clinical Information Models and Terminology for Sharing Clinical Information," presented at the eTELEMED 2016, The Eighth International Conference on eHealth, Telemedicine, and Social Medicine, 2016, pp. 153–159.

[16] "SNOMED Home page," *SNOMED*. [Online]. Available: /. [Accessed: 18-Mar-2020].

[17] L. Marco-Ruiz, K. Nicolaisen, R. Pedersen, A. Makhlysheva, and P. A. Bakkevoll, *Ontology-based terminologies for healthcare - Impact assessment and transitional consequences for implementation*, 1st ed., vol. 1. Tromsø: Norwegian Centre for E-health Research, 2017.

[18] A. Makhlysheva, K. Malm-Nicolaisen, L. Marco-Ruiz, P. A. Bakkevoll, and R. Pedersen, *Klinisk beslutningsstøtte - Vurdering av standard og arkitektur*, vol. 1, 37 vols. Tromsø, Norway: Norwegian Centre for E-health Research, 2017.

[19] "OpenCDS." [Online]. Available: http://www.opencds.org/. [Accessed: 18-Mar-2020].

[20] J. C. Mandel, D. A. Kreda, K. D. Mandl, I. S. Kohane, and R. B. Ramoni, "SMART on FHIR: a standards-based, interoperable apps platform for electronic health records," *J. Am. Med. Inform. Assoc. JAMIA*, vol. 23, no. 5, pp. 899–908, Sep. 2016, doi: 10.1093/jamia/ocv189.

[21] "CDS Hooks." [Online]. Available: https://cds-hooks.org/. [Accessed: 19-Mar-2020].

[22] Direktoratet for e-helse, "Anbefaling om bruk av HL7 FHIR for datadeling," Oslo, Norway, Mar. 2019.

[23] L. Marco-Ruiz and R. Pedersen, "Challenges in Archetypes Terminology Binding Using SNOMED-CT Compositional Grammar: The Norwegian Patient Summary Case," *Stud. Health Technol. Inform.*, vol. 245, p. 1332, 2017.

[24] L. Marco-Ruiz and R. Pedersen, "The Patient Summary Case: Challenges in Archetypes Terminology Binding Using SNOMED-CT Compositional Grammar," in *eTELEMED 2019 : The Eleventh International Conference on eHealth, Telemedicine, and Social Medicine*, Athens (Greece), 2019, pp. 49–55.

[25] K. Kawamoto, T. Hongsermeier, A. Wright, J. Lewis, D. S. Bell, and B. Middleton, "Key principles for a national clinical decision support knowledge sharing framework: synthesis of insights from leading subject matter experts," *J. Am. Med. Inform. Assoc. JAMIA*, vol. 20, no. 1, pp. 199–207, 2013, doi: 10.1136/amiajnl-2012-000887.

[26] N. Anani, R. Chen, T. P. Moreira, and S. Koch, "Retrospective checking of compliance with practice guidelines for acute stroke care: a novel experiment using openEHR's Guideline Definition Language," *BMC Med. Inform. Decis. Mak.*, vol. 14, no. 1, p. 39, May 2014, doi: 10.1186/1472-6947-14-39.

[27] A. A. Boxwala *et al.*, "A multi-layered framework for disseminating knowledge for computer-based decision support," *J. Am. Med. Inform. Assoc.*, vol. 18, no. Supplement_1, pp. i132–i139, Dec. 2011, doi: 10.1136/amiajnl-2011-000334.

[28] J. Fox, "Cognitive systems at the point of care: The CREDO program," *J. Biomed. Inform.*, vol. 68, pp. 83–95, Apr. 2017, doi: 10.1016/j.jbi.2017.02.008.

[29] J. A. Maldonado *et al.*, "Concept-Based Exchange of Healthcare Information: The LinkEHR Approach," in *2011 First IEEE International Conference on Healthcare Informatics, Imaging and Systems Biology (HISB)*, 2011, pp. 150–157, doi: 10.1109/HISB.2011.18.

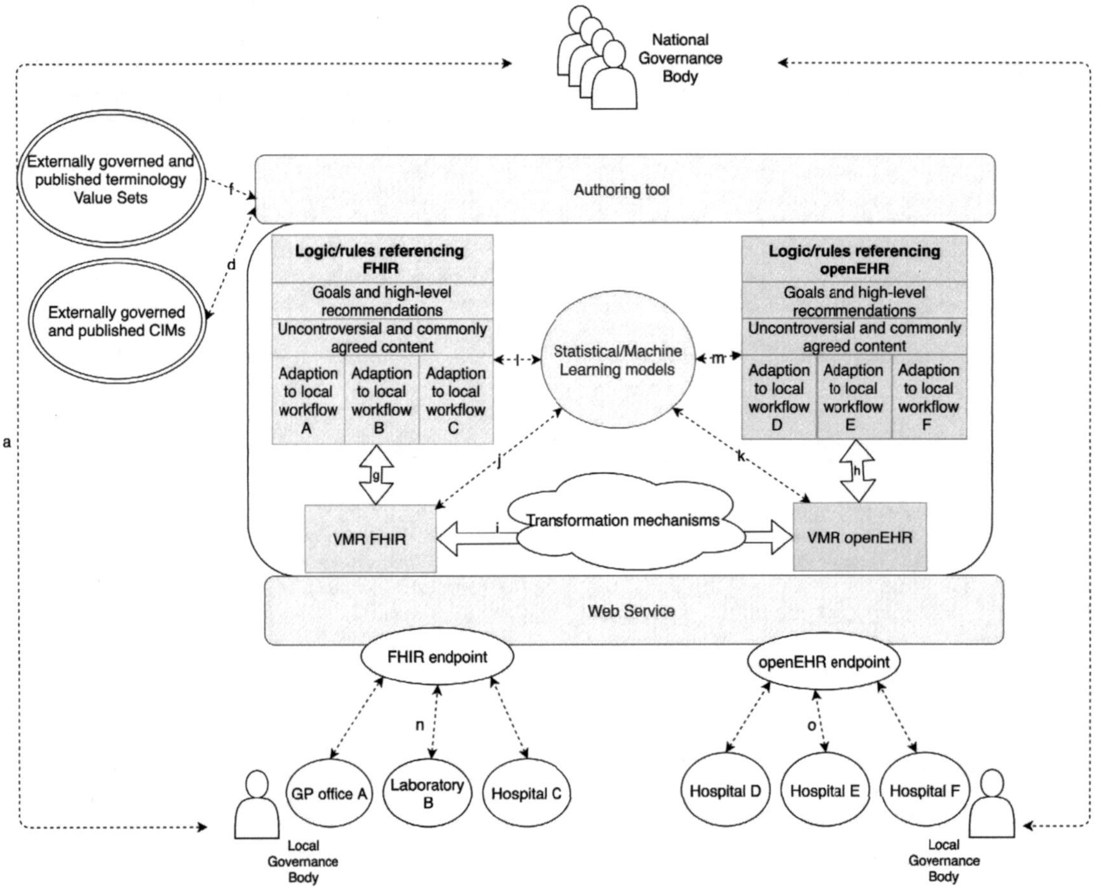

Figure 1. Architecture for the edition, governance, and deployment of CCDS in the Norwegian context.

Investigation on the Use of the PE873 Conductive Ink for Surface EMG Measurements

Andrea Federico, Marco Belcastro, Pasqualino Torchia, Salvatore Tedesco, Brendan O'Flynn
Wireless Sensor Networks Group, Micro & Nano System Centre
Tyndall National Institute, University College Cork
Cork, Ireland
e-mail: andrea.federico@tyndall.ie, marco.belcastro@tyndall.ie, pasqualino.torchia@tyndall.ie, salvatoretedesco@tyndall.ie, brendan.oflynn@tyndall.ie

***Abstract*—Nowadays, wearable devices are part of everyone's life and their popularity is constantly increasing. With diverse applications, spanning from healthcare to fitness tracking, more and more wearable devices are being developed which can send and receive information in real-time.**

To date, electric cables represent the most stable form of communication in terms of reliability and resistance. However, for wearable systems, cables restrict movement and introduce additional noise and movement artefacts on wearable sensing systems. Wireless devices, on the other hand, can be comparatively complicated in design, manufacturing and use. A possible strategy, to improve communications in wearable systems, is the adoption of conductive inks able to conduct electrical signals, these can be printed on fabric without the movement restriction normally associated with traditional wired systems. The use of such conductive inks in wearable sensors may, therefore, lead to a more comfortable method of monitoring health data (heart rate, muscle contraction etc.) throughout the day. In this paper, the properties of the promising conductive ink PE873 (manufactured by DuPont) are tested and analysed. The conductive ink electrical properties are studied in relation to stretching, folding and washing tests. The electrical performance of the ink printed onto the selected fabric is assessed and presented. Furthermore, an optimized printing procedure, aiming at improving the connection performances, is suggested and the development of a novel system able to read muscle contractions, based on PE873, is demonstrated, thus showing that this conductive ink is a promising solution for stretchable electrical connections in the wearable field.

Keywords—Wearable systems; textile electronics integration; conductive ink; interconnections; DuPont.

I. INTRODUCTION

Many of the wearable products currently available on the market have been developed for monitoring of health, well-being and fitness. Such systems, e.g. smart watches and wrist bands, can provide information about physical activity (e.g., steps, calories burned, etc.) and physiological parameters (pulse rate, blood oxygen levels, Blood pressure etc.). The constant monitoring of a subject's health status is a desirable feature for clinical and health-related applications, such as rehabilitation or the management of chronic diseases. This can be clearly observed by the amount of research outputs produced in the wearable field on health monitoring [1]-[3].

In particular, the interest in sensor-embedded clothing is constantly increasing enabled by recent advancements in materials and technology, in particular flexible and printed electronics. For example, biosignals, such as surface electromyography (EMG) or electrocardiography (ECG), can be directly collected from the location of interest (e.g., upper or lower limbs) by means of smart garments, such as fully sensorized shirts or shorts with integrated electrodes [4]-[6].

One of the most attractive features of those wearable devices is the capability to transmit data in real-time without the need of physical connections. On the other hand, wireless communication (e.g., ZigBee, Bluetooth, etc.) can be comparatively unreliable [7], relatively expensive and more complex in terms of design, manufacturing and use. Thus, simpler and less expensive solutions are required to make such systems universally accessible and affordable. To date, electric cables and copper wires have represented the best form of connection in terms of reliability and resistance; however, the presence of cables on wearable devices might restrict body movements and prevent the execution of specific physical tasks, as well as introduce additional noise and movement artefacts.

One of the possible solutions to this problem might come from the new generation of conductive ink [8] which are coming on the market and which are increasingly accessible to system designers. This type of ink can be directly printed on fabrics, and it can be washed and stretched several times while maintaining appropriate electric resistance. Generally, conductive ink is composed by a conductive material, thermoplastic polyvinylbutyral terpolymer binder and a glycol ether solvent [9] and, typically, conductive ink based on silver metal particles, which shows the best conductivity [10].

Although modern inks have an electrical conductivity comparable to copper [11], the reliability of conductive inks in wearable devices is still a critical issue, due to the change in electrical resistance caused by mechanical strain [12].

To address this problem, specific technical aspects need to be considered, such as the relationship between the viscosity of the ink and the porosity of the fabric. For example, viscous inks do not permeate well into porous textile structures [13]. Moreover, the electrical performance is also affected by the different direction of the fabric wires. As shown in [14], when comparing the electrical features under large strains performed in two different directions (e.g., parallel and perpendicular to the fibre bundles), the resistance of the printed line differs of a factor up to 20 between the two scenarios.

At present, the market of conductive inks is still limited, and performances reported by the manufacturers do not take into consideration the electrical properties of such materials under specific mechanical stress, e.g., ink tested on stretchable substrate or textile materials. Therefore, the present work aims at investigating the suitability of a conductive ink available on the market to be used as an electrical connection on both stretchable and unstretchable fabric by testing the electrical performance of the material in both conditions. Furthermore, the opportunity to adopt the conductive ink in a system able to monitor muscle contractions has been also assessed in a practical scenario. In order to investigate the suitability of a conductive ink in creating robust electric connections for surface EMG acquisition with a wearable device, the conductive ink was used to create an electrical connection on a stretchable strip of fabric between a conductive electrode (point A), which transduces the myoelectric signal from the skin, and "snap" connectors (point B) mounted on a custom printed circuit board (PCB), which connects the ink to the EMG measuring device. Moreover, a printing process and associated performance assessment procedures were developed to assess the system.

The manuscript is organized as follows. The methodologies adopted for the system development (ink selection, connection process, 3D printing, etc.) are described in Section II, while Section III shows the results obtained over different test conditions. Section IV presents a discussion of these results and conclusions are drawn in Section V.

II. METHODS

A. Conductive Ink Characteristics

Despite the market of conductive ink being still in its infancy, a number of companies have already produced conductive inks and pastes for photovoltaics, power electronics, displays, automotive, antenna design and others, but few products on the market have been developed for stretchable e-textile [15]. For this work, the conductive ink selected for analysis is the Intexar PE873 produced by DuPont (DuPont de Nemours, Inc., Wilmington, Delaware, USA). The selection was mainly made based on the waterproof capability of such ink which, by the time of this work, was a feature available only on a very limited range of similar product. Its physical and composition properties are shown in Tables I-II.

TABLE I. PHYSICAL PROPERTIES

Test	*Properties*
Sheet Resistivity (mΩsq/25μm) (5μm Dried Print Thickness on ST505 PET Film)	< 75
Resistivity After Crease (ASTM F1683, 180 deg, 1 cycle, 2kg)	< 5%
Abrasion Resistance (ASTM D3359 Pencil Hardness)	1 H
Adhesion (Tape Cross Hatch) (ASTM D3359 w/3M Scotch Tape 600)	No transfer
Clean-Up Solvent	Ethylene Diacetate
Encapsulant	PE771 / PE773

TABLE II. COMPOSITION PROPERTIES

Test	*Properties*
Solid (%) @ 150°C	60 - 65
Viscosity (PaS) Brookfield RVT, #14 spline, 10rpm, 25°C	50 - 80
Density (g/cc)	2.0
Coverage (cm^2/g @ 5 μm) Coverage (cm^2/g @ 10 μm)	350 175
Dried Print Thickness (microns)	8 - 12
Thinner	DuPont™ 8260

B. Ink-Sensor Connection

Point A of the overall system is represented by the electrode made via a conductive and biocompatible fabric (produced by Swift Textile Metalizing LLC, Bloomfield, CT) and placed on the bottom side of the fabric, which is in contact with the skin, and on the top side. Regarding the connection between the top and bottom side of the fabric electrodes, to ensure a robust connection the conductive ink was sandwiched between the two conductive layers, acting as a glue (Figure 1).

On the other hand, point B are snaps integrated in to the PCB design so as to connect with the MyoWare Muscle Sensor (Advancer Technologies LLC, Raleigh, USA), placed on the top side to monitor muscle contractions. To connect the strips of conductive ink with the reading electrodes of the EMG sensor, a PCB layout with three pads on the top and the bottom layer was designed. The top pads connect the PCB to the MyoWare by the means of snap buttons, soldered on top. The bottom pads, instead, connect the PCB with the ink paths (Figure 2a). As soldering on fabric is not possible, a chemical compound has been identified to connect the PCB and the ink paths (Figure 2b). The conductive epoxy 8331-B (produced by MG Chemicals, Burlington, Ontario, Canada) was chosen due to its suitable characteristics of low electric resistance, high water resistance, and short working life.

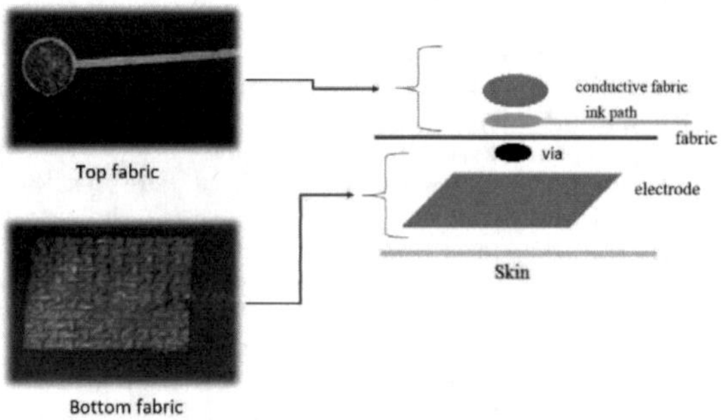

Figure 1. Sandwich structure for collection of EMG signal

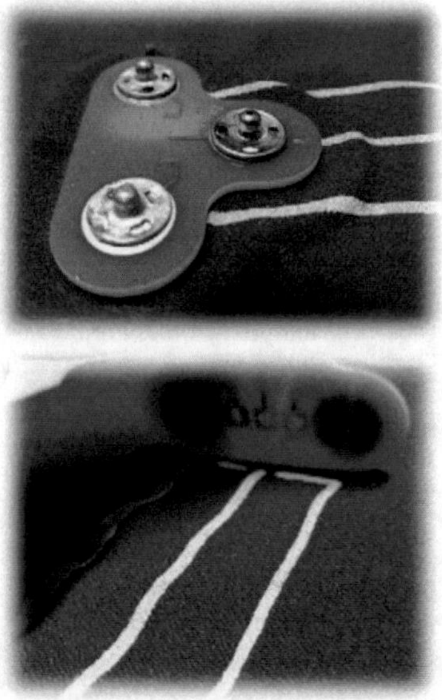

Figure 2. a) PCB with snap buttons soldered on top (top). b) Connection between PCB and fabric via 8331-B epoxy (bottom)

C. 3D Printed Support

From a design methodology perspective, a system designer needs to consider that electronics embedded into clothing can be under significant strain due to body physical movements, thus the use of epoxy on its own does not guarantee a reliable strong grip due to flexing and cracking. To ensure the stability of the communication, a mechanical enclosure was designed to support the electronics. It consists of two 3D printed parts, obtained via fuse deposition model technique using polylactic acid (PLA), which can be screwed together to work as a clamp (red components of Figure 3).

D. *Ink-Printing Process*

In order to evenly spread the conductive ink only across the conductive paths, namely the narrow strips that connect the fabric electrodes with the PCB, a custom mask was 3D printed. The mask was design to act as a "stencil" to confine the ink only within the designed geometry. The final design of the mask is shown in (Figure 4a). An important element of the 3D printed mask is a small step added on the bottom face (Figure 4b), which allowed more pressure on the fabric surface in order to avoid possible leakage of the ink. The final system is shown in Figure 5.

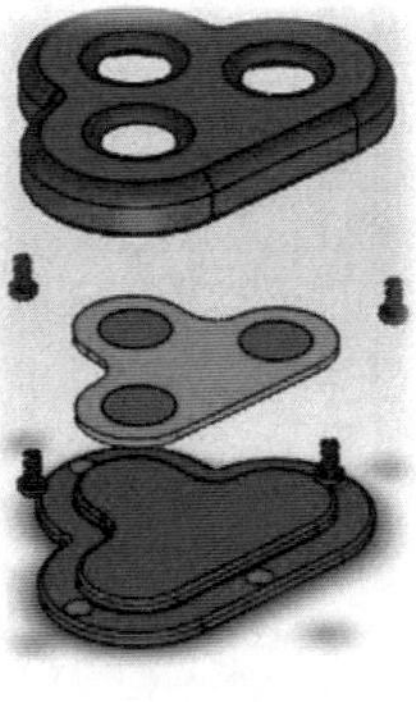

Figure 3. 3D printed support (exploded view)

Figure 4. 3D printed mask

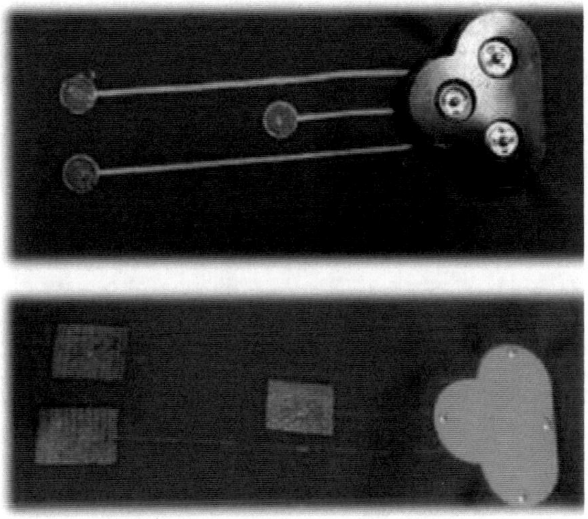

Figure 5. Developed system (top and bottom view)

E. *Test Protocol*

Three different tests have been carried out on the conductive material once deposited and cured on the textile material in the present investigation: stretching, folding, and washing.

The stretching test was divided in three steps: resistivity measurements in rest position, in a stretched position, and finally in rest position again. The second test investigated the impact of folding the garment along the conductive trace, in which the resistivity was measured after each folding procedure. The washing test consisted in testing the electrical performance of the material after washing at a selected program in a standard domestic washing machine: synthetic garment (2 hours) with a temperature of 30°C and a detergent for coloured garments.

III. RESULTS

To test the performance of the conductive ink, several test samples were made for analysis. The tests were made on two types of fabrics: unstretchable and stretchable materials. Resistance measurements in the following tests were conducted 10 times and the average value is presented with the related standard deviation shown on the graphs as error bars. The ideal condition was to obtain a resistivity near to 1 Ohm, which is comparable to standard cables resistivity.

Stretching results are shown in Figures 6-7, while folding results are illustrated in Figures 8-9, and washing test in Figure 10.

The present work aimed at investigating the suitability of the DuPont PE873 conductive ink to be used as an electrical connection on stretchable and unstretchable fabric by testing the electrical performance in both conditions across different test. This section briefly discusses the results obtained with the implemented experimental protocol.

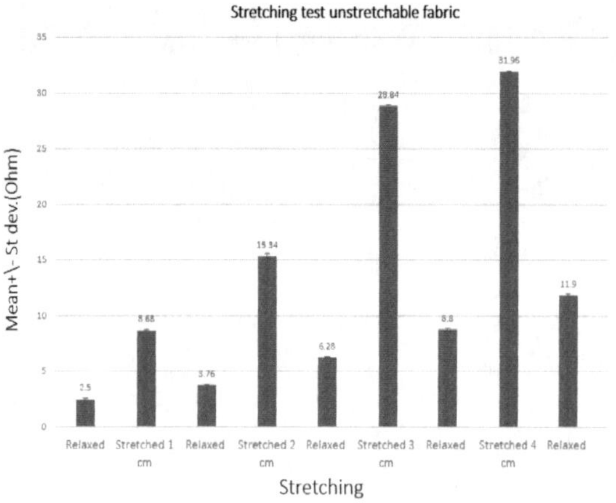

Figure 6. Stretching test results for unstretchable fabric

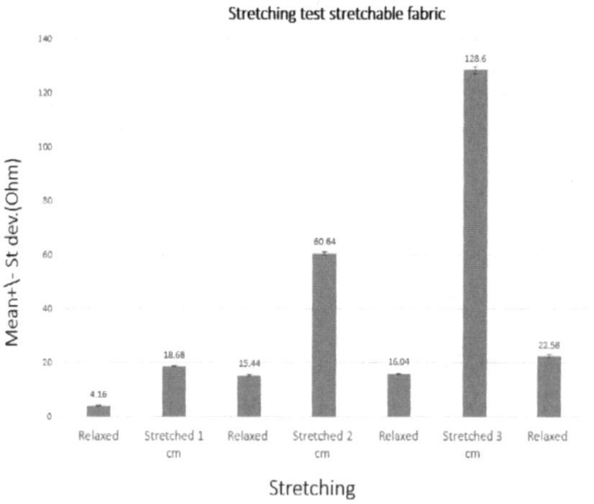

Figure 7. Stretching test results for stretchable fabric

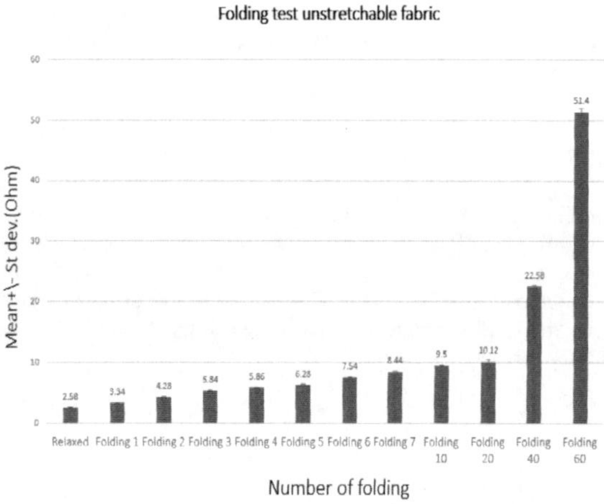

Figure 8. Folding test results for unstretchable fabric

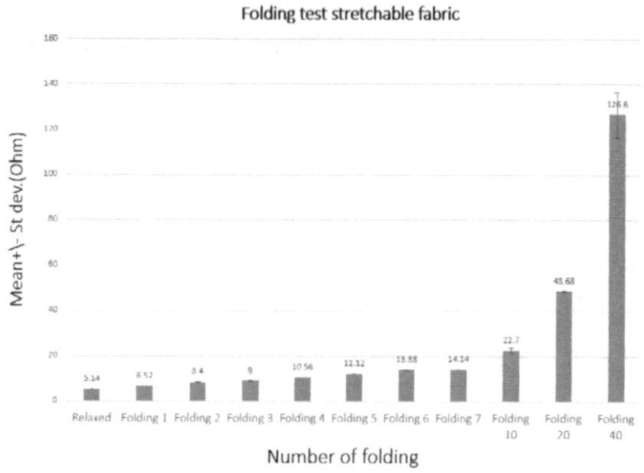

Figure 9. Folding test results for stretchable fabric

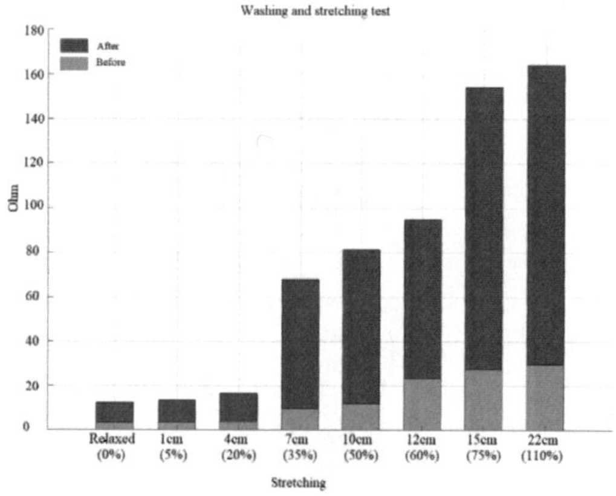

Figure 10. Washing and stretching test results for stretchable fabric

IV. DISCUSSION

A. Stretching Test

Figure 6 illustrates the resistance changes for the stretching test for an unstretchable fabric. Initially, the measured resistivity is 2.5 Ohms in the rest position, which increases to 8 Ohms after stretching the fabric of 5% its length (e.g., 1 cm). However, when returning to a rest position, the resistivity measured is slightly larger than initially obtained. This behaviour is amplified when longer stretching are applied and the measured electrical resistance when applying the maximum stretching of 20% (e.g., 4 cm) is over 30 Ohms.

The same behaviour is shown when performing the test on the stretchable fabric (Figure 7), even though the resistivity values measured are generally higher compared to the previous fabric. It is worth noticing the behaviour of the ink over one day in rest position. Indeed, after 24 hours, the resistivity decreases of 1.08 Ohms and 3.26 Ohms for the unstretchable and stretchable fabrics, respectively, thus showing an adaptation of the textile fibres.

B. Folding Test

The results for the folding test are displayed in Figures 8-9. In both cases, the resistivity measured grows with the number of foldings; however, the unstretchable fabric shows better results. As an example, after the 60th folding, its resistivity (51.4 Ohms) is less than half than the resistivity measured from stretchable fabric after the 40th folding (126.6 Ohms).

C. Washing Test

After washing, the stretching test was carried out on the sample according to the previously described procedure.

At the first attempt, the mechanical stress of the washing procedure was enough to cause an open circuit on the conductive track. To this purpose, it is important to specify that every sample analysed was developed with only one layer of ink; therefore, as a possible solution, samples with more than one layer of conductive ink were made [16]. Moreover, to increase the conductivity of the stretchable fabric, it was decided to apply the ink while the fabric was stretched. The results achieved from using this methodology are shown in Figure 10. Using this new ink-printing method, the resistivity value of the conductive trace after washing is still unsatisfactory. However, the resistivity before and after washing is lower than the results obtained in the stretching test. This test was carried out only on the stretchable fabric, which was the worst between the two cases.

Overall, it can be concluded that the performance of the PE873 conductive ink is acceptable only in a small range of mechanical strain, thus conductive ink is still not able to provide results comparable to standard cables in a wearable solution requiring significant stretching capability (such as lycra shorts for instance with integrated sensors for example). One of the biggest issues is the ink penetration in the textile fibres, which deeply impact on the conductivity [16]. DuPont has now made available a new series of conductive inks, which promise better performance, as shown in Figure 11.

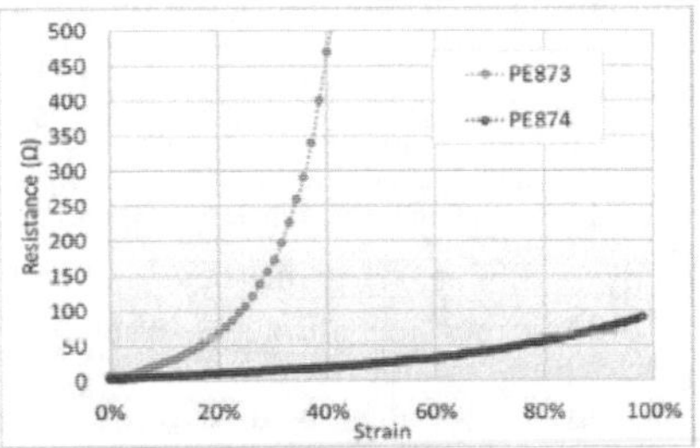

Figure 11. PE874 and PE873 mechanical and electrical performance compared

In the picture, the characteristics of the PE873 and of the new generation ink (PE874) are compared. Furthermore, DuPont has also produced a stretchable TPU (thermoplastic polyurethane) film, which works as a printing surface and is characterized by high recovery and composed of a melted adhesive layer for bonding to fabric.

V. Conclusion

Conductive inks have the potential to represent a possible efficient and inexpensive solution for solving the interconnection issues typical of smart wearable garments with embedded sensors and electronics. In this paper, the features of the promising conductive ink PE873 (manufactured by DuPont) were tested and analysed across several experiments (stretching, folding and washing) by measuring the electrical performance of the ink. Furthermore, the methodologies adopted for implementing an optimized printing procedure were discussed.

Even though this investigation presented mixed results, it can be concluded that the performance of the PE873 conductive ink was acceptable in a limited range of mechanical strain. The next-generation conductive inks, and improved printing procedures, can potentially ensure better performance of printed connections, thus making it a desirable technology in the field of wearable devices in the coming years.

Acknowledgment

This publication has emanated from research supported by a research grant from the Enterprise Ireland (EI) funded project SKYRE under grant number CF-2015-0031-P. Aspects of this publication have emanated from research conducted with the financial support of Science Foundation Ireland under Grant number 12/RC/2289-P2, and 13/RC/2077-CONNECT, which are co-funded under the European Regional Development Fund.

References

[1] R.K.Y. Tong, Wearable technology in medicine and health care, Academic Press, Elsevier, 2018.

[2] T. Loncar-Turukalo, E. Zdravevski, J. Machado da Silva, I. Chouvarda, and V. Trajkovik, "Literature on wearable technology for connected health: Scoping review of research trends, advances, and barriers," J Med Internet Res, vol. 21, no. 9, e14017, 2019.

[3] S. Tedesco, J. Barton, and B. O'Flynn, "A review of activity trackers for senior citizens: Research perspectives, commercial landscape and the role of the insurance industry," Sensors, vol. 17, no. 6, pp. 1277-1316, 2017.

[4] C. Chen et al., "Novel flexible material-based unobtrusive and wearable body sensor networks for vital sign monitoring," IEEE Sensors Journal, vol. 19, no. 19, pp. 8502-8513, 2019.

[5] J. Shi et al., "Smart textile-integrated microelectronic systems for wearable applications," Adv Mater, e1901958, 2019.

[6] S. Tedesco et al., "A multi-sensors wearable system for remote assessment of physiotherapy exercises during ACL rehabilitation," IEEE ICECS, Genoa, Italy, 27-29 Nov 2019.

[7] A. Kos, V. Milutinović, and A. Umek, "Challenges in wireless communication for connected sensors and wearable devices used in sport biofeedback applications," Future Generation Computer Systems, vol. 92, pp. 582-592, 2019.

[8] H. Yan, W. Cai, Y. Zhang, H. Wang, and Y. Liu, "Review of conductive inks in printed electronics," Beijing Gongye Daxue Xuebao/Journal of Beijing University of Technology, vol. 43, pp. 1117-1122, 2017.

[9] A. Vena, E. Perret, and S. Tedjini, Chipless RFID based on RF encoding particle, 1st edition, ISTE Press – Elsevier, 2016.

[10] K.R.R Venkata, A.K. Venkata, P.S. Karthik, and P.S. Surya, "Conductive Silver Inks and its Applications in Printed and Flexible Electronics," RSC Advances, vol. 5, pp. 77760-77790, 2015.

[11] C. Maxfield, Bebop to the Boolean Boogie, 3rd edition, O'Reilly, 2009.

[12] A. Christopoulos and G. Tsamasphyros, "Strain measurements using electrically conductive silver paint," Sensor Letters, vol. 12, 2014.

[13] S. Hasan, K. Inhwan, H. Soewardiman, and J.S. Jur, "Inkjet Printing of Reactive Silver Ink on Textiles," ACS Applied Materials & Interfaces, vol. 11, pp. 6208-6216, 2019.

[14] Jin H. et al., "Enhancing the performance of stretchable conductors for e-textiles by controlled ink permeation," Adv Mater, vol. 29, no. 21, 1605848, 2017.

[15] https://www.idtechex.com/en/research-report/conductive-ink-markets-2019-2029-forecasts-technologies-players/665 [retrived: Dec, 2019].

[16] M. Stoppa and A. Chiolerio, "Testing and evaluation of wearable electronic textiles and assessment thereof," in Performance testing of textiles, Woodhead Publishing Series in Textiles, pp. 65-101, 2016.

Development of a Program for Analytical Systems of Personal Diagnostics of People and Animals Based on the Piezoelectric Sensors Array

Tatiana Kuchmenko
Physical and Analytical Chemistry Department
Voronezh State University of Engineering Technologies
Voronezh, Russia
e-mail: tak1907@mail.ru

Ruslan Umarkhanov
Physical and Analytical Chemistry Department
Voronezh State University of Engineering Technologies
Voronezh, Russia
e-mail: rus_270487@mail.ru

Ivan Murakhovsky
Physical and Analytical Chemistry Department
Voronezh State University of Engineering Technologies
Voronezh, Russia
e-mail: ivan335522@mail.ru

Daria Kuchmenko
Endocrinology department
Voronezh Children's Clinical Hospital at the Voronezh State Medical University named after N.N. Burdenko
Voronezh, Russia
e-mail: darinochka_08@mail.ru

Abstract— **Modern methods of analytical chemistry are based on the measurement and processing of two- up to multidimensional signals. Multidimensional analytical signals include the responses of multisensory systems, such as artificial "nose, tongue, eyes". The purpose of this work is to develop software for personal diagnostic devices "electronic nose" with the possibility of clinical diagnosis of humans and animals by easily renewable selected biosamples. A program has been developed as a native Android application written in a high-level Java language, designed to interact with the electronic nose device. For the analysis of signals received from an array of 8 sensors, two algorithms have been developed. Analysis of the measurement is carried out on a complete data matrix which is generated when registering sensor signals over the skin of a person's hand or over a biosample. When processing the output data of the sensor, the most informative parameters about the health state of the body, individual organs and systems are obtained. Based on these parameters, visual smell traces are constructed by the sensor signals. The developed software personalizes the diagnosis of the health state of the human or animal using multisensory systems like electronic nose, and displays information in an understandable form to an untrained user. This makes it possible to use such devices for personal purposes, on small farms, therefore, to prevent the development of complex diseases, the death of livestock, and to improve the quality of life.**

Keywords- program; Java; non-invasive diagnostics; screening; sensors; metabolism; eHealth.

I. INTRODUCTION

Modern methods of analytical chemistry are based on the measurement and processing of two- up to multidimensional signals. Multidimensional analytical signals include the responses of multisensor systems, such as artificial "nose, tongue, eyes".

The purpose of this work is to develop software for personal diagnostic devices "electronic nose" with the possibility of clinical diagnosis of humans and animals by volatile metabolome of body or biosample [1].

Section 2 presents some state-of-the-art methods for processing and visualizing the signals from the multisensory systems. Section 3 describes the processing and visualizing features of sensor signals and the code of the proposed program. In Section 4, the conclusions and perspectives of the developed program application are provided.

II. STATE-OF-THE-ART METHODS

Currently available scientific developments of mobile applications for processing different information regarding health problems can be found in [2]-[4]. All such applications have basic diagnostic information, but there is no online connecting with a device based on sensors for non-invasive screening of health state. The final stage of our development will be a cloud service with artificial intelligence technology to provide complete information to customers about their health condition and tips for improving it.

III. DESCRIPTION OF PROGRAM

This program is written in a high-level Java language (Figure 1) as Android application, designed to interact with the electronic nose (e-nose) device and it is based on previous our works [5][6]. For the analysis of signals received from an array of 8 sensors, two algorithms have been developed. The first one allows getting the measurement results in a way that is understandable to any user. Initial processing of the measurement results is carried out by the maximum sensor responses, which forms a health state diagram (sphere), constructed based on the calculated data. Figure 1 shows a fragment of a software algorithm for assigning color to a sector in accordance with the values of sorption efficiency parameters calculated from the maximum sensor responses.

```
public class ResultAl_2 extends BaseResult {

  public ResultAl_2(double A, DataByMaxParcelable
inputData, Context context) {

    supei(A, inputData, context);

    setLegend("1_2");
}

//todo add "dry skin" If the skin is dry - color - red,
comments – Inflammation

 public void setResult(){

  if(getA() >=0.9&& getA() <= 1.14){

   if(getInputData(). isPractice()){
```

Figure 1. Fragment of a program about classes that work with the model data

The numerical boundaries of the reference values were determined experimentally for each sorption efficiency parameter. Depending on the interval in which the calculated parameters fall, certain comments about the health state of the body or about the volatile substances contained in the analyzed sample, are displayed. The program allows using the simplest responses of the sensor array in electronic nose to calculate a set of parameters that reflects the health state of both individual organs in the human body and the psycho-emotional sphere. Each calculation parameter is assigned numerical thresholds, text decoding, and color gamut, when the calculated value corresponds to certain semantic numerical ranges. It is possible to manually enter the values of the sensor signals (Figure 2a), as well as from the database of the e-nose for 2 measurements simultaneously with the possibility of averaging the calculated parameters. These 2 measurements may be the analysis results for one sample or for different ones.

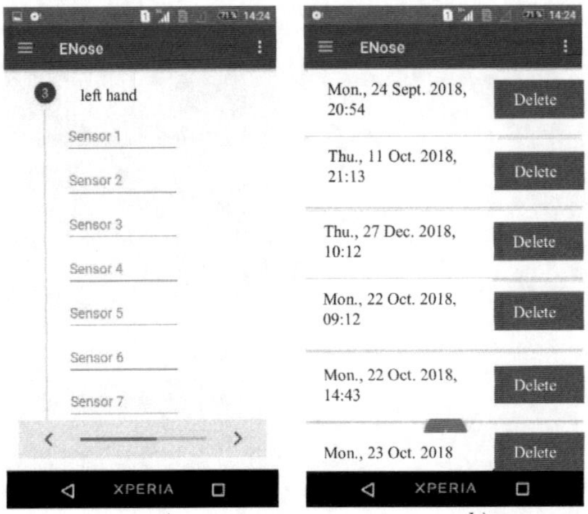

a) *b)*

Figure 2. The dialog windows of program to input of sensor responses (*a*) and save the mesurements (*b*).

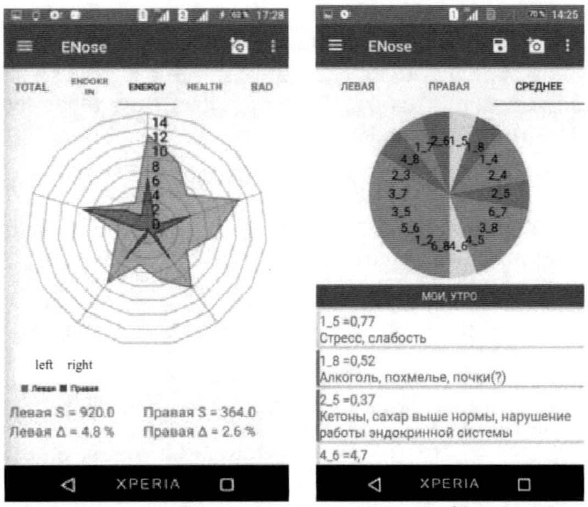

a) *b)*

Figure 3. Program dialog windows with text and graphical information as well as results of the comparison of two measurements for left and right forearm (*a*) and for average measurement with norm (*b*) – The green sector represents the normal health values for parameters included in the diapazon, while yellow and red represent deviations from norm.

The program provides processing as well as presentation of data for individual measurement and for averages as separate recordings in the dialog window of the program (Figure 2b). The decoding of the health state of the organism corresponding to the color scheme of each parameter is displayed on the screen together with a full set of calculated parameters in the form of a sphere, where each parameter corresponds to a separate sector (Figure 3 *b*). It is envisaged to save all data entered, calculated, and visualized (data measurement) to the database on a personal device.

A complete set of data is generated when registering sensor signals for 80 s of sorption and 120 s for desorption volatile substances excreted by the skin of a person's hand or by a biosample (for example, nasal mucus of cattle). When processing the output data of the sensor, the most informative parameters about the health state of the body, individual organs and systems are obtained. Based on these parameters, visual smell traces are constructed by the sensor signals.

The use of adaptive, high-level Java language provides high availability for users via Google Play Market. The proposed mobile application can describe more than 17 health conditions, including tiredness, stress, weakness, endocrine gland disorders, non-numeric level of glucose etc., due to connecting with portable e-nose based on piezoelectric sensors. Thus, the developed software surpasses modern world analogues in many parameters.

IV. CONCLUSION AND FUTURE WORK

The developed software personalizes the diagnosis of the health state of the human or animal using multisensory systems and displays the information in an understandable form to an untrained user. This makes it possible to use such devices for personal purposes, on small farms, therefore, to prevent the development of complex diseases, the death of livestock, and to improve the quality of life.

This approach to creation of software can be useful for processing and visualizing of output data from other sensor devices with 2-8 transducers.

ACKNOWLEDGMENT

This work was financially supported by Russian Scientific Found (grant number 18-76-10015) and grant UMNIK, № 13141GU/2018.

REFERENCES

[1] B. De Lacy Costello et al., "A review of the volatiles from the healthy human body," J Breath Res., vol. 8, p. 29, Feb. 2014, doi: 10.1088/1752-7155/8/1/014001.

[2] A. R. Lucas et al., "Development of an eHealth System to Capture and Analyze Patient Sensor and Self-Report Data: Mixed-Methods Assessment of Potential Applications to Improve Cancer Care Delivery", Jmir Medical Informatics, vol. 6, pp. 138-150, Oct.-Dec. 2018, doi: 10.2196/medinform. 9525.

[3] J. J. Liu, Z. X. Geng, Z. Y. Fan, J. Liu, and H. D. Chen, "Point-of-care testing based on smartphone: The current state-of-the-art (2017-2018)", Biosensors & Bioelectronics, vol. 132, pp. 17-37, May 2019, doi: 10.1016/j.bios.2019.01.068.

[4] N. Mohaghegh, E. Ghafar-Zadeh, and S. Magierowski, "NGRID: A novel platform for detection and progress assessment of visual distortion caused by macular disorders", Computers in Biology and Medicine, vol. 111, UNSP 103340, Aug. 2019, doi: 10.1016/j.compbiomed.2019.103340.

[5] T. A. Kuchmenko, R. U. Umarkhanov, and T. Chikake, "Server for reading, calculation and storage of geometric parameters of the elastic e-nose server sensor response matrix," Certificate of registration of a computer program RU 2019616314, 05.22.2019. Application No. 2019615346 dated 05.13.2019. (Т.А. Кучменко, Р.У. Умарханов, Т. Чикаке, "Сервер для чтения, расчета и хранения геометрических параметров матрицы откликов сенсоров elastic e-nose server" Свидетельство о регистрации программы для ЭВМ RU 2019616314, 22.05.2019. Заявка № 2019615346 от 13.05.2019)

[6] T. Chikake, D. A. Kuchmenko, T. A. Kuchmenko, and R.U. Umarkhanov, "Cloud service for the collection and processing of information systems piezoelectron nose," In the collection: Problems and innovative solutions in chemical technology (PIIRT-2019), Oct. 2019, pp. 23-24. (Т. Чикаке, Д.А. Кучменко, Т.А. Кучменко, Р.У. Умарханов, "Облачный сервис для сбора и обработки информации систем пьезоэлектронный нос," В сборнике: Проблемы и инновационные решения в химической технологии (ПИРХТ-2019), Октябрь 2019, ст. 23-24.)

Emotional Self-Awareness System for Mental Health

(SAMBEDS – Lifestyle management model for depression and anxiety)

Sakirulai Olufemi Isiaq
Department of Computing
Solent University
Southampton, United Kingdom
email: femi.isiaq@solent.ac.uk

Masooma Saeed
Department of Computing
Solent University
Southampton, United Kingdom
email: 4saeem44@solent.ac.uk

Abstract— **A considerable proportion of the human population experience a form of mental health disorder but estimated are about 350 million people suffering from depression according to the World Health Organisation. Now considered one of the leading health conditions, depression contributes significantly to the recorded health-related work absences in the western world. Although mental health associated issues exacerbate for various shortcomings, practitioners and patients have been exploring different methods to conquer this exacerbating circumstance. A notable potent strategy in recent times is the adoption of lifestyle management through emotional self-awareness. While the concept of emotional self-awareness is linked to emotional intelligence, accomplishing a hallmark of realistic self-assessment can be challenging for the potential difficulty of comprehension (i.e., signs and symptoms may be ambiguous or incorrectly comprehended), which may cause misdiagnosis and maltreatment. This work acknowledges the surge in eHealth and presents SAMBEDS emotional self-aware model for managing anxiety and depression through lifestyle management. Thirty-four lived experience candidates participated in the co-created emotional self-aware system prototype, which considers lifestyle factors for managing depression and anxiety. We adopted quantitative and qualitative methods (questionnaires, informal and formal feedback sessions) to co-create the system requirements, design and evaluation. Besides, the user experience study conducted appraises the user activities using a 360-degree camera and eye-tracking device. Although only 50% of the service users are depression and anxiety diagnosed candidates, 70% of the overall participants (including other types of diagnoses) showed high fulfilment in terms of design, usability and efficacy of the for managing system mental health conditions.**

Keywords- emotional intelligence; emotional self-awareness; lifestyle changes; mHeath; Software as a Medical Device (SaMD); eHealth; mental health systems; diagnosis; prevention; alleviation; SAMBEDS.

I. INTRODUCTION

Mental Health is defined as a state of wellbeing by which an individual can cope with the normal stress of life, recognises own abilities, work productively and contribute to own society [1]. While many authors subscribe to distinctive definitions, the lack of a consensus definition is attributed to the differences in values, cultural and social backgrounds among other attributes, although the notion of mental health not being solely the absence of mental illnesses is consented by most authors [2]. Therefore, a less-restrictive or cultural-bound definition that accommodates emotional state variation and imperfect functioning is promoted. Consequently, factors including the dynamic state of internal equilibrium, cognitive and social skills, empathy, flexibility to cope, harmonious body-mind relationship, and emotional regulation are prescribed [2].

Nowadays, depression increasingly becomes a popular mental health condition, which about 300 million people suffer globally [3]. Not only is it considered one of the leading condition responsible for health-related work absences in the western world, but a UK study also showed 595,000 cases of workers suffered from depression, stress and anxiety between 2017 and 2018. A recent Labour Force Survey (LFS) in a similar period revealed 15.4 million workdays (57.3% of total days lost to ill-health) was lost to mental health-related issues [4]. Furthermore, 43.4 million adults (17.9%) were estimated to suffer a form of mental illness in 2015 in the United States [5].

Although various alleviation procedures are being proposed, a significant proportion of mentally ill candidates yet lack appropriate interventions or supports. Factors hindering such support range from a shortage of medical practitioners particularly, psychiatrists in rural areas [6], the lack of progress monitoring tools, and deficit in mental health budgets - for instance, 78.9% of African countries spend less than 1% and only 54% of European countries spend over 5% of their health budgets on mental health [7] - among other factors.

Beyond these limitations, stakeholders are now taking different turns to improve this escalating health condition. One of such shifts is alleviation through self-monitoring of thoughts and behavioural activities to improve depressive symptoms [8]. Subsequently, self-monitoring is a concept of self-awareness, which is one of the five major components (including self-regulation, empathy, motivation, social skill) of emotional intelligence. Its major hallmark of realistic self-assessments harmonises the capacity to recognise one's moods, emotions, strengths, weaknesses, drives, values, and goals and their corresponding impacts [9]. A realistic assessment will not only help understand one's emotional state and imperfect functions but can help adopt an appropriate lifestyle for any realised imperfect functioning.

Nonetheless, achieving a realistic assessment (i.e., the management of own emotional awareness) can be a challenge without appropriate support and adequate guiding tools. For instance, without a rigorously tested tool, how do individuals measure corresponding lifestyle activities in relations to their emotions? Yet, the advances in technology incorporate innovations such as automation even with further investments. For instance, the mHealth industry is expected to hit a hundred billion dollars by 2025 thus, these challenges are becoming surmountable. Nowadays, more individuals own a smartphone and are likely to use established apps to monitor changes in their lifestyle activities including physical and diet than those without a smartphone [10]. In additions, users have been seeking interventions through apps in recent times. For instance, the Kooth counselling app - an online counselling and wellbeing platform - where many teens turn to seek support with mental health issues has recorded a surge in recent years [11]. Considering the increasing access to technology, which brews the surge in the adoption of Software as a Medical Device (SaMD) and medical apps, this work proposes SAMBEDS model, a co-created depression and anxiety management system based on the Framework for Lifestyle Management pro Mental Health, FLMMHS [12].

The rest of the paper is structured as follows; Section II discusses the methodology involving co-creation and data analysis techniques; Section III describes the development of SAMBEDS model-based system, an emotional self-aware system for anxiety and depression; Section IV presents the evaluation and results of the system, while conclusions and future work were manifested in Section V.

II. METHODOLOGY AND DATA ANALYSIS

Taking that a single method may not cover all areas of research and system development, we adopted mixed methods involving techniques of qualitative, quantitative and engineering approaches. Although the system development emulates the Framework for Lifestyle Management pro Mental Health Systems (FLMMHS) [12], which the concept of Diagnosis, Prevention and Alleviation (DPA) is preeminent, we utilised a co-creation strategy to develop an effective emotional self-aware system. Adopting mixed methods will not only support the understanding of areas uncovered by a singular method [13], it also allows a provision of suitable coverage for any discovered aspects. Therefore, we utilised techniques including questionnaire surveys, prototyping and interviewing and feedback following a co-creation approach.

A. *Co-creating emotional self-aware elements*

Although FLHMMS framework emphasises lifestyle management approach for mental health oversight, the procedures of diagnosis, prevention and alleviation adopt digital tools to manage the associated lifestyle factors of the intended mental health disorder. Hence, this work utilises a document analysis technique to establish corresponding lifestyle elements necessary for emotional self-awareness in managing depression and anxiety.

B. *SAMBEDS derivation and Document Analysis*

Similar to other qualitative methods, document analysis allows the review of specific field data and perhaps, develop empirical knowledge from the examined documents or data [14]. Hence, this work examines published data about anxiety and depression from different publication platforms including IEEE Library, NCBI/PubMed (National Center for Biotechnology Information), American Journal of Psychiatry and BMC Public Health among others. Then, we utilised the examined information imperatively for the derivation of the system knowledge and its design and development. For instance, monitoring mood changes and its corresponding symptoms are rarely monitored outside a clinical setting, but such information could help improve clinical care and intervention [15]. Also, a study by Robertson et al, indicated walking can improve symptoms of depression [16], but Meyer and Broocks concluded that a good amount of aerobics exercises will improve mood against depression [17]. In terms of behaviour and thought, not only can note-taking help to enhance the decluttering of the mind, but patterns identified through the logs of negative thoughts and feelings can be enhanced through emotional self-awareness. Similarly, sleep quality has been highlighted to contribute to depressive symptoms [18]. While low vitamin D level is identified to be associated with depression and anxiety [19], a reduced blood flow of key stress-reducing hormones is also linked to dehydration [20].

Based on the different established findings, this work devises a seven lifestyle factors, SAMBEDS to enhance emotional self-awareness for mental health management. Highlighted SAMBED factors include *"Sleep Pattern"*, *"Activity and Social"*, *"Mood Change"*, *"Behaviour and Thought"*, *"Exercise and Fitness"*, *"Diet and Hydration"*, and *"Spiritual and Meditation"*. We further examined these seven core elements as an area of emotional self-awareness for mental health in the data collection section.

C. *Data collection*

In addition to the document analysis exercise that establishes relative emotional self-awareness factors of mental health disorder, the derived survey questions were also validated by service users in interactive sections. Thirty-four lived experience candidates were randomly selected and the nature of their disorders were established based on the eight-point question survey as shown in Table I.

TABLE I. QUESTIONS TO EVALUATE CANDIDATES' UNDERSTANDING OF CURRENT SUPPORT SYSTEM

No	Questions
1	Have you ever been diagnosed with any mental health conditions?
2	If Yes, what kind of support do you receive?
3	Do you get support immediately available when you need it?
4	Generally, rank the order of importance of keeping track and monitoring the following in relation to mental health: mood, sleep, social activities, thought/behavior, diet, exercise, spiritual/meditation.

5	Chose the three activities that you consider most relevant to you (manage own life, seek professional advice, be in control own life, keep diary, prefer face-2-face contact, dislike discussing own mental health and prefer to find own solution).
6	Do you feel regular self-monitoring of your lifestyle will enhance mental health and general wellbeing?
7	Do you ever or currently use any healthcare management app or SaMD?
8	What is your opinion on the current mental health care system?

An indication of mental health cases and corresponding percentages of service users are as shown in Table II.

TABLE II. MENTAL HEALTH DISORDER AND CORRESPONDING CASES

Mental Health Disorder	No of cases	Percentage of cases (%)
Depression	22	64.7
Anxiety	22	64.7
Substance use Disorder	1	2.9
Stress-related Disorder	6	17.6
Bipolar Disorder	1	2.9
Post-traumatic stress Disorder	1	2.9
Schizophrenia	1	2.9
Don't want to disclose	1	2.9

Table II reveals that in more than a singular case, a candidate may have been diagnosed with more than one disorder. We recorded fifty-five diagnosed disorder cases from the thirty-four participating service users. Emphatically, we expect that the survey questions will thoroughly establish the participants' perception of the current mental health system particularly, how the existing mental health system supports their conditions, i.e., in terms of usability, expectations and efficacy. The survey questions are further explained in the following section, survey questions explained.

1) Survey questions explained: While the survey proposes to validate the defined seven core elements of emotional self-awareness for mental health management, the impacts of the questions vary in diverse perspectives. While the first question aims to establish the candidates' type of diagnosis as highlighted in Table I, we establish the nature of support they received and the immediate frequency of getting these supports with second and third questions, respectively. Although question four confirms the significance level of these core elements, question five identifies the participants' preferred approach for dealing with their associated mental health conditions. Questions six measures the respondents' perception of the concepts of emotional self-awareness to manage mental health and general wellbeing. Subsequently, questions seven and eight examine the participants' ability to use health devices and their opinions on the existing methods of dealing with mental health, respectively. The results of questions conjoining with the formal feedback exercise effectively aid the development of an "emotional-awareness" system for depression and anxiety. The survey results were further elaborated in the survey result section as follows.

2) Survey results explained: Regarding the supports received by candidates, only about 35% of the participants responded to this question. While these candidates may have received varied support types perhaps, on multiple occasions, at about 75% of these occasions were these supports provided by families and friends as shown in Figure 1. Conversely, 41.7% of each occasion were candidates seeking to manage the situation by themselves or go for therapy or other medical options. Interestingly, none of the participants has chosen not to have had any form of support perhaps, such situations may not have arisen in cases of mild mental health situations. However, it is particularly unsurprising that a significant proportion of candidates do not get immediate support when needed; only 29% of the candidates get immediate support while 71% do not, hence an explanation the rapid mental health cases recorded in recent times. Ranking the order of importance of SAMBEDS elements by the participants reveals behavioural and thought element as the most important.

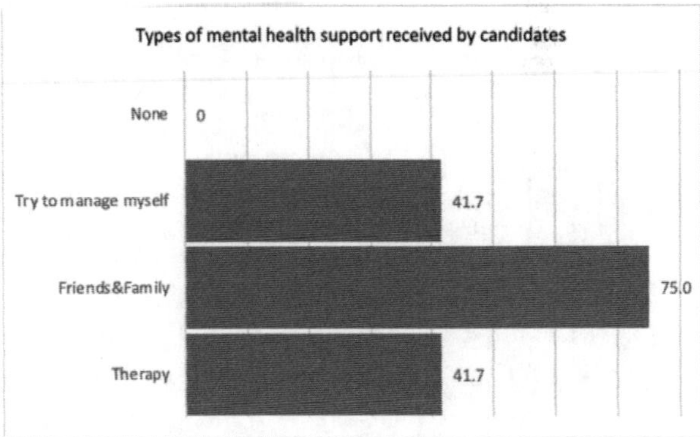

Figure 1. Percentage (%) of candidate and types of support received.

Followed in the order of importance are *sleep pattern*, *mood change*, *spiritual/meditation* and *activity/social*, respectively. Bottom ranked elements are *diet/hydration* and *exercise/fitness* with the same rating score. However, considering that the ranking of these elements is perceptual, it is interesting to note all elements recorded same overall rating value (68%) except for *sleep pattern* (70%) and *mood changes* (84%) which are slightly ranked higher, respectively (see Figure 2). Not only does this evidence the importance of these identified elements in mental health management, but it also exhibits mood changes as a very important self-aware element in managing depression and anxiety.

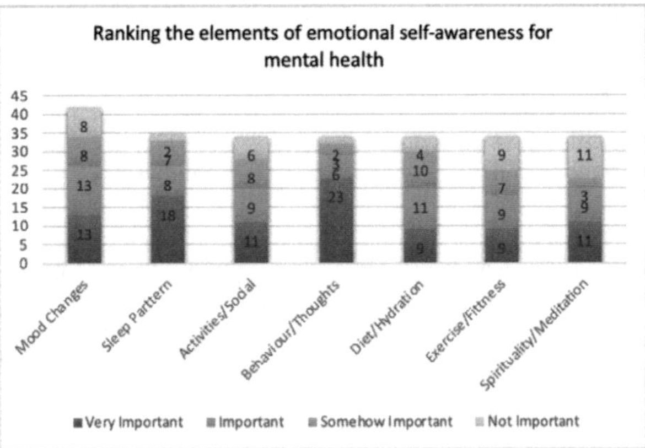

Figure 2. Ranking (n) the importance of emotional self-aware elements.

Not only do the participation of lived-experienced candidates better reflect the preferred approaches of managing mental health situations, but also provides a depth understanding of preeminent options habitually adopted by mental health candidates. For instance, as shown in Figure 3, a significant proportion of the candidates prefer to manage their own life, preferred a face-to-face conversation and perhaps, be in control of things by themselves. Not only does this result evidence own freedom characteristics of humans as a requirement of any mental health management system, but it also explains the reluctance of mental health candidates in seeking professional supports.

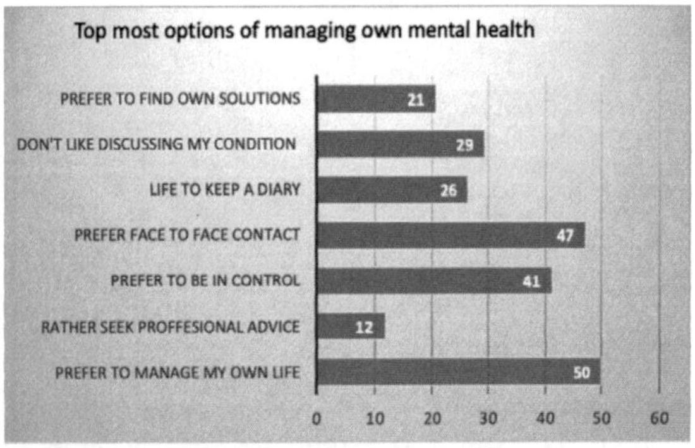

Figure 3. Participants rating the topmost options in numbers for managing mental health.

Trailing behind are options of candidates preferring not to discuss own disorder or conditions but will keep a diary or prefer finding own solutions. It is bolstering that most candidates are averse to seeking professional advice, yet this negatively correlates the reality of candidates' willing to manage their own life and being in control of their life. However, a significant proportion of candidates feel self-monitoring of their emotions 'will' or 'maybe' help improve mental health conditions. Contrary to the 88% who believe emotional self-monitoring could improve mental health symptoms (see Figure 4), only 24% have ever used health apps or any forms of Software as a Medical Device, SaMD.

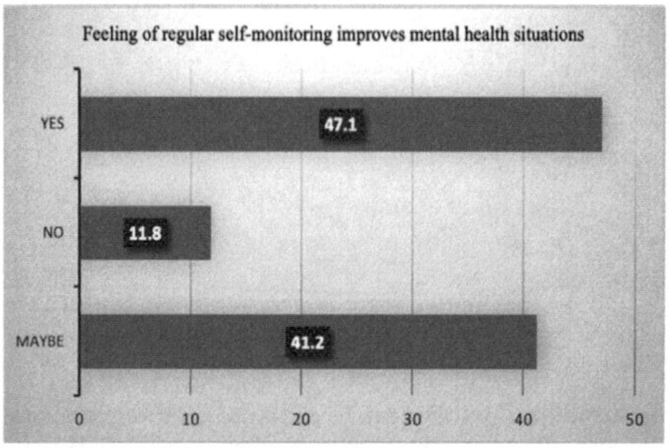

Figure 4. Rating in percentages (%) of a feel of regular self-monitoring improves mental health

Finally, participants were able to provide free-text answers to question 8 by presenting their overall perception of the current mental health system. Interestingly, some of the responses include "*pills are not the answer*"; "*dislike going into therapy*"; "*more investments in providing therapy and being able to see doctors within a day instead, of having 4-8 weeks wait*"; "*more individual and group supports*"; *feeling of therapist not being interested in users' problems*"; "*keeping track of thoughts, activities and mood swings will help to better understanding what bothers candidates*" among other comments. While these comments provided further justification for the results of the preceding questions, they also create opportunities to analyse the limitations of the existing mental health system. Not only do these results aid the derivation of robust requirements but also promote an avenue for the co-creation of an effective SAMBEDS model-based emotional self-awareness system, as discussed in the development section as follows.

III. Developing SAMBEDS based System to Manage Anxiety and Depression

While a User-Centered-Design, UCD approach is adopted for the system development, the results derived from survey aids the specified requirements of SAMBEDS emotional self-awareness system for mental health. The system is partitioned into multiple segments to manage different activities including *mood assessment*, *progress check*, *therapist support*, *general information*, *exercise*, *diet*, *sleep pattern* and *journal* as shown in Figure 5. It is important to note that the DPA aspect of FLHMMS framework is covered in a different partition of the system. While the diagnosis component is accomplished via the mood assessment section, i.e., "*assess yourself*", users are able to rate their feelings at intervals using a corresponding smiley (very happy, happy, indifferent, sad and very sad).

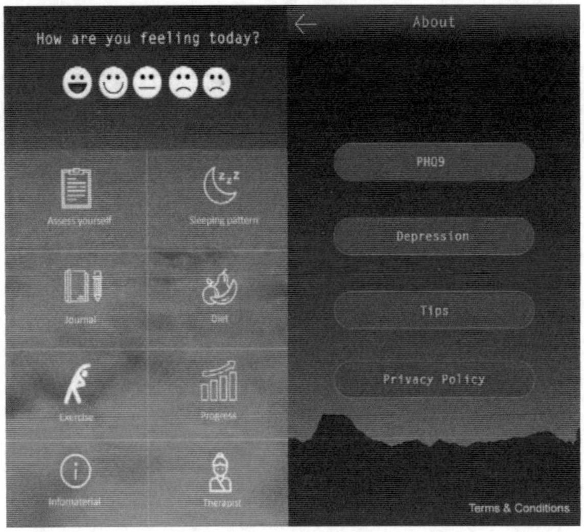

Figure 5. SAMBEDS-based system - Home and About screen.

Furthermore, a PHQ9 instrument was adopted to measure severity due to its concise diagnostic criteria for depression [21] and perhaps, its accuracy when compared to the Beck Depression Inventory, which tends to assign higher scores for severity [22]. Besides, prevention and alleviation processes were based on lifestyle changes with the intention of mood upliftment through these changes. For instance, understanding the correlation of recorded sleep patterns with other lifestyle factors are made easy through suitable life entities association such as family, health, school and studies among others via the journal (see Figure 6).

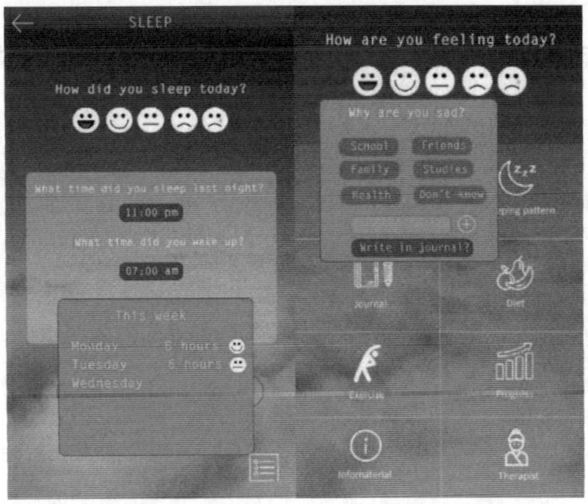

Figure 6. SAMBEDS-based system - Screens for Sleep and Journal

In additions are the *progress* screen, which presents a graphical representation of individual progress of the SAMBEDS elements. While the *Stats* page as shown in figure 7 provides statistical awareness about the SAMBEDS elements, the therapist page allows connection with a setup therapist. Hence, users can easily share their lifestyle records and in real-time with therapists to get appropriate advice or treatment (see Figure 7). Not only has the co-creation approach helps with an understanding of the system requirements, but it also influences the development of a usable system for the intended users. The following section discusses the results of user experience study conducted for the developed system.

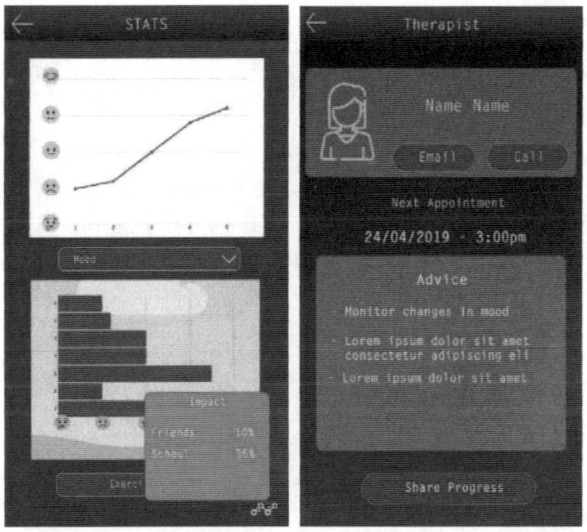

Figure 7. SAMBEDS-based system - Stats and therapist screen.

IV. RESULTS AND EVALUATION

Besides the survey and interview exercises by lived-experienced candidates, a usability study was conducted in a usability lab using the eye-tracking device and a 360-degree camera to understand the impact of the system in aiding emotional self-awareness for mental health conditions. We adopted a five-factor usability approach including learnability, memorability, errors, satisfaction and efficiency [23] with a rank scale questionnaire. The rank scale ranges from 1 to 7, with 1 being the lowest and 7 being the highest rating value. We aggregated and evaluated the users' results, laying emphasis on the aforementioned five factors.

TABLE III. LIST OF USABILITY TASKS WITH TARGET TIME AND OBJECTIVES

Task No	Task Description	No of Screens	Time (S)	System Objective
1	Navigate to the registration page	3	20	Efficiency
2	Take the assessment, return score	4	20	Memorability
3	Click mood and go to Journal	2	10	Efficiency
4	Go to	2	20	Memorability

	exercise screen, click goal and go back to home screen			
5	Access PHQ9 and return to information screen	2	10	Learnability
6	Check progress	1	10	Learnability
7	Check sleep hours	2	20	Efficiency
8	Contact therapist and read advice	1	10	Efficiency

Figure 8 shows the ease of completing all activities as highlighted in Table III. Specifically, 50% of the participants give the highest rate value, 7 for interface simplicity, while 40% rate with value 6, therefore, 90% of participants consider the interface to be simple enough for emotional self-awareness.

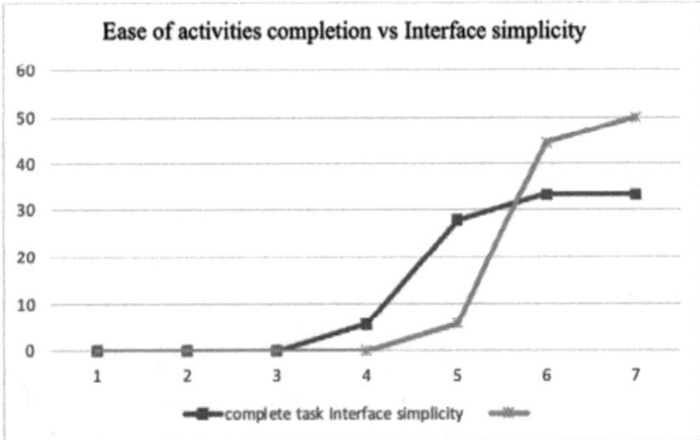

Figure 8. Perceived simplicity of interface versus ease of completion.

Also, 66.6% of participants rated the ease of completing as high, i.e., 33.3% rated value 7 and value 6 (33%). While the ease of completion only recorded an instance (5.6%) of value 4 with no instance recorded below 4 for simplicity.

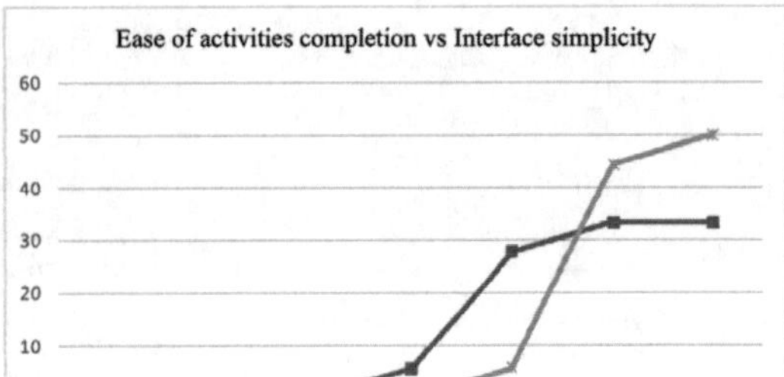

Figure 9. The ease of access to information versus learnability.

Additionally, the ease of accessing information and learnability were evaluated as shown in figure 9. A total of 83% of users rated the high values, i.e., 50% and 33.3% of users rated the learnability of system 7 and 6, respectively. Similarly, 77.8% of users rated the access to all information with the high values within the targeted time (i.e., 38.9% each rated the values 7 and 6).

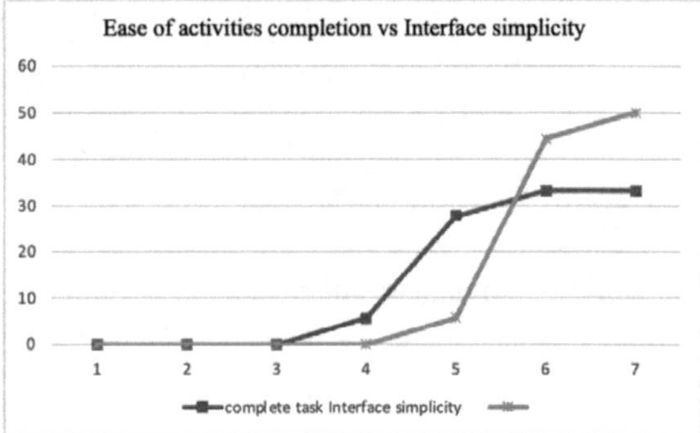

Figure 10. Overall users' satisfaction versus difficulty in navigation.

Besides, the difficulty in navigation and overall satisfaction evaluated indicated 77.8% of the participants perceived the difficulty in navigation (values 1 and 2) very low (as shown in Figure 10), thus, participants find the system easy to use in accomplishing self-awareness for mental health. Also, the recorded rating for the overall satisfaction appears between the values between 5 and 7, i.e., all the service users (100%) were highly satisfied with the system for the accomplishment of emotional self-awareness for mental health. Finally, the participants' comments strengthen the satisfaction recorded on the system. These comments include "*motivating, easy navigation, appealing layout and interactive mood monitoring*"; "*the included PHQ9 is a really good idea*"; "*icons are very useful and appealing"; "colour scheme is appealing, gave no eye strain*"; "*clear design balanced with image, interface clean, calming and relaxing*"; "*sharing progress with a therapist*" among other comments. Not only do these comments validate the quantitative results, but they open further research on the efficacy of SAMBEDS as emotion self-awareness tool for mental health management.

V. Conclusion and Future work

Lifestyle factors including diet and exercise among other factors have been found to improve the quality of life and depression particularly, in overweight people. Also, a strong association has been established between improved mood and exercising. Similarly, a robust correlation between exercises and sleep has been noted in lifestyle psychiatry. This work presents a SAMBEDS model for emotional self-awareness in depression and anxiety management with seven core lifestyle elements. While a co-creation of a prototype system allows a true evaluation of these elements, all elements were considered equally important except for sleep patterns and mood changes, elements which are considered slightly more eminent. Independently, the behavioural and thought element is rated the highest among other SAMBEDS elements, but adequate management of all elements prove contributory to emotional self-awareness in relations to mental health states. While a limitation of the system is the digital characterisation of most activities, a mitigation approach is the contraction of users' interactivity through automation. Future work must devise automatic means of extracting these core lifestyle data with minimal or no hindrance to the users' daily activities. A further interesting research challenge is the development of automatic behavioural and thoughts algorithms that will synchronise mental health state with mood changes, exercises and sleep and other SAMBEDS elements.

REFERENCES

[1] "Promoting mental health: concepts, emerging evidence, practice", World Health Organization, 2005. [Online]. Available: https://www.who.int/mental_health/evidence/MH_Promotion_Book.pdf. [retrieved: February, 2020].

[2] S. Galderisi, A. Heinz, M. Kastrup, J. Beezhold and N. Sartorius, "Toward a new definition of mental health", *World Psychiatry*, vol. 14, no. 2, pp. 231-233, 2015.

[3] "World Health Organization WHO (2017). Depression and Other Common Mental Disorders, Global Health Estimates. - References - Scientific Research Publishing", *Scirp.org*, 2017. [Online]. Available: https://www.scirp.org/reference/ReferencesPapers.aspx?ReferenceID=2360252. [retrieved: January, 2020].

[4] "Work-related stress, depression or anxiety statistics in Great Britain", HSE (2018). [online]. Available: https://www.hse.gov.uk/statistics/causdis/stress.pdf [retrieved: February, 2020].

[5] *Samhsa.gov*, 2015. [Online]. Available: https://www.samhsa.gov/data/sites/default/files/cbhsq-reports/NSDUHNationalFindingsReport2018/NSDUHNationalFindingsReport2018.pdf. [retrieved: January, 2020].

[6] T. Butryn, L. Bryant, C. Marchionni and F. Sholevar, "The shortage of psychiatrists and other mental health providers: Causes, current state, and potential solutions", *Ijam-web.org*, 2017. vol. 3, pp. 5-9, 2018.

[7] "WHO | New WHO mental health atlas shows global mental health resources remain inadequate", *Who.int*, 2005. [Online]. Available: https://www.who.int/mediacentre/news/notes/2005/np21/en/. [retrieved: January, 2020].

[8] P. Chandrashekar, "Do mental health mobile apps work: evidence and recommendations for designing high-efficacy mental health mobile apps", *mHealth*, vol. 4, pp. 6-6, 2018.

[9] D. Goleman, *Emotional intelligence*. New York, NY: Bantam Books, 2006.

[10] J. Carroll, A. Moorhead, R. Bond, W. LeBlanc, R. Petrella and K. Fiscella, "Who Uses Mobile Phone Health Apps and Does Use Matter? A Secondary Data Analytics Approach", *Journal of Medical Internet Research*, vol. 19, no. 4, p. e125, 2017. Available: 10.2196/jmir.5604.

[11] D.Campbell, Mental health - UK teenagers turn to mobile apps to help with mental ill health, Guardian News and Media Limited, 2018 [Online] Available: https://www.theguardian.com/society/2018/nov/21/uk-teenagers-turn-to-mobile-apps-to-help-with-mental-ill-health. [retrieved: February, 2020].

[12] S. Isiaq, and D. Sobnath. A Collective Intelligence Framework for Lifestyle and Mental Health Management Systems (FLiMMS), International Journal on Advances in Life Sciences, vol, 11 no. 3 and 4 pp. 138-150, 2019.

[13] J. Creswell and V. Plano Clark, *Designing and conducting mixed methods research*. Thousand Oaks: Sage, 2009.

[14] R. Service, "Book Review: Corbin, J., and Strauss, A. (2008). Basics of Qualitative Research: Techniques and Procedures for Developing Grounded Theory (3rd ed.). Thousand Oaks, CA: Sage", *Organizational Research Methods*, vol. 12, no. 3, pp. 614-617, 2008. Available: 10.1177/1094428108324514.

[15] M. Nahum et al., "Immediate Mood Scaler: Tracking Symptoms of Depression and Anxiety Using a Novel Mobile Mood Scale", *JMIR mHealth and uHealth*, vol. 5, no. 4, p. e44, 2017. Available: 10.2196/mhealth.6544.

[16] R. Robertson, A. Robertson, R. Jepson and M. Maxwell, "Walking for depression or depressive symptoms: A systematic review and meta-analysis", *Mental Health and Physical Activity*, vol. 5, no. 1, pp. 66-75, 2012.

[17] T. Meyer and A. Broocks, "Therapeutic Impact of Exercise on Psychiatric Diseases", *Sports Medicine*, vol. 30, no. 4, pp. 269-279, 2000.

[18] H. Tanaka, Y. Sasazawa, S. Suzuki, M. Nakazawa and H. Koyama, "Health status and lifestyle factors as predictors of depression in middle-aged and elderly Japanese adults: a seven-year follow-up of the Komo-Ise cohort study", *BMC Psychiatry*, vol. 11, no. 1, 2011.

[19] P. Bongiorno, "Put Anxiety Behind You eBook by Peter Bongiorno ND, LAc - 9781609259280 | Rakuten Kobo", 2015. [Online]. Available: https://www.kobo.com/us/en/ebook/put-anxiety-behind-you-2. [Accessed: 21 Feb 2020].

[20] "Can Anxiety Be Caused by Dehydration?", Calm Clinic, 2018. [online] Available: https://www.calmclinic.com/anxiety/causes/water-dehydration, [retrieved: January, 2020].

[21] B. Dear et al., "The Pain Course: 12- and 24-Month Outcomes From a Randomized Controlled Trial of an Internet-Delivered Pain Management Program Provided With Different Levels of Clinician Support", *The Journal of Pain*, vol. 19, no. 12, pp. 1491-1503, 2018. Available: 10.1016/j.jpain.2018.07.005.

[22] B. Weobong, H. Weiss, I. Cameron, S. Kung, V. Patel and S. Hollon, "Measuring depression severity in global mental health: comparing the PHQ-9 and the BDI-II", *Wellcome Open Research*, vol. 3, p. 165, 2018. Available: 10.12688/wellcomeopenres.14978.1.

[23] R. Dermawi, H. Tolle and I. Aknuranda, "Design and Usability Evaluation of Communication Board for Deaf People with User-Centered Design Approach", International Journal of Interactive Mobile Technologies (iJIM), vol. 12, no. 2, p. 197, 2018. Available: 10.3991/ijim.v12i2.810

Leveraging Machine Learning and Natural Language Processing for Monitoring E-health Publications

Andrius Budrionis, Rune Pedersen, Torbjørn Torsvik, Karianne Lind, Omid Saadatfard
Norwegian Centre for E-health Research
University Hospital of North Norway
Tromsø, Norway
e-mail: Andrius.Budrionis@ehealthresearch.no, Rune.Pedersen@ehealthresearch.no, Torbjorn.Torsvik@ehealthresearch.no, Karianne.Lind@ehealthresearch.no, Omid.Saadatfard@ehealthresearch.no

Abstract—**E-health is a rapidly developing field governed by national and international guidelines. These guidelines are often generic, making it problematic to monitor the development of the field with regards to the expected directions. To address this shortcoming, we present a data analytics pipeline for continuous monitoring of e-health publications in Norway with regards to the national e-health strategy. The pipeline contains PubMed data import module, machine learning, natural language processing modules and a visualization component. The potential of the proposed approach is illustrated by identifying publication trends in Norway for the last ten years. These trends show how well focus areas of the Norwegian e-health strategy are represented in scientific publications. The pipeline is customizable and can be extended to support other countries, e-health strategies and publication channels.**

Keywords-e-health strategy; publications; machine learning; natural languge processing.

I. INTRODUCTION

In recent years, e-health has become an important topic in political agendas of both developing and developed countries. According to the World Health Organization, 58% of member states have developed national e-health strategies and 87% reported one or more national mobile health (m-health) initiatives in 2016 [1]. These initiatives are often connected to ongoing research activities disseminating achievements and lessons learned through scientific publication channels.

In Norway, e-health has undergone a strong national foundation process, especially from 2016 up till now. Through the establishment of a Directorate for E-health (E-Dir) and the Norwegian Center for E-health Research (NSE), as well as national initiatives and strategies, such as One Citizen – One Journal [2], the National E-health Strategy [3], and the National Action Plan for E-health 2020-22 [4], the national development and focus on e-health has escalated. In light of these initiatives, monitoring how well national development of e-health corresponds to the goals defined in the aforementioned documents is an important feedback to the decisionmakers.

Considering the multifaceted nature of e-health [5], it is difficult to draw a boundary between e-health, telemedicine, health technology, medicine and other closely related fields. This uncertainty makes it difficult to isolate and measure achievements within e-health and, thus, monitor compliance of national e-health development against strategic documents, such as National E-health Strategy [3] and the National Action Plan for E-health 2020-22 [4].

Publication of scientific papers could be considered as a proxy for research and development in the field [6]. Building on our previous publication [7] where we manually searched for e-health papers and classified them into several groups, we present improvements allowing continuous monitoring of scientific publications in Norway with regards to the National E-health Strategy [3].

Our previous work showed that it is problematic to classify publications into the focus areas of the National E-health Strategy [3] with high accuracy, however, achieved performance was considered sufficient for revealing publication trends. These trends showed value without being completely accurate at a single publication level [7]. This paper presents continuation of our work on e-health publication monitoring in Norway and classification of the identified manuscripts into the six underlying focus areas: 1) digitization of work processes, 2) seamless/coherent patient pathways, 3) improved use of health data, 4) new ways to provide healthcare, 5) common foundation for digital services and 6) national e-health management and increased implementation [3].

The reminder of this paper is organized as follows. Section II provides a summary of methods used to produce results, which are presented in Section III. Section IV discusses the key findings and limitations of this work, while Section V concludes the paper.

II. METHOD

To provide continuous monitoring of e-health publications, a data analytics pipeline, covering data collection analysis and visualization was developed. Considering the multifaceted definition of e-health, a neural network-based approach, encompassing language representation was selected for differentiating e-health publications from other irrelevant content. To be specific, a Bidirectional Encoder Representations from Transformers (BERT-base) model pretrained on large Wikipedia corpus [8] was adapted to differentiate between e-health and not e-health publications. BERT is a general-purpose language representation model trained in an unsupervised manner. Unlike simpler Natural Language Processing (NLP) models based on word counts, BERT takes context of tokens into consideration. This way, the model captures language semantics and is able to differentiate between tokens that would be considered the same by word-count models.

To learn contextual relationships between words (or text tokens), BERT utilizes a Transformer neural network architecture and an attention mechanism. BERT is trained using two strategies: masked word prediction in a sequence of tokens and next sentence prediction. Both strategies are trained together minimizing combines loss function. BERT model, trained on a big language corpus, learns language representation and relationships between text tokens. Such general-purpose model can be later finetuned for specific NLP tasks without the need to retrain entire model [8]. In comparison to model pretraining, finetuning is considerably less demanding computationally and can be performed on much smaller datasets.

To adapt the general purpose BERT-based model to a specific task, it was finetuned on a manually labelled publication corpus presented earlier (e-health publication dataset) [7]. This corpus contains 1891 publications (816 e-health, 1075 not e-health) papers. Text from title, keywords and abstract fields was used for model finetuning. Finetuning was performed in Google Colab environment. The e-health publication dataset was split into training (60%), validation (20%) and testing (20%). The model was evaluated using a random test set (20% of the e-health publication dataset) after model finetuning.

Due to manual data collection, the e-health publication dataset was skewed and contained a much larger proportion of e-health publications than data available in PubMed [7]. To ensure that the model generalizes for data available in PubMed, an additional evaluation step was included. A PubMed dataset, containing 924 publications (25 e-health and 899 not-e-health) was manually labelled and used for validating the model's performance. The PubMed dataset represents e-health and not e-health class distribution in data extracted from PubMed.

To map e-health publications to the focus area of the National E-health Strategy, a machine learning model developed previously was employed [7]. This model is based on token count values (Term Frequency – Inverse Document Frequency, TF-IDF) and was trained on the e-health publication dataset to differentiate between the following publication classes:

1) Digitization of work processes.
2) Seamless/coherent patient pathways.
3) Improved use of health data.
4) New ways to deliver healthcare.
5) Common foundation for digital services.
6) National e-health management and increased implementation.

III. Results

The architecture and validation of the developed models are summarized in this section.

A. System design

To support up-to-date monitoring of e-health publications, a data analytics pipeline was set up. The pipeline contains three major components (Figure 1):

1. Data import. This component handles queries to the data providers (PubMed in the current setup) and returns metadata for every publication meeting inclusion criteria. When monitoring publications from Norway, inclusion criteria was limited to at least one coauthor affiliated with Norway and publication dates (01-01-2010 – 01-04-2020). These data are stored in a relational database for further analysis.
2. The data analytics component hosts pretrained machine learning models for classifying publications into specific groups. The two-class classifier performs dataset denoising, discarding irrelevant (not e-health) publications. Since our focus is monitoring production e-health related papers, only these manuscripts are considered in further analysis. The six-class classifier classifies e-health publications into 6 focus areas of the Norwegian E-health Strategy [3].
3. The visualization module performs data aggregations and presents the results of the data analytics step in a visual way.

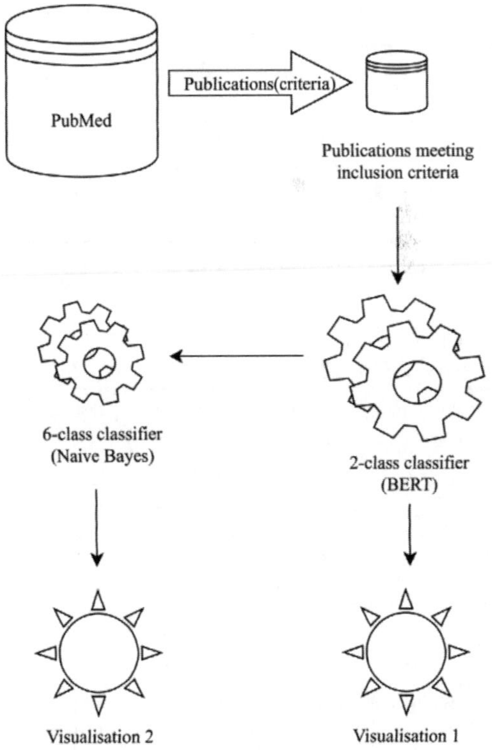

Figure 1. Data analytics pipeline

B. Performance of the classification models

The performance of the 2-class model was tested on two datasets containing different ratios of positive class (e-health) publications. The e-health publication dataset contains the same ratio of positive and negative class examples as the dataset used for model finetuning. The PubMed dataset represents a realistic ratio of positive and negative class examples observed in PubMed (Table I).

TABLE I. PERFORMANCE OF THE 2-CLASS MODEL

Dataset	Class	Precision	Recall	f-1 score	AUC
E-health publication dataset	Not e-health	0.92	0.88	0.9	0.888
	E-health	0.85	0.90	0.87	
PubMed dataset	Not e-health	0.99	0.99	0.99	0.858
	E-health	0.82	0.72	0.77	

A trained 6-class classifier, reported in an earlier publication, was used for classifying e-health publications into the focus areas of the Norwegian E-health Strategy [7]. The performance of this model is available in Table II.

TABLE II. PERFORMANCE OF THE 6-CLASS CLASSIFIER [7]

Class	Precision	Recall	f1-score
1. Digitization of work processes	0.70	0.58	0.63
2. Better continuity of care	0.61	0.62	0.62
3. Improved use of health data	0.62	0.71	0.67
4. New methods to provide healthcare	0.74	0.77	0.75
5. Common foundation for digital services	0.53	0.62	0.57
6. National e-health management and increased implementation	0.66	0.64	0.65

C. *Visualizations*

To illustrate how the trained model could be used for monitoring scientific publications in e-health, all publications containing "Norway" in author affiliation and published during the last 10 years (01-01-2010 – 01-04-2020) were extracted using PubMed API. More than 70 000 publications were published in PubMed by authors affiliated to Norway. The aforementioned 2-class model was used to filter out irrelevant papers based on their title, keywords, and abstract. The number of e-health publications stratified yearly are visualized in Figure 2.

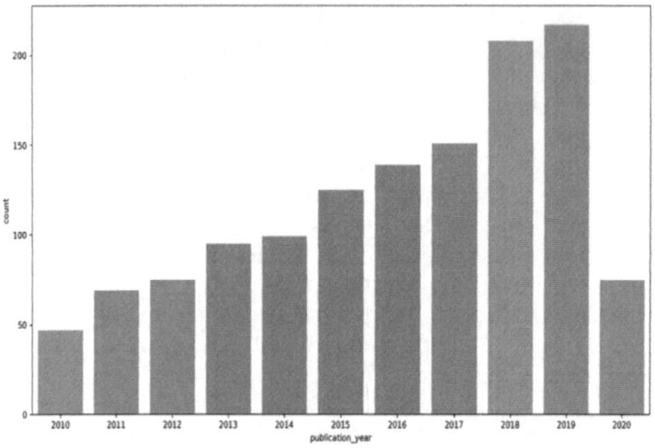

Figure 2. E-health publications authored by researchers affiliated to Norway. Publication period 01-01-2010 – 01-04-2020

To map the identified e-health publications to the focus areas of the National E-health Strategy, they were classified using a 6-class classifier (Figure 3, Figure 4) [3].

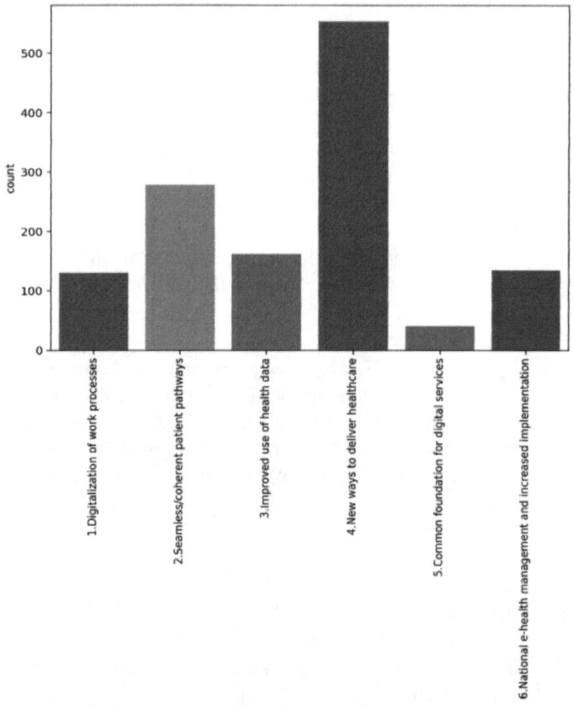

Figure 3. Classification of e-health publications into focus areas of a National E-health Strategy

To show how number of publications in each class developed in time, they were stratified on yearly basis and visualized in Figure 4.

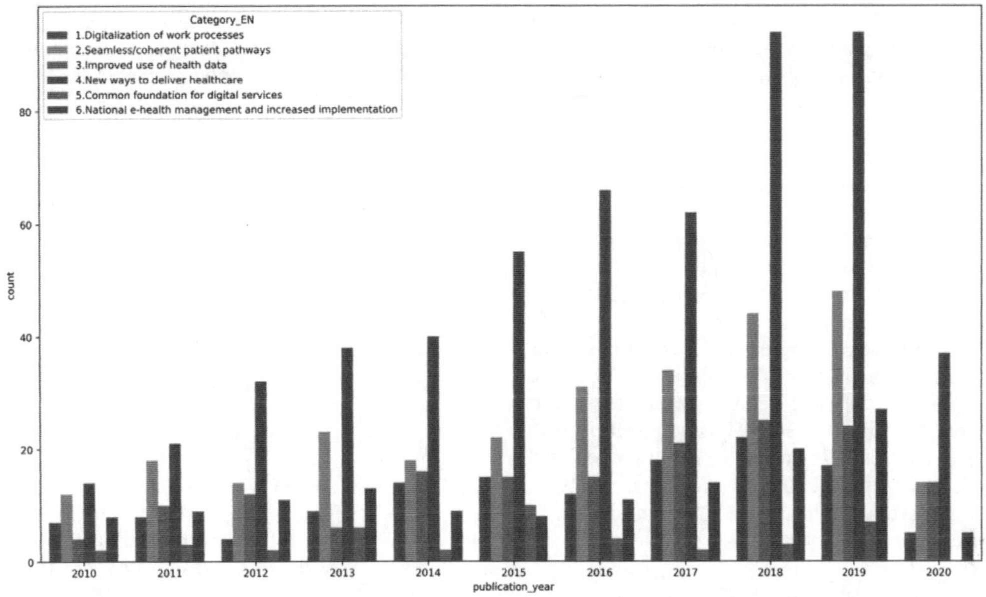

Figure 4. Classification of e-health publications into focus areas of a National E-health Strategy stratified yearly

IV. Discussion

In this paper, we demonstrated how to leverage a data analytics pipeline for monitoring the production of scientific publications. This pipeline is flexible and easily extendable to other fields, data sources and visualizations.

A. Key findings

Monitoring of e-health publications shows an increasing pace of publishing scientific contributions in the field of e-health in Norway (Figure 2). It may not be surprising, considering similar trends are observable in other fields of research. However, it is an important result, demonstrating the feasibility of the presented monitoring approach. We previously performed a similar experiment manually querying selected research databases with a set of keywords specifically combined to capture e-health publications [7]. Considering that manual search included more data sources, it is natural that some differences in yearly publication counts were observed when comparing these methods. Regardless of these deviations, both methods captured the same publication trend.

B. Model performance

Classification of e-health and not e-health publications has been attempted previously using more traditional approaches [7]. These data processing pipelines consist of two major steps: transformation of free text into numeric representations (for instance, TF-IDF) that are later used for training machine learning models. This approach is rather simple computationally, however, does not take language semantics into account. Advanced NLP models, such as BERT can address language specifics much better, however, are computationally intensive and require more data to deliver satisfactory performance. Even though pretrained BERT models are less dependent on the amount of data (they have learned language representation during pretraining), generating sufficient amount of data for model finetuning could be problematic in some fields. In our case, finetuning BERT for 2-class classification showed performance increase with regards to the baseline model based on TF-IDF [7]. However, finetuning BERT for the 6-class classification resulted in poor performance, indicating data insufficiency. The traditional NLP model based on TF-IDF features and Naive Bayes classifier performed better for this task.

C. Limitations

Accuracy of the machine learning model has to be taken into consideration when looking at the absolute numbers of e-health publications. Even though the model generalizes reasonably well, some performance decline is observable in the PubMed dataset (Table 1). It is caused by the different class label distribution in the e-health publication dataset used for model finetuning and PubMed dataset used for validation. While the absolute numbers presented in Figure 1 should be interpreted with caution, the identified publishing trend is not affected by these discrepancies.

Due to the difficulties of accessing other research publication databases, only papers indexed in PubMed were included in this study. Unfortunately, PubMed lacks publications indexed elsewhere or only published in non-indexed conference proceedings. However, it is fair to assume that the major part of e-health related publications is available in PubMed. E-health is often considered as an intersection of health, technology and social sciences. Focus on health topics makes PubMed the preferred database for e-health publications.

D. Future work

In this proof-of-concept phase, automatic data collection was implemented using PubMed API for querying relevant publication metadata. Even though the number of publications in PubMed shows general trends in scientific paper production, this number could be considered misleading in terms of absolute publication count. Previous research shows that some e-health publications are published in channels that are not indexed by PubMed [7]. Other major research databases were not included in data collection due their limitations and costs associated with consumption of metadata APIs. Inclusion of publication data from Scopus and Web of Science databases is planned for the future.

Publications from other countries could contribute to the insights delivered by this system by contrasting global e-health research and development trends. Even though the 6-class model is optimized for the Norwegian context, it could be useful for other countries. Replacing the 6-class model with another one, addressing specific use case requirements better (for instance, focus areas of e-health implementation strategy in other countries) is straightforward.

V. Conclusion

National strategies are complex and often generic documents that are difficult to map to the scientific development in a field. Academic publishing could be used as a proxy hinting to the maturity of a specific field and a direction it may take in the future. In this paper, we presented how novel data science methods could be leveraged to map production of research papers into focus areas of the Norwegian E-health Strategy. This mapping shows the coverage of various focus areas in the e-health strategy by scientific publications and provides insights to the decision makers about underresearched topics.

References

[1] WHO Global Observatory for eHealth and World Health Organization, Global diffusion of eHealth: making universal health coverage achievable : report of the third global survey on eHealth. 2016.

[2] Direktoratet for E-helse, "Utredning av «Én innbygger – én journal»," [Report "One Citizen - One Health Record"] Dec. 2015. [Online]. Available from: http://www.webcitation.org/73nstSwCb. [Accessed: 09-Nov-2018].

[3] Direktoratet for E-helse, "Nasjonal e-helsestrategi 2017-2022," [National e-health strategy 2017-2022] [Online]. Available from: http://www.webcitation.org/73ntEa1LR. [Accessed: 09-Nov- 2018]

[4] Direktoratet for E-helse, "Nasjonal handlingsplan for e-helse 2017-2022," [National action plan for e-health 2017- 2022] [Online]. Available from: http://www.webcitation.org/73ntLCceo. [Accessed: 09-Nov-2018].

[5] G. Eysenbach, "What is e-health?," J. Med. Internet Res., vol. 3, no. 2, p. e20, 2001, doi: 10.2196/jmir.3.2.e20.

[6] C. W. Belter, "Bibliometric indicators: opportunities and limits," J. Med. Libr. Assoc. JMLA, vol. 103, no. 4, pp. 219–221, Oct. 2015, doi: 10.3163/1536-5050.103.4.014.

[7] A. Budrionis, K. Lind, I. M. Holm, O. Saadatfard, and R. Pedersen, "Establishing Baseline in the Status of E-health Research in Norway," presented at the eTELEMED 2019, The Eleventh International Conference on eHealth, Telemedicine, and Social Medicine, Feb. 2019, pp. 85–89, Accessed: Mar. 01, 2019. [Online]. Available: https://www.thinkmind.org/index.php?view=article&articleid=etelemed_2019_5_40_40070.

[8] J. Devlin, M.-W. Chang, K. Lee, and K. Toutanova, "BERT: Pre-training of Deep Bidirectional Transformers for Language Understanding," ArXiv181004805 Cs, May 2019, Accessed: Apr. 28, 2020. [Online]. Available: http://arxiv.org/abs/1810.04805.

Comparison of Bed-Sensors for Nocturnal Behaviour Assessment

Bruce Wallace[1,2,3], Frank Knoefel[1,2,3,4,5], Rafik Goubran[1,2,3], Jeffrey Kaye[6,7], Neil Thomas[2,3,4,5]
[1]Systems and Computer Engineering,
Carleton University, Ottawa, Canada;
[2]Bruyere Research Institute,
Ottawa, Canada;
[3]AGE-WELL SAM3 Innovation Hub,
Ottawa, Canada;
[4]Bruyere Continuing Care,
Ottawa, Canada;
[5]Faculty of Medicine,
University of Ottawa, Ottawa, Canada;
[6]NIA-Layton Aging & Alzheimer's Disease Center,
Portland, OR, USA;
[7]Oregon Center for Aging & Technology (ORCATECH),
Portland, OR, USA;
Emails: wally@sce.carleton.ca, fknoefel@bruyere.org, goubran@sce.carleton.ca,
kaye@ohsu.edu, nthomas@bruyere.org

Abstract— **Most older adults wish to age in place. Numerous home-based monitoring technologies are being developed to help provide solutions to the human resource challenge created by the aging population. Bed-based monitoring systems can measure something as simple as frequency of bed exits and as complex as sleep quality. This work compares the abilities of three available consumer bed mats from Best Buy, Withings and Emfit to identify bed exit, sleep time and sleep quality. The results show that the Best Buy Canada Assured Living platform provides the most accurate measurement of bed exit/entry with only a single error over 54 nights of testing. The estimates of sleep quality, based on sleep scores derived using commercial algorithms, did not align well between the Withing and Emfit sensors. However, these mats provide more detailed information on estimated time asleep and sleep state compared to the Best Buy sensor. Analysis of the raw data provided by the sensors may provide a more useful method to assess sleep related outcome measures in the research setting and to eventually guide clinical decision making.**

Keywords- IoT; cloud processing; bed-sensing; pressure sensing; aging-in-place.

I. INTRODUCTION

Aging in place and independence for aging adults is important for their wellbeing and also to avoid the costs and demands that an alternative of communal care [1] places on the healthcare system or the family. For older adults to continue to live independently, they are frequently supported by family care-givers [1] and this represents a burden affecting the care-giver especially when care is provided by an elderly partner.

The potential for sensors, smart homes and Internet of Things (IoT) technologies to provide an alternative source of information to support the aging adult, their care-giver or formal care providers is an emerging area of research including proposals for systems to assist in the ongoing assessment [2]–[5] of the aging adult.

The sensors used within these systems include wearables [6][7] and passive sensing systems for assessment of very specific behaviours such as urination [8]. IoT sensors within a residence can now be easily connected to cloud services [9][10] and the advanced processing of the combined information from many sensors can then lead to insight into behaviours and behavioural changes by the aging adult [11] [12].

Sleep is important to the well-being of older adults, and changes in sleep pattern can be associated with numerous health conditions. For instance, Alzheimer's disease can lead to disturbances in the circadian rhythm [12]. Nocturia can be associated with a number of clinical conditions and the resulting impact on sleep can be significant. Medical issues as simple as an infection can lead to frequent need urinate. Respiration related illness can also lead to sleep interruption [13] while sleep quality [14], often assessed by time spent in various sleep states, is also important.

Bed sensing in conjunction with smart home technology has been applied and trialed for a supportive smart home system to support care-givers of persons with dementia specifically associated with the challenge of nighttime disorientation to time and/or place that can lead to wandering and elopement [15][16]. These works specifically used a pressure sensitive mat (Ideal Security Pressure Mat SK630) located between the mattress and box spring. This mat was identified as the source of many errors associated with incorrect reporting of bed occupancy, such as detection of status for a specific occupant in a double occupancy bed [16].

This report explores alternative bed mat sensors that have recently entered the consumer market to evaluate their performance to accurately detect overnight bed exits for a specific subject within a double occupancy bed, assess time in bed, and quality of sleep (if provided by the manufacturer). The Ideal Security Pressure Mat was not included in this assessment as two different versions of this mat with different sensitivities were tested in the study bed and they were either closed by the mattress weight alone (indicating someone always in bed) or never closed continuing some of the issues identified in the previous works [15][16].

This paper presents the methodology for the data collection and study in Section II, Section III presents the results from the bed sensors within the study with Section IV discussing the results and potential areas of further exploration.

II. METHOD

A. Research Subjects

The participants in the project are summarized in Table I and included a spousal pair. The research project had research ethics approved by the Bruyère Continuing Care and Carleton University Research Ethics Boards. Both of the participants are adults with post graduate education and no known issues associated with general, physical or cognitive health.

TABLE I. DEMOGRAPHIC INFORMATION FOR THE TWO BED OCCUPANTS. SUBJECT 1 SLEPT ON THE SENSORS WHILE SUBJECT 2 WAS IN THE BED ADJACENT TO THE SENSORS.

Subject	*Gender*	*Age (yr)*	*Height (m)*	*Weight (kg)*
1	M	55	1.85	85
2	F	50	1.65	60

B. Sensors Systems and Set-up

The sensors were placed in the queen sized (North American sizing) bed as shown in Figure 1. Each of the sensors was placed on subject 1's side of the bed in the positions shown in the figure. Each sensor was placed according to the respective installation instructions between the box spring and mattress. The three sensors had differing placement instructions allowing for them to be placed side by side without overlap. The selected sensors included two sensors used by the team in current trials (E) and (W) and a sensor just entering the market (B). There is a practical limit on the number of sensors that can be used by one subject without them overlapping or being installed in an incorrect location.

The Withings Sleep (sensor W) [17], is a 62 x 18.8 x 0.5 cm sensor that is placed under the chest area of the subject. The sensor consists of a single air bladder that occupies the full sensor which is then connected to a pneumatic sensor. The sensor inflates and when placed under the mattress, assesses the variations in the pressure in air chamber associated with the movements of the subject.

The Emfit QS (sensor E) [18], is a 56 x 6 x 0.5 cm sensor that uses a proprietary plastic technology to create a pressure sensitive capacitor and from this assesses the subjects motion through ballistocardiography. This sensor is placed at the upper chest region.

The bed sensor mat from Best Buy Canada's Assured Living solution (sensor B) is a 76cm x 28cm x 1cm sensor from Telehealth Sensors [19] that is provided as a component of the Assured Living platform and the sensor is located near the hip region. This sensor is also based on a pressure sensitive capacitor but the unit includes post processing to determine an adaptive threshold leading to a binary output of occupied or not for the bed. The Assured Living platform also includes other sensors such as motion sensors that were placed throughout the house. For this work, the motion sensors located in the bathroom and a flight of stairs are reported as these represent the two potential directions for an overnight bed exit.

The sensors were each placed in the bed per their respective installation instructions and each was connected through the residential Wi-Fi network to the Internet. Sensor B, uses the Z-Wave wireless protocol to connect to a base station that then connects to cloud based Internet services within the Assured Living portal for data logging and review. The W and E sensors directly connect to the residential Wi-Fi network and through that to cloud based services and portals provided by each of the vendors.

The B sensor system was not available at the start of the study and was introduced 21 days into the study leading to 75 consecutive nights of study for the E and W sensors and 54 nights for the B sensor. The data presented represents a sample of convenience consisting of all nights that had all data available including API portal data allowing for direct comparison between the sensors.

C. Data Collection and Analysis

This work reports on the first 75 consecutive nights within a longitudinal study. During the study, the research subject (Subject 1 – author BW) maintained a paper log for their bed entry and exit times and also the times associated with any overnight exits from the bed. The subject also made note of additional details associated with each night such as the perceived sleep quality based on a simple rating of poor, average or above average.

Each of the sensors independently captured their information within respective cloud-based services provided by each of the sensor vendors. The data for each sensor is stored within the respective vendor's cloud and the sleep summary is available each morning. The E sensor includes a real-time monitoring tool within its portal. The cloud systems for each of the sensors presents that sensor information to the research subject. An end-user, such as the aging adult, their family or a health care provider, would access the data in this manner. The systems had options for the download of raw data that were not readily available to a typical end user. This data was accessed to determine if there were any differences between the raw data and the Web browser presented summary.

The comparison of the times recorded for each of the sensors to the recorded times within the logs was assessed such that if the log time and sensor reported time was within two minutes of each other, they were accepted as the same time to account for any drift or error in the reference time sources for each of the sensors and the log. The actual results indicated that sensor logs were very similar to the paper log reported time and there were no cases of sensor events that were just outside of the two-minute window.

Time asleep, sleep state and sleep quality were assessed by comparing the results reported by the sensors that assessed these attributes and through comparison to the user log notes regarding subjective assessment of sleep quality. As this was a longitudinal study, it was not practical to use physiological sensors to directly assess sleep state and time.

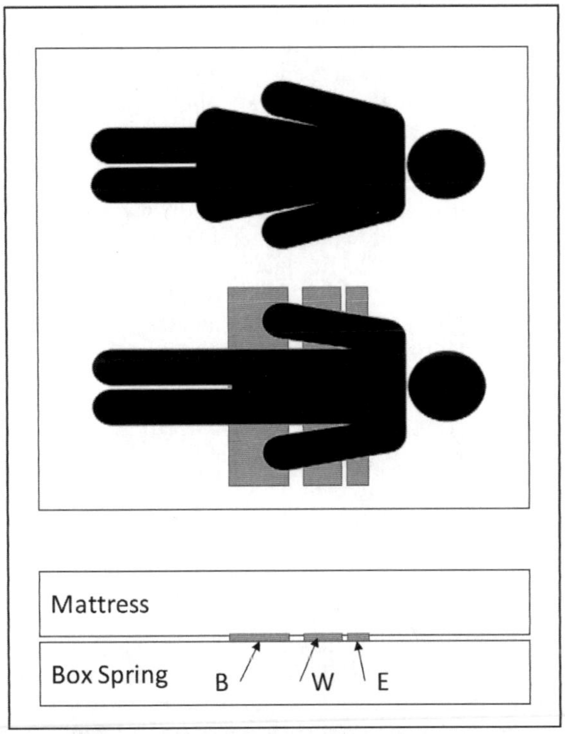

Figure 1. Image showing the placement of the sensors within the bed. Upper portion of the figure shows a top view while the lower portion shows a side view of the bed.

III. RESULTS

The three sensors under evaluation within this longitudinal assessment each provided a measure of presence in bed that would allow assessment of overall time in bed and also for the detection of nocturnal exits during the overnight period. Two of the sensors (W and E) also claim to be able to identify the 4 distinct sleep states (awake, light sleep, deep sleep, REM sleep) purporting to be able to determine time asleep and sleep quality.

A. Measuring Time in-Bed

The performance of the three sensors to assess time in bed is summarized in Table II where the results for the sensors are shown for the 75 consecutive night study period. The W and E sensors were present for the complete study period while the B sensor was added when it arrived and was present for the last 54 days of the study period.

Comparison of the results for the three sensors to the detailed log maintained by the subject show that all three sensors were able to measure the mean Time in Bed with sensor B providing the best performance compared to the research subject logs. Sensor W assessment is lower than the log reference values as this appears to be associated with cropping of the overnight period on some nights where nocturnal exits happened just after the initial entry to bed or just prior to the final bed exit in the morning.

During the study, subject 1 typically went to bed and rose from bed prior to subject 2 and the results for sensor E shown in Table II are higher than the log data for subject 1. The additional time is associated with the sensor reporting a later rise time, which turned out to specifically align with the noted rise time for subject 2.

Table II shows the number of nights that each of the sensors reported the incorrect Start or End time for the night and the errors for each of the W and E sensors are associated with the above described cases. The one error for sensor B (also reflected in sensor E) occurred on one morning where subject 2 was more centered in the bed when subject 1 rose leading to the sensors not detecting this rise. This was the only occurrence of this in the reported period.

TABLE II. PRESENCE IN BED BASED ON TIME OF ENTRY AND TIME OF EXIT COMPARISON OF EVENTS LOGGED BY EACH OF THE SENSORS IN COMPARISON TO A WRITTEN LOG MAINTAINED BY THE RESEARCH SUBJECT

Sensor	*Nights (count)*	*Start Time Error (count)*	*End Time Error (count)*	*Mean Time in Bed (min)*	*St. Dev. Time in Bed (min)*
Log	75	na	na	562.4	39.6
W	75	7	9	550.9	44.2
E	75	4	19	580.2	36.9
B	54	0	1	566.2	39.4

B. Measuring Nocturnal Exits

Each of the sensors detects and reports nocturnal exits for the subject and these are summarized in Table III for all three of the sensors in comparison to a count of exits derived from the log maintained by the research subject. Again, the W and E sensors were used for a 75-day longitudinal period while the B sensors were assessed for the last 54 days of that period. In addition to the B sensor deployed within the bed, motion sensors from the B system were also deployed within the bathroom and flight of stairs. This allowed for a second measure of nocturnal exits as these were either for washroom use or to let a family dog outside during the night.

The results show that B bed sensor detected all of the bed exits during the study period with no errors, the motion sensors from the B system did have two errors that are directly attributable to a failed battery in a motion sensor. Although this issue was identified by the B system for correction by the subject, it was not corrected until the two errors occurred associated within a single overnight period.

The W sensor performed extremely well during the period and it only had 3 missed bed exits (false negatives - FN) while also reporting three exits that did not occur (false positives - FP). The cause of the latter are not known while the cause of the FN appears to also be associated with the cropping of the overnight period noted previously for this sensor.

The results for the E sensor show that this sensor frequently misses bed exits. The bed exits during the study in almost all cases were associated with the bed having double occupancy and so although subject 1 had left the bed, subject 2 remained in the bed.

TABLE III. NOCTURNAL EXIT DETECTION BY THE SYSTEMS OVER THE STUDY PERIOD IN COMPARISON TO THE LOG MAINTAINED BY SUBJECT 1.

Sensor	*Actual Exits (n)*	*Exits Detected (TP)*	*Exits Missed (FN)*	*Extra Exits Reported (FP)*
W	92	89	3	3
E	92	30	62	0
B bed	73	73	0	0
B motion	73	71	2	0

C. *Assessment of Sleep State and Quality*

In addition to being able to assess time in bed and nocturnal bed exits, two of the sensors (W and E) provide information regarding sleep state and time asleep with the results summarized in Table IV. These results show a large difference between the values reported for the two sensors. For instance, sensor W suggested mean time asleep was 7 hours and 17 minutes, while sensor E provided a number of 8 hours and 41 minutes. Causes of this difference include the inclusion of sleep within the score for sensor E from participant 2, such as in example night 1 in Figure 3 vs 4, showing participant 1 as asleep during and around missed overnight exits and lastly missed bed exits resulting in the inclusion of sleep from participant 2 after participant 1 had risen.

Further analysis of these data from the two sensors was performed by downloading the raw data for each sensor and doing subsequent analysis on the raw data in comparison to the known data within the sleep log. Two example nights are reported and for the first night, the results presented within the Web portal for the W and E sensors is shown in Figure 2 while the raw data for the W sensor is shown in Figure 3 and the E sensor is shown in Figure 4. This particular night, subject 1 reported poor sleep and specifically was unable to get to sleep at the start of the night leading to a number of hours of wakefulness, including two bed exits around 1 hour and 2 hours after entry to bed before the initial occurrence of sleep. There was a noted additional overnight bed exit around 6 hours after entry within the sleep period.

TABLE IV. TOTAL SLEEP TIME AND SLEEP STATE TIMES AS REPORTED BY THE W AND E SENSORS.

Sensor	*Nights (n)*	*Mean Time Asleep (min)*	*St. Dev. Time Asleep (min)*	*Mean Light Sleep (min)*	*Mean Deep Sleep (min)*	*Mean REM Sleep (min)*
W	75	437.1	50.5	350.7	64.8	21.5
E	75	520.8	47.3	291.4	92.9	136.6

The results for the W sensor shown in Figure 2, suggest that it incorrectly identified a reentry after one of the bed exits (the one about 2 hours after initial entry to bed) during the wakeful period as the first entry to bed and start on the night. The W sensor does correctly report the subsequent overnight bed exit. The raw data for the W sensor in Figure 4 provides a more accurate portrayal of the night as it does show the correct entry time and the first two bed exits. The documentation for the W sensor notes that it determines and reports the overnight period automatically and this is an example of the effect of this algorithm leading to the portal not including two bed exits and the period of wakefulness not being reported.

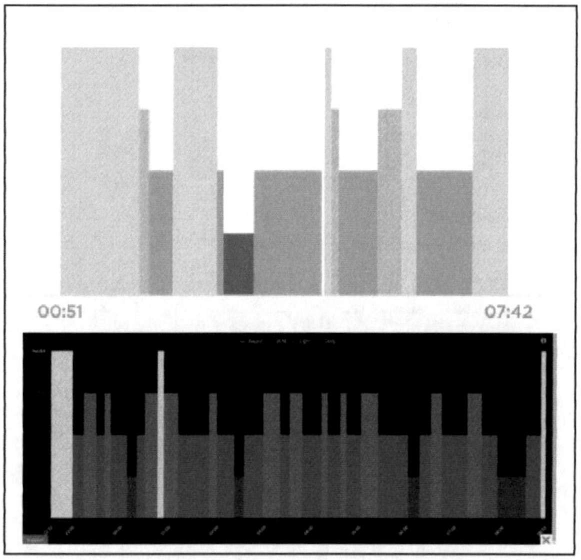

Figure 2. Sleep State information as presented through the user Web portal for the W(upper) and E(lower) sensor example night 1. Legend Upper: grey - awake; cyan - REM, light blue - light sleep, dark blue - deep sleep, white gap – out of bed
Lower: cyan - awake, light purple - REM, medium purple - light sleep, dark purple - deep sleep

The portal results for the E sensor are shown in Figure 2 while Figure 4 shows presents the raw data. The raw data does not differ from the portal presented data and in both cases show the correct entry to bed time. The E sensor appears to show the onset of sleep much earlier in the night than shown by the W sensor and during the actual prolonged wakeful period reported by the subject. This onset of sleep appears to be associated with subject 2 that was adjacent to the sensor while subject 1 was directly on the sensor.

Further analysis for the results for the end of night bed exit time show that the exit time reported for the E sensor matches the time in the subject log while the results for the W sensor match the bed exit time for subject 2. This is the night where it was noted that subject 2 was centered in the bed in the morning leading to the potential for this subject to "seen" by both sensors.

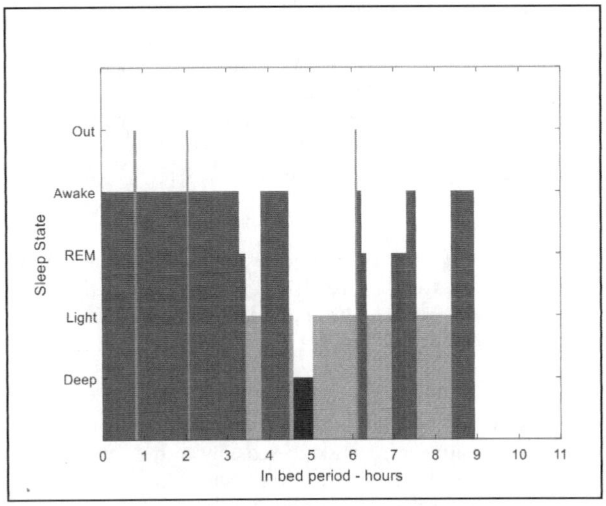

Figure 3. Sleep State reports for an example night per the information available only through detailed raw data download from the portal for the W sensor for example night 1.

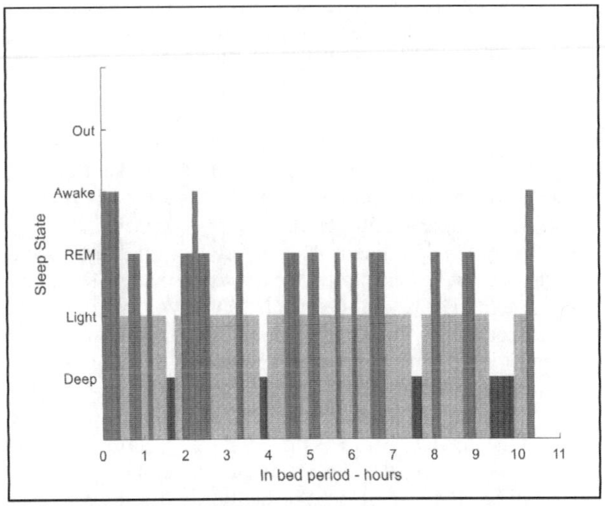

Figure 4. Sleep State reports for an example night per the information available only through detailed raw data download from the portal for the E sensor for example night 1.

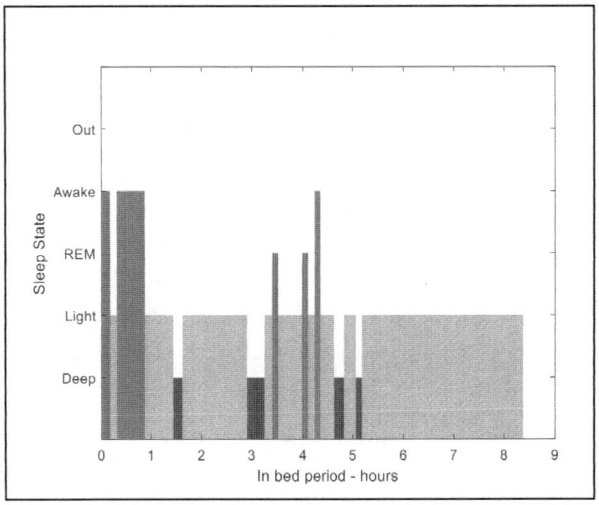

Figure 5. Sleep State reports for an example night per the information available only through detailed raw data download from the portal for the W sensor for example night 2.

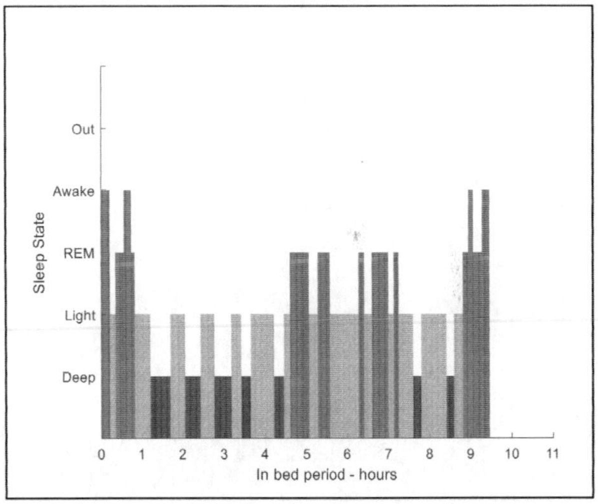

Figure 6. Sleep State reports for an example night per the information available only through detailed raw data download from the portal for the E sensor for example night 2.

Figures 5 and 6 show the raw data for another example night for the two sensors. In this case there is a lot of similarity in the periods reported for the subject being asleep although there is variation in the actual sleep state being reported. In both cases, the sensors correctly reported the start of the night, but the data for the W sensor extends longer than the E sensor. This end of night reporting for the E sensor is correct while again the W sensor results match those for subject 2 that was in a normal position in the bed and not on top of the W sensor in any way, as noted for example night 1.

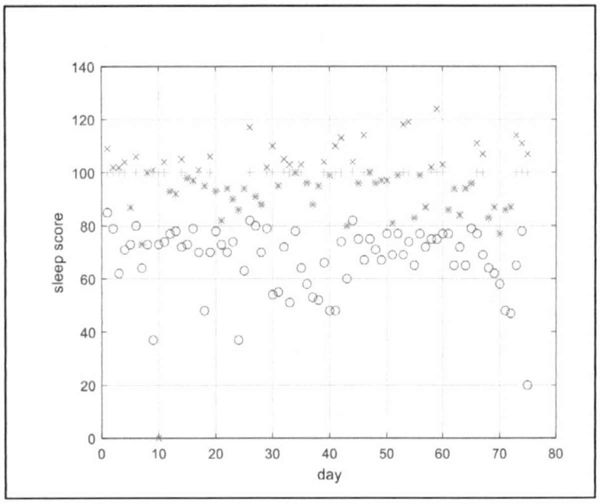

Figure 7. Overnight sleep score are reported by the sensors over the duration of the study. Legend: Blue o – W, Red x – E through API, Orange + – E through Web portal.

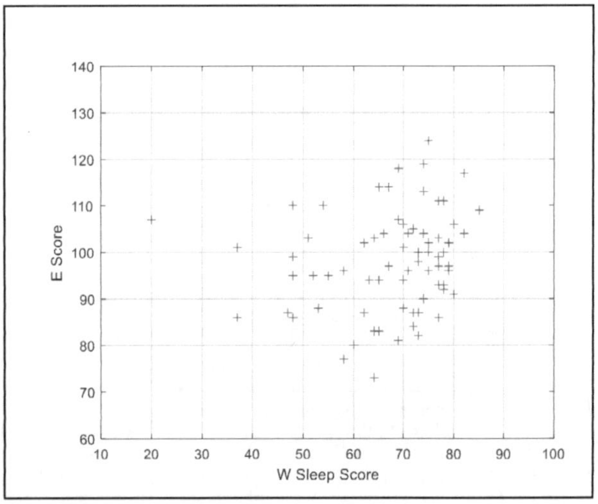

Figure 8. Comparison of the Sleep Score reports for Sensor W in comparison to Sensor E as reported through API for each of the nights within the study.

In addition to measuring sleep state, both the E and W sensors report a sleep score for each night that is shown within the Web portal and available within the raw data. Neither sensor manufacturer provides details on the methodology or algorithm used for this assessment. The sleep score for the E and W sensors is shown in Figure 7. The E sensor data is shown twice as the Web portal presents a sleep score to the user that appears to be capitated at a score of 100 and when the hidden raw data was reviewed through an API, it was found that the scores were no longer capitated at 100, and results up to 124 were found. The minimum reported value was 73, so 100 is now almost the middle of the range and not the maximum.

The comparison of the results between the E and W sensors is shown in Figure 8 where the E sensor data for each night is compared the non-capitated W sensor data. Since these two systems were measuring the sleep quality of the same subject, it should be expected that there would be a high correlation in the results (i.e. a straight line). Figure 7 clearly shows that this is not the case and the correlation coefficient for the best-fit straight line through the data is 0.077, indicating the poor correlation of the values.

The results for the two example nights provide some contrast between the two sensors. The subject reported poor sleep quality on night 1 with above-average sleep quality on night 2. The W sensor correctly scored sleep quality as 20 and 80, respectively. In contrast, sensor E scored sleep quality as 107 and 106, respectively.

IV. DISCUSSION

The challenge of well-being sensing associated with time in bed, nocturnal exits, overall sleep time and quality is of significant interest to care providers that are supporting adults wishing to age in place. As many of these are supported through care partners that are frequently a spouse, the ability to assess the well-being of a specific subject within a shared bed is highly relevant and also presents a significant challenge as passive sensing within the bed has the potential to confuse the two bed occupants.

The results of this work show that the B sensor is able to provide a highly accurate assessment of the time in bed for the specific subject. This sensor greatly improves on the performance of a previous occupancy sensor used by the researchers [15][16]. However, the B sensor by design is limited to only the assessment of bed occupancy.

The E and W sensors provide the potential for significantly more knowledge to be obtained for the overnight periods although each was shown to have limitations. The W sensor was extremely effective at detecting entry and exits from the bed for the research subject with only a few errors but it did report shorter than expected nights on a few occasions. A detailed review of available documentation for this sensor identified that the sensor reports only a single sleep period in each 24 hours and the start and end of the period are not based on first bed entry and last exit. The result is that this sensor will not report on sleep associated with a nap (second sleep period in a day) and can ignore time at the start or end of the night such as removing a period of wakefulness at the start or end of night.

The E sensor performed well for the study and does not include the limitation to a single reporting during any 24-hour period leading it to be better and more appropriate for use within a study where naps or multiple daily sleep periods are expected such as with aging adults. The sensor appears to have had more difficulties associated with confusion between the bed occupants. The differences in sleep stage and sleep scores between the E and W sensors suggests more work needs to be done on these algorithms before assumptions can be made about reliability.

V. Conclusion and Future Work

The goal of this work was to understand the capability of emerging consumer bed sensing devices to provide accurate assessment of time in bed, bed entry and bed exits. The work also explored the sleep quality assessments provided by some of the sensors. The work has shown the potential for passive bed sensing using consumer bed sensors while also showing the challenges that occur within a double occupancy bed. The results show that studies should consider the use of multiple sensor technologies including wearable sensors and perhaps the use of sensors on both sides of a shared bed.

One area for additional exploration is the effect of mattress structure and size on the performance of mats as this structure could have variable effect on any cross detection by the sensors. Future work could include the creation of a fusion based algorithm that can leverage the more accurate assessment of time in bed provided by the B sensor with the more detailed knowledge of sleep related measures provided by the W, E sensors or other sensors located through-out the home to provide better understanding of behaviour, such as the reasons for bed exit leading to potential for treatment if amenable.

Acknowledgment

The work was supported in part by the CABHI Spark program and AGE_WELL NCE. BestBuy Canada Inc provided their Assured Living platform for the study. It was also supported in part by the Natural Sciences and Engineering Research Council (NSERC) Discovery Grant, and by the National Institute on Aging (NIA) grants: P30AG008017; P30AG066518; P30AG024978.

References

[1] D. G. Manuel *et al.*, "Alzheimer's and other dementias in Canada, 2011 to 2031: a microsimulation Population Health Modeling (POHEM) study of projected prevalence, health burden, health services, and caregiving use," *Popul Health Metrics*, vol. 14, no. 1, pp. 1-10, Dec. 2016.

[2] N. W. Thomas *et al.*, "Assessing Everyday Cognition in Older Adults with MCI or Alzheimer's Disease Using a Home-Based Sensing and Computing System," *Alzheimer's & Dementia: The Journal of the Alzheimer's Association*, vol. 14, no. 7, pp. P1466–P1467, Jul. 2018.

[3] S. Casaccia *et al.*, "Assistive sensor-based technology driven self-management for building resilience among people with early stage cognitive impairment," in *2019 IEEE International Symposium on Measurements Networking (M N)*, pp. 1–5, Jul. 2019.

[4] S. Massie, G. Forbes, S. Craw, L. Fraser, and G. Hamilton, "Monitoring Health in Smart Homes using Simple Sensors," *Proceedings of the 3rd International workshop on knowledge discovery in healthcare data co-located with the 27th Intenational joint conference on artificial intelligence and the 23rd European conference on artificial intelligence (IJCAI-ECAI 2018)*, p. 1-6, 2018.

[5] J. Kaye *et al.*, "Methodology for Establishing a Community-Wide Life Laboratory for Capturing Unobtrusive and Continuous Remote Activity and Health Data," *JoVE (Journal of Visualized Experiments)*, no. 137, p. e56942, Jul. 2018.

[6] E. Como and F. Lucca, "Improving the Impact of Wearable Devices in Health Promotion and Wellbeing: the WEHMIX Project," in *eTELEMED 2018: The Tenth International Conference on eHealth, Telemedicine, and Social Medicine*, p. 5, 2018.

[7] R. Wettstein and C. Fegeler, "Evaluation of Machine Learning Algorithms to Detect Irregular Health States in Wearable Sensor Generated Data," in *eTELEMED 2019: The Eleventh International Conference on eHealth, Telemedicine, and Social Medicine*, p. 1-2, 2019.

[8] T. Tsukiyama, "Ambient Monitoring System for Urination," in *eTELEMED 2019: The Eleventh International Conference on eHealth, Telemedicine, and Social Medicine*, p. 1-4, 2019.

[9] Y. Hu, S. Eriksén, and J. Lundberg, "Future Directions of Applying Healthcare Cloud for Home-based Chronic Disease Care," in *eTELEMED 2017: The Ninth International Conference on eHealth, Telemedicine, and Social Medicine*, p. 1-4, 2017.

[10] R. B. Wallace, F. Horsfall, R. Goubran, A. El-Haraki, and F. Knoefel, "The Challenges of Connecting Smart Home Health Sensors to Cloud Analytics," in *2019 IEEE Sensors Applications Symposium (SAS)*, Sophia Antipolis, France, pp. 1–5, Mar. 2019.

[11] A. Akl, B. Chikhaoui, N. Mattek, J. Kaye, D. Austin, and A. Mihailidis, "Clustering home activity distributions for automatic detection of mild cognitive impairment in older adults," *AIS*, vol. 8, no. 4, pp. 437–451, Jul. 2016.

[12] G. L. Sprint, D. J. Cook, and R. Fritz, "Behavioral Differences Between Subject Groups Identified Using Smart Homes and Change Point Detection," *IEEE J. Biomed. Health Inform.*, pp. 1–1, 2020.

[13] E. Dzięciołowska-Baran, A. Gawlikowska-Sroka, and J. Szczurowski, "Diagnosis of Sleep-Disordered Breathing in the Home Environment," *Adv. Exp. Med. Biol.*, pp. 1-6, Feb. 2020.

[14] M. Baran Pouyan, M. Nourani, and M. Pompeo, "Sleep state classification using pressure sensor mats," in *2015 37th Annual International Conference of the IEEE Engineering in Medicine and Biology Society (EMBC)*, Milan, pp. 1207–1210, Aug. 2015.

[15] B. Wallace, T. N. E. Harake, R. Goubran, N. Valech, and F. Knoefel, "Preliminary results for measurement and classification of overnight wandering by dementia patient using multi-sensors," in *2018 IEEE International Instrumentation and Measurement Technology Conference (I2MTC)*, Houston, TX, pp. 1–6, May 2018.

[16] L. Ault, R. Goubran, B. Wallace, H. Lowden, and F. Knoefel, "Smart Home Technology Solution for Nighttime Wandering in Persons with Dementia," *Journal of Rehabilitation and Assistive Technologies Engineering*, pp. 1-8, Jan. 2020.

[17] A. Gruwez, A.-V. Bruyneel, and M. Bruyneel, "The validity of two commercially-available sleep trackers and actigraphy for assessment of sleep parameters in obstructive sleep apnea patients," *PLoS ONE*, vol. 14, no. 1, pp. 1-11, Jan. 2019.

[18] J. M. Perez-Macias, J. Viik, A. Varri, S.-L. Himanen, and M. Tenhunen, "Spectral analysis of snoring events from an Emfit mattress," *Physiol. Meas.*, vol. 37, no. 12, p. 2130, Nov. 2016.

[19] S. J. Hinterlong and M. S. Buck, "Electronic mattress or chair sensor for patient monitoring," US20100228516A1, Sep. 09, 2010.

User Experience Design for Persons Living With Dementia- Current Methods and Experimental Experience

Sabrina Knappe[1], Bruce Wallace[1,2,3], Laura Ault[2,3], Rafik Goubran[1,2,3], Neil Thomas[2,3,4,5], and Frank Knoefel[1,2,3,4,5]

[1]Systems and Computer Engineering, Carleton University, Ottawa, Canada
[2]Bruyère Research Institute, Ottawa, Canada
[3]SAM[3] Innovation Hub, Ottawa, Canada
[4]Bruyère Continuing Care, Ottawa, Canada
[5]Faculty of Medicine, University of Ottawa, Ottawa, Canada
Email: *sabrinaknappe@cunet.carleton.ca*
Email: {*wally, goubran*}*@sce.carleton.ca*
Email: {*lault, nthomas, fknoefel*}*@bruyere.org*

***Abstract*—Persons Living With Dementia (PLWD) present both a challenge and opportunity for ambient monitoring within smart homes. Persons Living With Dementia could benefit greatly from smart home technologies that monitor their well being and enable them to live at home longer. This potential has not been fully realized, partly because of the difficulty in creating interfaces that individuals with dementia can interpret and use. User experience design and user centred design aim to produce technology that is intuitive and accessible to target users. While traditional user experience design methods are difficult to carry out with Persons Living With Dementia, they can be adapted to better suit this population. By involving the families of Persons Living With Dementia in the design process, adapting the prototyping process, and encouraging empathy on the part of designers, smart home technology for ambient monitoring can be developed for Persons Living With Dementia in an inclusive way. The paper presents results from research projects where Persons Living With Dementia and those with less severe mild cognitive impairment have engaged with technology user interfaces. The results show that users were much more successful with simpler interfaces such as a touch screen over mouse and keyboard. They were also more successful with concepts that drew on early learnings. Our results also show that Persons Living With Dementia that experience night time wandering, can be supported with visual and audio cues from a supportive smart home system. This is a new user interface model for interaction with smart home systems that will require the involvement of PLWD in the design process.**

Keywords–Ambient monitoring; Dementia; User Experience Design; Supportive Smart Home.

I. INTRODUCTION

Dementia is a progressive disease of the brain causing cognitive decline and resulting in an impaired performance of activities of daily living. Some of the cognitive domains affected by dementia include memory, language, and abstract thinking. PLWD, especially moderate or severe dementia, require constant supervision, placing a significant burden on caregivers who are often close relatives. PLWD may struggle with independent living as a result of their cognitive difficulties and many caregivers opt to place loved ones in care homes when overwhelmed by the responsibilities associated with caring for an individual with dementia. Remote assessment and support using in-home technology are a potential avenue to allow PLWD to live at home longer by alleviating a caregiver's responsibilities [1][2]. A variety of projects have been undertaken in an effort to integrate home assessment into the lives of PLWD [3][4].

Projects, such as the AGE-WELL SAM[3] (Sensors and Analytics to Monitor Mobility and Memory) collaboration between Bruyère Research Institute and Hospital, Carleton University and AGE-WELL NCE Inc., seek to facilitate life at home for Persons Living With Dementia using technology based solutions. The process of designing supportive ambient assessment solutions for PLWD is challenging due to the cognitive impairments of the target users [5]. The classic User Experience Design (UXD) pipeline includes the end user in most parts of the design process through interviews and testing which are used to gather insights into how to improve a product's usability.

In this paper, we discuss the current best practices for UXD for PLWD according to previous work in the field. In Section II, we discuss current best practices in designing for PLWD. Section II, A covers working with the families of PLWD to facilitate the UXD process. Section II B, covers adaptations that have been made to prototyping methods for use with PLWD. Section II C, covers the role of empathy in designing for PLWD as well as methods that have been used to leverage it. We also outline the challenges and insights we have gathered from past development of user interfaces for PLWD in Section III. In Section III A, we outline an experiment using word search and sudoku games. In Section III B, we cover an experiment where PLWD played whack-a-mole on a digital tablet. In Section III C, we go over an experiment where we had older adults, some with PLWD, use a balance tracker. In Section III D, we discuss a project where we designed and tested a system to reduce nighttime wandering in PLWD. In Section IV we outline a framework for future UXD for continuing work within the AGE-WELL SAM[3] collaboration in section IV. In Section IV A, we list some rules for designing PLWD-friendly interfaces. In Section IV B, we summarize the importance of prototyping and iterative design in designing for PLWD. Finally, we present discuss possibilities for further work in Section V.

II. Current Best Practices

Don Norman, who coined the term User Experience, described the process as "think(ing) through all of the stages of a product or service, from initial intentions through final reflections, from first usage to help, service, and maintenance. Make them all work together seamlessly" [6]. Current best practice methods in user experience include integrating the end user in the design process early and often. Iterative prototyping is one of the most common methods used to involve end users. It consists of the production of multiple prototypes with increasing fidelity, each influenced by the testing of the previous prototype with a sample of end users. For example, an initial prototype of a phone application might consist of a series of paper cutouts representing the various screens of the application. The end user must be able to interpret the paper cutouts for what they represent, which requires abstract thinking.

When working with PLWD, traditional user experience methods must be adapted to better suit the cognitive difficulties this group of end users experiences. The impairment in abstract thinking in PLWD makes traditional prototype testing difficult [7]. In prior work, groups of older adults with low digital literacy and moderate to advanced cognitive impairment were involved in the prototyping of a TV remote [8]. The group of participants in the study represent one of the more impaired subgroups of PLWD. They had "little to no experience with digital technology" [8] and most had an educational level equivalent to 4th grade.

When afflicted with dementia, an individual's memory deteriorates starting with short term memory and eventually affecting more long term memory. Individuals with both significantly progressed dementia and very little education represent a group with significant cognitive impairments. There is emerging evidence that education has some protective effects against dementia, so this group with low education would be doubly challenged, for instance in the development of mental maps. The remote controls the participants designed during the prototyping exercise researchers led "would not have been fully functional, and the participants had difficulties or were unable to describe or explain their designs" [8]. This type of challenge is common when trying to include PLWD in the design process and has led to the adaptation of traditional UXD methods. Though there has been limited work in adapting design methods for PLWD, a few strategies have emerged.

A. *Working with families of PLWD to facilitate UXD*

PLWD often live with their families who act as informal caregivers. The bulk of the caregiving responsibility often falls on the spouse or a child of the PLWD [1][2]. These caregivers are familiar to the PLWD and are familiar with the challenges the PLWD faces. Because of this insight, they have frequently either assisted PLWD in communicating with researchers and designers or acted as a stand-in for the interests of their loved ones. Previous projects have had caregivers in the long term care home where the study was held assist the PLWD with prototyping activities when they were unable to grasp what was required of them [8].

In other work, families were used as units for participation in the design of a board game [9]. The families presented their opinions together, allowing the families to assist the PLWD in communicating their thoughts. Including caregivers when designing for PLWD is useful because they communicate more easily with researchers. However, it is important to recognize the distinction in the experiences of the PLWD and the experiences of the caregiver. While the caregiver may have empathy for the individual with dementia they care for, they cannot have the same perspective. Relying solely on caregivers to inform development can lead to missing important insights on the part of the end user.

B. *Adapting Prototype Methods for PLWD*

Through trial and error, researchers have found ways of adapting traditional prototyping methods to make them accessible to Persons Living With Dementia. The prototypes themselves can be made more accessible by being explicit and higher fidelity in order to reduce the amount of abstract thinking PLWD must engage in. Persons Living With Dementia also can find themselves unable to describe their thoughts and feelings about a prototype to researchers. One study found that in spite of moderate to severe cognitive impairment, PLWD are able to engage in discussion about photographs [8]. In this study, participants were asked to use various paper "components" to make paper prototypes of remotes. The participants found this task difficult and were unable to explain what their prototypes represented. Despite the difficulties PLWD found in creating traditional protoypes in this study, they found discussing photographs natural. While the prototypes in the study were abstracted (the researchers used paper to represent different pieces of a television remote), photographs require very little abstraction on the part of PLWD to make them interpretable. A later study used photographs as a discussion tool for individuals with dementia to express their feelings about their circumstances [10]. PLWD often find it difficult to recall specific words to describe their situation. The photographs helped prompt them and communicate their state of mind to the researchers.

Another method that has been used to aid in vocabulary recall is a modified semantic differential, "a scaling tool which has been used frequently for measuring social attitudes, particularly in the fields of linguistics and social psychology" [11]. Semantic differentials usually involve subjects choosing a number to represent a level of intensity for a particular characteristic. A modified version has been used with PLWD, allowing the individual with dementia to select from a variety of adjectives on cards and place them on scales of 'a little bit', 'rather', or 'very' [9]. This simplified version of the scale was interpretable to the PLWD and every participant was able to use it in spite of varying levels of cognitive impairment. These two methods present themselves as ways to facilitate discussion about prototypes when PLWD find it difficult to produce criticism unprompted.

C. *Empathy as a Design Strategy*

Designers and researchers are often able bodied and neurotypical, and have not had experiences similar to those of PLWD. "Young and healthy design team members often find it difficult to collaborate with users who have different abilities from them and live in difficult situations" [10]. Empathy has been posited as a strategy to bridge this gap in experiences. "In empathic design, designers attempt to get closer to the users' experiences and circumstances" [10]. In empathic co-design, one researcher is able to impart insights from meetings with PLWD to the rest of their team. This is done through a variety of exercises that encourages team members to consider the perspective and experiences of PLWD. One activity explored is a role-play exercise where the design team re-enacts scenes with some team members acting as individuals with dementia. After the exercise there was discussion to encourage participants to relate the role-play to their own experiences, particularly emotions associated with the frustrations of living with dementia.

A project detailing the design of assistive technology for PLWD highlighted the utility of empathy on the part of designers [12]. A few designers had consistent contact with PLWD and deliberated with them over design choices. This relationship fostered empathy in the "point of contact" for the individuals with dementia and a deeper understanding of their life experiences. Empathy lends itself well to personally tailored designs. While not always feasible, tailoring designs specifically for individuals can be empowering for PLWD. Researchers personally tailored two tracking devices to help participants' families locate them should they become lost. One participant remarked "they did not want to have to carry a device that made them 'feel disabled'" [12]. This sentiment reflects the importance of empathy in preserving the personhood and dignity of Persons Living With Dementia.

Another method to help researchers understand the lives of PLWD is the "diary interview method," where participants keep a diary that outlines their daily activities for a period of time. While this method would be impractical for individuals with advanced dementia, it can be an effective way to bridge the gap between the experiences of researchers, designers, and PLWD. One project involved the collection of diary interviews with PLWD involved in dementia activism [13]. In the diaries, participants explained what was and wasn't difficult for them and expressed feelings of frustration and helplessness. They were able to do so at their own pace, encouraged to record their thoughts daily but not required to. The easy-going nature of this method lends itself well to PLWD who become frustrated with their inability to express themselves in conversation. "The diary interview method, when modified, can engage people with dementia as equal partners in the data-gathering process" [13], and gathering qualitative data on the lives of PLWD in this manner could provide researchers with significant insights.

III. Insights from Previous Work

Previous collaborations between Bruyère Research Institute and Hospital and Carleton University have focused on the development of technology to longitudinally assess the cognitive decline of PLWD. While formal User Experience Methods were not used, insights into how PLWD interact with interfaces were gleaned.

A. *Word Search and Sudoku Computer Games*

We designed two simple computer games to measure the cognitive impairment of PLWD. The first game was a word search game, shown in Figure 1. The second was a sudoku game, shown in Figure 2 [14].

Eleven participants played the games over a six month period and their results for speed, performance, errors, and hints used were measured and tabulated. For the word search game, the user interface was shown on a computer screen and users interacted with the game using a mouse and clicking to highlight letters. Some users understood the game easily, likely because of the similarity to physical pen and paper word search games they have experienced for many years. Some participants had difficulty associating the movements they made with their mouse to the movement of the cursor on the screen. Clicking the mouse was not always cognitively associated with the ability to select a letter on the screen. Some additionally felt the need or desire to watch their hand, which led them to ignore the screen.

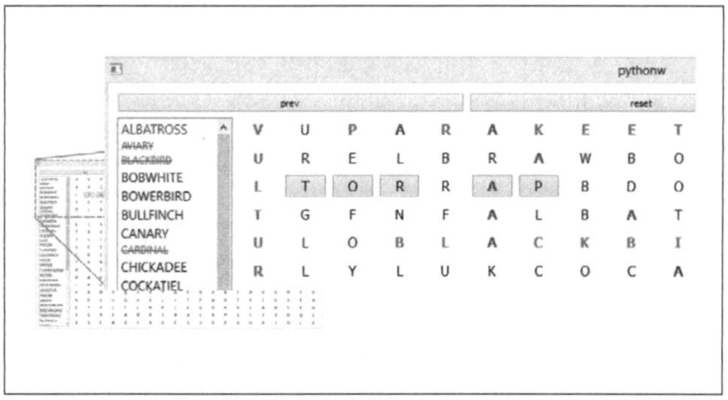

Figure 1. The word search game interface [14].

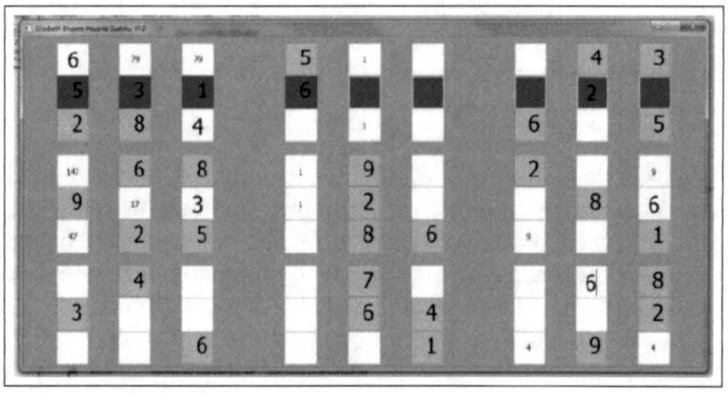

Figure 2. The sudoku game interface [14].

Particpants' prior experience with sudoku was more limited than their experience with word search type games. The visual interface in the game was also dissimilar to what you might encounter in the physical world. As a result, they demonstrated less interest and had less success playing it. In addition, the interface was significantly more complex. Participants not only had to navigate to the cell they wanted to interact with and click on it to engage with it, they also needed to use a keyboard to enter the numerical value they wanted to use and hit enter to confirm the value they chose. In spite of these challenges, some participants were able to successfully play the game. Relabeling keys, such as the "enter" key, helped some participants better understand how to use the keyboard in relation to the game.

B. *Wack-a-mole Tablet Game*

In another study, we presented twelve PLWD with a whack-a-mole game and their performance was measured "to determine the potential for a game-based instrument to provide an indication of cognitive change and to study the suitability of game-based techniques with patients" [15]. Based on our prior work, we chose wack-a-mole, a game that users would be likely to have early life experiences with. We also chose a

Figure 3. The whack-a-mole game interface [15].

touch-screen interface, as users in our previous work had struggled with the mouse and keyboard setup. Visually, the interface was similar to what participants may have seen at fairgrounds in their childhood, with near photographic renderings of the elements, shown in Figure 3.

To add some additional challenge, we added "bunnies" to the game, which participants were supposed to avoid hitting. It has been conventionally accepted that PLWD do not learn new information because of impairments to episodic memory, however, the results of the study indicate that PLWD can learn an interface and a game. Several participants' results improved not only within sessions but also between weekly sessions. "Analysis of the results indicates that they improve their play over the first few weeks and then eventually are able to achieve a steady level of play. This indicates that participants are showing an ability to learn the game," and its interface [15]. The participants even understood that they should hit the moles and avoid hitting the bunnies. The possibility that PLWD can learn an interface over time is a hopeful one, suggesting that even if an interface initially presents a struggle to users with dementia it may eventually be learned.

C. Balance and Cognitive Decline

We have also done work using more traditional digital interfaces not specifically designed for PLWD. In a study done using the Biodex Balance System^{TM}shown in Figure 4, we tracked the balance of older adults [16]. The Biodex Balance System^{TM}was designed with generally physically and mentally well users in mind. For example, it is used to evaluate the balance of high performance athletes after a concussion. The system's interface consists of a plate participants stand on and a screen that presents the participant with information on their body position and motion using a dot representing their centre of pressure overlaid on a target.

Users were positioned so that the dot aligns with the center of the target and then asked not to move and to keep the dot at the center of the target. Because the participants had dementia, "reminding them to keep their feet in exactly the same position [was] not always effective" [16], and they would often shift their position even between the initial positioning and the actual test. This indicates that they may not have made the connection between the movement of the dot on the screen and their shifting of their centre of pressure, similar to how some users playing the word search game did not understand

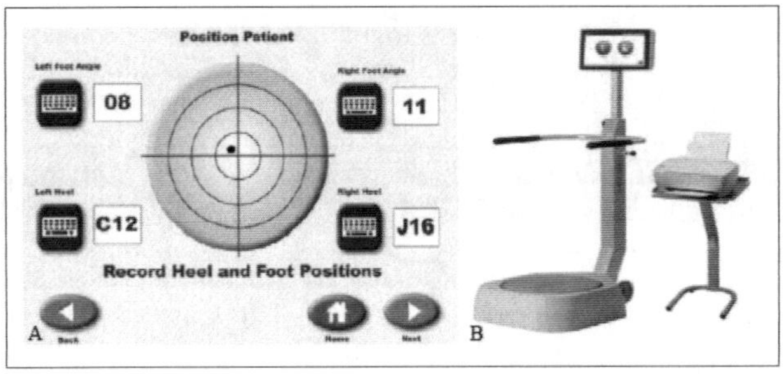

Figure 4. The Biodex Balance System^{TM}[16]

the association between the movement of their mouse and the movement of the cursor on the screen. A modified interface for the balance system may have helped the participants stay in place and resulted in more accurate readings. For instance, outlines of feet on the board may have helped remind PLWD that they were not to move their feet. More research in this area is required to know how best this interface could be adapted.

D. Nighttime Wandering

Our most recent effort has been the design and testing of a system to aid in the reduction of nighttime wandering and its impacts. Nighttime wandering is a common expression of dementia symptoms in those with advanced cognitive decline. It consists of the PLWD awakening at night and finding themselves disoriented, sometimes leading them to leave the home [17][18]. Our smart home solution consisted of a pressure mat embedded in the bed of the participant which is linked to a smart speaker and smart lights. This is a marked break from previous interfaces used in studies by PLWD, which have been much more conventional, using a screen that users interact with. Our interface for this project has no explicit interface for users to interact with, it instead reacts to activities that users might not instinctively understand to be connected to an interface.

When the participant got out of bed at night, the pressure mat in their bed would trigger a series of lights leading to the bathroom, "as bathroom need is typically the cause for wakening during the night hours" [17][18]. If the participants strayed away further into another room an audio message was triggered telling the participant that it is nighttime and they should return to their bed. The message was typically in the voice of caregivers, a choice made after consulting families about what would suit them best. Fifteen individuals participated in the study, dyads consisting of a Person Living With Dementia and a caretaker. Because of the advanced nature of the cognitive decline of individuals who exhibit nighttime wandering behavior, it is possible that they interpreted the message as actually being their caregiver speaking. We found this system effective both in convincing the PLWD to return to bed and in alleviating the stress caregivers felt about the nighttime wandering of their loved ones. Future work could explore how variations in the design of the audio and light interfaces change the efficacity of the system.

IV. Framework for UXD for PLWD

A. "Rules of thumb" for PLWD-Friendly Interfaces

Heuristics for the design of PLWD-Friendly interfaces can be drawn from expert knowledge on the condition, as well as previous work on interfaces from PLWD. Outlines on how user interfaces should be adapted based on expert knowledge of the condition have been created. The initial impacts of dementia are on the cognitive domains such as memory, orientation, and abstract thinking, however more advanced forms of dementia also affect visuo-spatial skills [19]. These outlines emphasize the difficulty in designing "one size fits all" interfaces, and the importance of making solutions that can be scaled to fit the progressive nature of dementia [20]. As their dementia progresses, the individual may need adjustments made to suit their more limited abilities [21]. While any interface designed for PLWD should be user tested, we have compiled a list of "rules of thumb" to use in initial designs for PLWD:

- Large font size
  As a general rule, larger font sizes are easier to read, although "care should be taken that the font is not increased to the point where scrollable windows are required" [19].
- Clear Navigation
  PLWD are easily disoriented and may be lost trying to navigate scrollable windows and other interfaces. The location of the user within the structure of a product should be clear at all times.
- Reduce hidden affordances
  PLWD often have difficulty associating elements with actions that can be taken. For instance, PLWD found using the touch-screen we used for the wack-a-mole game easier than the mouse and keyboard with the word search game, in part because they had difficulty associating the movements of their mouse with the movement of the cursor on the screen. Interfaces should be designed in a way that PLWD understand how they can interact with what they see.
- Accommodate limited motor skills
  Complex motor movements can also be difficult for PLWD, so interfaces that do not require advanced motor skill are preferable. In our work using the Biodex Balance System™we asked participants not to move, which is a motor skill they found difficult. While this may have been due in part to memory, discomfort with maintaining their position may also have played a role [22] [23].
- Avoid the need for information recall
  Problems with short term memory can make information recall difficult for PLWD. Reiterating information across multiple parts of the interface or separating it into smaller, easier to remember chunks, are possible alternatives [20].
- Provide Hints
  In our prior work with PLWD, we found the inclusion of a hint function to be useful to PLWD. Providing them with information when confused is important. PLWD should always have some form of support system to help them orient themselves [20]. The simple hints we provided in the word search game were used, but those given in the sudoku game were largely ignored. It is possible that the display method used in the sudoku game was too complex and so users did not understand that hints were available.
- Avoid Iconography
  PLWD struggle with abstract thinking. Using icons to indicate important information can be problematic because PLWD may not associate it with what it indicates correctly, and may not learn the association over time.

Overall, interfaces for PLWD should be characterizable as "structured," and "simple" [24].

B. *Prototyping and Iterative Design*

The rules above cannot replace testing and iterative design, where designers and engineers engage with stakeholders during the development of a product. The particulars of dementia mean that the best designs for PLWD may be designs that designers find counter-intuitive. These may only be discovered through the testing of a wide variety of possibilities [25]. This necessitates the involvement of PLWD in the iterative design process. The adapted prototyping methods discussed above may be appropriate for some populations with dementia. In the case of severe dementia, teams may need an alternative way to connect with their end user. Teams could consult with less impaired PLWD to gain an understanding of how they view and interact with the world. Alternatively, family caregivers can be consulted or aid PLWD in communicating with researchers, as discussed in Best Practices. Whatever the method, it is crucial that individuals with dementia are involved in the design process when they are the ultimate end users of any piece of technology.

V. Discussion and Conclusion

While it can be difficult, PLWD can be integrated into the UXD process. By using adapted prototyping methods, design heuristics, and involving caregivers in discussions, designers and researchers can gain insights into the lives of PLWD and how the products they are creating can best serve them. Our past work with PLWD has highlighted the difficulties they encounter when interacting with interfaces the cognitively able would consider obvious. Limitations in memory, orientation, and abstract thinking are all barriers for PLWD when trying to use interfaces. The future of the AGE-WELL SAM^3 collaboration will include further work on non-traditional visual user interfaces, interfaces in non-visual modalities such as audio interfaces, and multi-modal interfaces such as the one implemented in our nighttime wandering project. The limited work done in these areas will necessitate further research into how non-traditional interfaces can be made less abstract and more interpretable to PLWD to allow communication of richer and perhaps more complex supportive cues and messages.

Testing interfaces in the prototype stage is crucial to identifying problems PLWD encounter while using them. By using the framework outlined in this paper, we will be able to make more stringent efforts at accessible and adaptable technological solutions for individuals living with dementia.

Acknowledgment

This work was supported in part by the AGE-WELL NCE Inc and National Innovation hub programs (Platform Project Program, AW-PP2019-PP5) and by the Natural Sciences and Engineering Research Council (NSERC) Discovery Grant.

REFERENCES

[1] K. J. Gilhooly et al., "A meta-review of stress, coping and interventions in dementia and dementia caregiving," BMC Geriatrics, vol. 16, no. 1, May 2016, p. 106.

[2] N. Farina et al., "Factors associated with the quality of life of family carers of people with dementia: A systematic review," Alzheimer's & Dementia, vol. 13, no. 5, 2017, pp. 572 – 581.

[3] M. Cash, "Assistive technology and people with dementia," Reviews in Clinical Gerontology; Cambridge, vol. 13, no. 4, Nov. 2003, pp. 313–319.

[4] D. J. Cook, M. Schmitter-Edgecombe, L. Jonsson, and A. V. Morant, "Technology-Enabled Assessment of Functional Health," IEEE Reviews in Biomedical Engineering, vol. 12, 2019, pp. 319–332.

[5] C. Ancient and A. Good, "Considering People Living with Dementia When Designing Interfaces," in Design, User Experience, and Usability. User Experience Design Practice, ser. Lecture Notes in Computer Science, A. Marcus, Ed. Cham: Springer International Publishing, 2014, pp. 113–123.

[6] D. Norman, "The way i see it - systems thinking: a product is more than the product." Interactions, vol. 16, 09 2009, pp. 52–54.

[7] E. Jokisuu, P. M. Langdon, and P. J. Clarkson, "A Framework for Studying Cognitive Impairment to Inform Inclusive Design," in Designing Inclusive Systems, P. Langdon, J. Clarkson, P. Robinson, J. Lazar, and A. Heylighen, Eds. London: Springer, 2012, pp. 115–124.

[8] L. M. Muriana and H. Hornung, "Towards Participatory Prototyping with Older Adults with and Without Cognitive Impairment: Challenges and Lessons Learned," in Human-Computer Interaction - INTERACT 2017, ser. Lecture Notes in Computer Science, R. Bernhaupt, G. Dalvi, A. Joshi, D. K. Balkrishan, J. O'Neill, and M. Winckler, Eds. Cham: Springer International Publishing, 2017, pp. 344–363.

[9] R. M. Branco, J. Quental, and O. Ribeiro, "Personalised participation: an approach to involve people with dementia and their families in a participatory design project," CoDesign, vol. 13, no. 2, Apr. 2017, pp. 127–143.

[10] W. Smeenk, J. Sturm, and B. Eggen, "Empathic handover: how would you feel? Handing over dementia experiences and feelings in empathic co-design," CoDesign, vol. 14, no. 4, Oct. 2018, pp. 259–274.

[11] J. Al-Hindawe, "Considerations when constructing a semantic differential scale," La Trobe papers in linguistics 9, no.7, 1996, pp. 1–9.

[12] S. Lindsay et al., "Empathy, participatory design and people with dementia," in Proceedings of the SIGCHI Conference on Human Factors in Computing Systems, ser. CHI '12. Austin, Texas, USA: Association for Computing Machinery, May 2012, pp. 521–530.

[13] R. Bartlett, "Modifying the Diary Interview Method to Research the Lives of People With Dementia," Qualitative Health Research, vol. 22, no. 12, Dec. 2012, pp. 1717–1726.

[14] B. Wallace, R. Goubran, F. Knoefel, M. Petriu, and A. McAvoy, "Design of games for measurement of cognitive impairment," in IEEE-EMBS International Conference on Biomedical and Health Informatics (BHI). Valencia, Spain: IEEE, Jun. 2014, pp. 117–120.

[15] B. Wallace, F. Knoefel, R. Goubran, P. Masson, A. Baker, B. Allard, V. Guana, and E. Stroulia, "Detecting cognitive ability changes in patients with moderate dementia using a modified "whack-a-mole" game," IEEE Transactions on Instrumentation and Measurement, vol. 67, no. 7, Jul. 2018, pp. 1521–1534.

[16] L. Ault et al., "Preliminary Results from Longitudinal Balance Assessment for Older Adults with Cognitive Decline," in 2019 IEEE International Symposium on Medical Measurements and Applications (MeMeA), Jun. 2019, pp. 1–6.

[17] L. Ault, R. Goubran, B. Wallace, H. Lowden, and F. Knoefel, "Smart home technology solution for night-time wandering in persons with dementia," Journal of Rehabilitation and Assistive Technologies Engineering, vol. 7, 2020, pp. 1–8.

[18] B. Wallace, T. N. E. Harake, R. Goubran, N. Valech, and F. Knoefel, "Preliminary results for measurement and classification of overnight wandering by dementia patient using multi-sensors," in 2018 IEEE International Instrumentation and Measurement Technology Conference (I2MTC), May 2018, pp. 1–6.

[19] C. Ancient and A. Good, "Issues with Designing Dementia-Friendly Interfaces," in HCI International 2013 - Posters' Extended Abstracts, ser. Communications in Computer and Information Science, C. Stephanidis, Ed. Berlin, Heidelberg: Springer, 2013, pp. 192–196.

[20] P. Gregor, A. F. Newell, and M. Zajicek, "Designing for dynamic diversity: Interfaces for older people," in Proceedings of the Fifth International ACM conference on Assistive Technologies, Jul. 2002, pp. 151–156.

[21] A. Astell, Using Technology in Dementia Care : a Guide to Technology Solutions for Everyday Living. London: Jessica Kingsley Publishers, 2019.

[22] A. Kluger et al., "Patterns of motor impairment in normal aging, mild cognitive decline, and early alzheimer'disease," The Journals of Gerontology Series B: Psychological Sciences and Social Sciences, vol. 52, no. 1, 1997, pp. 28–39.

[23] A. Seelye et al., "Computer mouse movement patterns: A potential marker of mild cognitive impairment," Alzheimer's & Dementia: Diagnosis, Assessment & Disease Monitoring, vol. 1, no. 4, 2015, pp. 472–480.

[24] C. Tamanini, M. Majewski, A. Wieland, C. Schlehuber, and F. Kamieth, "Graphical User Interface for an Elderly Person with Dementia," in Constructing Ambient Intelligence, ser. Communications in Computer and Information Science, R. Wichert, K. Van Laerhoven, and J. Gelissen, Eds. Berlin, Heidelberg: Springer, 2012, pp. 157–161.

[25] L.-H. Chen and Y.-C. Liu, "Affordance and Intuitive Interface Design for Elder Users with Dementia," Procedia CIRP, vol. 60, 2017, pp. 470–475.

Spatial Awareness for the Deafblind in Natural Language Presentation using SPIN Rules: A Use Case in the SUITCEYES Platform

Vasileios Kassiano, Thanos G. Stavropoulos, Spiros Nikolopoulos and Ioannis Kompatsiaris
Information Technologies Institute (ITI)
Centre for Research & Technology Hellas (CERTH)
Thessaloniki, Greece
e-mail: {vaskass, athstavr, nikolopo, ikom}@iti.gr

Marina Riga
Department of Mechanical Engineering, Aristotle University of Thessaloniki, Thessaloniki, Greece
e-mail: mriga@isag.meng.auth.gr

Abstract— **This paper presents a rule-based approach towards spatial awareness for the deafblind through natural language constructs. The approach entails two components, a novel ontology for the interoperable representation of data pertaining the domain (objects, space, etc.) and a rule set to derive the natural language constructs for spatial awareness and answer related user queries. The rule set is expressed in SPARQL Inferencing Notation (SPIN), which enables simplicity and flexibility in rule definition as opposed to other frameworks. Both are applied in a use case scenario of the SUITCEYES platform for the deafblind, extending it with the ability to answer spatial awareness queries. More specifically, the ontology component uses rules to provide the users of the platform answers to queries regarding their environment. We present those rules and show how they inform the user of their surroundings, using natural language. Furthermore, we provide a differential population solution to avoid overloading the ontology with unnecessary data.**

Keywords- ontology; rules; natural language; SPIN; deafblindness.

I. INTRODUCTION

Communication with and between users with deafblindness is constrained by the medical nature of this disability, ranging from congenital to acquired deafblindness, including worsening sight or worsening hearing or both over time, plus, ultimately, symptoms of ageing as well.

This paper presets an approach towards spatial awareness for the deafblind using an ontology and a rule set to provide the user with information about their environment expressed in natural language and to dynamically update the spatial information by only keeping the most recent and relevant information provided to the ontology. Also, we apply a use-case of our approach in the SUITCEYES platform [1], by testing it with incoming data from the SUITCEYES system. The purpose of our work is to provide a way to represent spatial context and enable spatial awareness using semantic web technologies and rules to form natural language constructs that can support the deafblind.

The rest of the paper is structured as follows: In Section II we present the most related work with our approach and ontologies regarding natural language. In Section III, we present the ontology that was developed, which extends already existing ontologies. In Section IV, we describe our method for providing spatial context in natural language and in Section V we describe our method for dynamically updating the ontology with differential population. Finally, in Section VI we offer a proof of concept, a use-case performed in the SUITCEYES platform.

II. RELATED WORK

More and more Internet of Things (IoT) applications are used for healthcare purposes due to their high interoperability and expressiveness [2][3][4]. These types of applications acquire knowledge from multiple sources and from continuous and heterogeneous data flows [5][6]. Semantic technologies provide comprehensive tools and methods for representing knowledge and producing new ones. IoT environments are increasingly found in home healthcare technologies in actions that create better living conditions for the elderly, through the use of IoT technologies, such as Active and Healthy Ageing (AHA) and Home Ambient Assisted Living (AAL).

Furthermore, in the case of deafblind people, the simple and accurate representation of their surrounding environment is one of the most basic needs for their quality of life. This can be achieved by providing the nature of their surroundings in natural language. Some of the most relevant works for language processing with ontologies are [7][8].

KnowSense [3] is an activity monitoring system for elderly with dementia, deployed in controlled and diffuse environments. Semantic Web technologies, such as OWL 2, are widely used in KnowSense to display sensor and specific application observations, as well as to implement solutions for identifying activities and identifying problems in everyday life activities with the aim of clinical evaluation in various stages of dementia. Description Logic Reasoning (DL reasoning) for activity detection and SPARQL questions are used to extract clinical problems.

ACTIVAGE [4] is a large-scale pilot project, with the purpose of developing Smart Living solutions that strengthens active and healthy aging. The ACTIVAGE IoT Ecosystem Suite (AIOTES) project, contains a set of techniques, tools, and methodologies (rule-based reasoning, interoperable ontologies, etc.) that increases semantic interoperability at different levels between heterogeneous IoT platforms. The approach uses multiple mechanisms of reasoning that can improve the understanding of patients' heterogeneous data and help generate new knowledge by providing services to end users.

Dem@Care [4] is a system based on heterogenous sensors that provides support for independent living for elder people with dementia or similar health problems. This approach incorporates a heterogeneous set of detection methods and technologies, including video, audio, in addition to normal, environmental, and other measurements. Semantic technologies (e.g., rule-based reasoning) are used to process and analyze sensor data according to user requirements. This leads to feedback and decision support, which is communicated to end users through appropriately designed user interfaces. The support includes various clinical scenarios, both short (trials in hospital settings) and long term (daily living at work), for independent living.

In [9], a system for healthcare in Smart Home environments is developed, which considers social relationship-based contexts to provide a fully personalized healthcare service.

An ontology-based sensor selection for real-world wearable activity recognition is presented in [10], in which the use of ontologies is proposed to thoroughly describe the wearable sensors available for the activity recognition process. This enables the semantic selection of sensors to support a continuity of recognition.

In [7], an extended version of a linguistic ontology is presented that works particularly with space. Language regarding space, spatial relationships and actions in space is covered and an ontological structure that relates such expressions with ontology classes is developed. Finally, examples of the ontology's results based on natural language examples are presented.

In [8], a project in which ontologies are part of the reasoning process used for information management and for the presentation of information is presented. Both accessing and presenting information are mediated via natural language and the ontologies are coupled with the lexicon used in the natural language component. This work, as well as [7], is related with the natural language aspect of our approach.

An approach to transform natural language sentences into SPARQL is proposed in [10], with the use of background knowledge from ontologies and lexicons. The results of this approach show that the diagnosis process and the data search for a broad range of users is improved.

In [11], an evaluation is made on how efficient the SPARQL query language is and the SPARQL Inferencing Notation (SPIN) when utilized to identify data quality problems in Semantic Web data automatically, and within the Semantic Web technology stack.

In the case of deafblind patients, the simple and accurate representation of their environment is one of the most important needs. In [7][8][9] even though forms of natural language processing through ontologies are proposed, they do not involve healthcare or wearable sensors. For the healthcare related work [2][3][4] and [9][12], the natural language presentation of information component is absent. In the SUITCEYES project, we combine techniques involving both semantic technologies for healthcare and representation of the environment of the deafblind patients using natural language. The SUITCEYES ontology [13] includes concepts for spaces, e.g., rooms, halls, stairways etc., and entities found in them, e.g., objects and persons. In this work, we reuse and extend this ontology with classes and properties, and use rules expressed in SPARQL and SPIN notation, to achieve spatial awareness with natural language constructs for the deafblind.

III. THE PROPOSED FRAMEWORK

A. The Proposed Ontology

1) Ontology Components

The SUITCEYES ontology [13] is extended for the purposes of this work can be found online [14]. One of its aims is to integrate heterogeneous, multimodal input from different sensors in a formal and semantically enriched basis, and thus to combine user's context-related information so as to provide enhanced situational awareness that can potentially augment users' navigation and communication capabilities. The ontology was developed to augment its semantic interoperability with other ontologies that exist in the domains of interest and to enrich its semantic representation capabilities for covering the additional concepts and functional requirements that may emerge through any system addressed to deafblind people.

In ontology engineering, it is common practice to reuse existing third-party models and vocabularies during the development of a custom ontology. This approach was also followed here, including the adoption of third-party vocabularies in order to rely on previously used and validated ontologies.

The semantic representation of objects and activities from the Dem@Care ontology [4] was adopted, which contains a set of descriptions of every-day activities and common objects used in an every-day context that are highly relevant to our goals. Moreover, the ontology is using SOSA/SSN [15] ontologies for representing sensors and the respective observations. The Friend-Of-A-Friend (FOAF) specification [16] is used for representing persons and social associations. Finally, the Smart Energy Aware Systems (SEAS) Building Ontology [17] was integrated, which is a schema for describing the core topological concepts of a building, such as buildings, building spaces and rooms. Figure 1, shows the basic classes imported from existing ontologies for interoperability.

2) Ontology Concepts

Some of the basic classes of the ontology represent objects, spaces and people that can be detected as raw data form sensors, such as cameras, or processed data, through a visual analysis component. For example, some of the objects that can be found in the ontology are computer, laptop, alarm clock, mug, table, chair and other everyday objects and some of the spaces include bedroom, bathroom, living room and other spaces that can usually be found in a home environment. We extend these concepts by adding further objects, spaces and rooms that could be useful to any deafblind user. In Figure 2, the main object entities of the combined ontologies are presented, while in Figure 3, the main spaces, such as rooms, are presented.

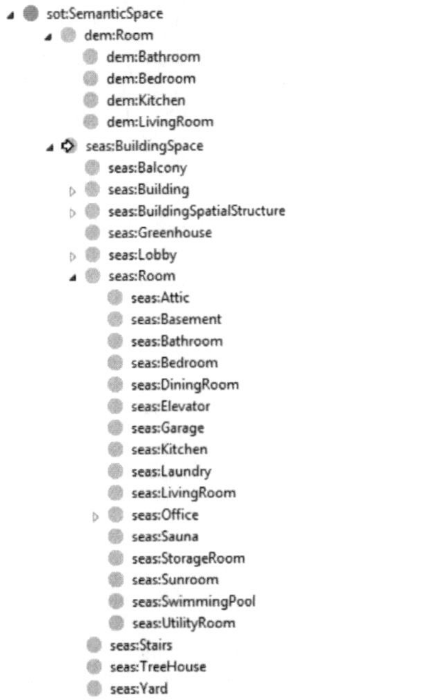

Figure 1. Classes imported from the Dem@Care and SEAS ontology for interoperability

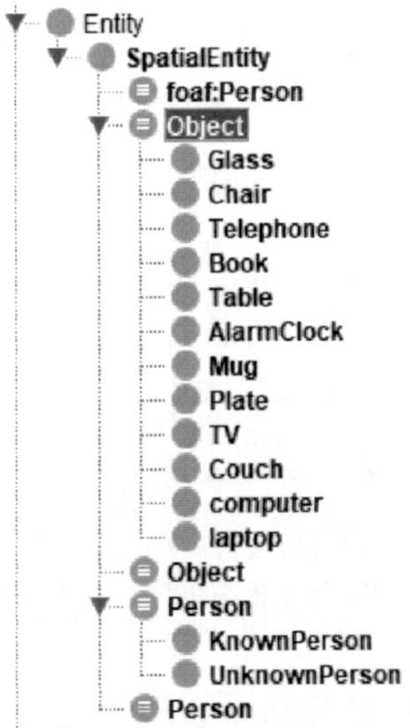

Figure 2.Object Class of the ontology

3) *Spatial Relations*

Topological relations of geometric objects have been widely described in literature and are generally utilized for navigation-, location- and context-based services. More specifically, the Egenhofer relations [18] or the DE-9IM topological model [19] can be used to specify how an object is located in space in relation to some reference object. For any two spatial objects, which can be points, lines and/or polygonal areas (represented by the definition of a bounding box), there are 9 relations derived from the model, which are:

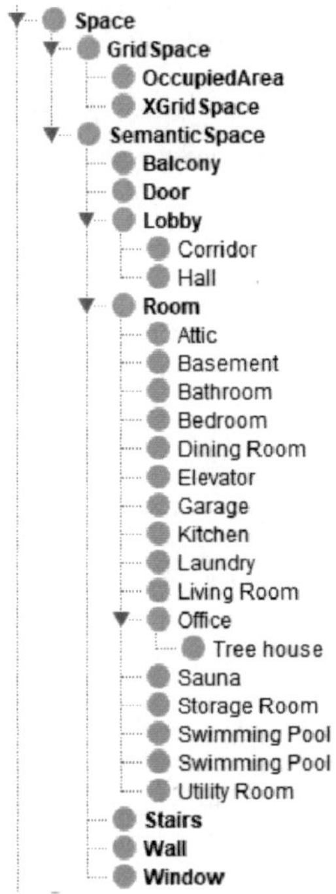

Figure 3. Space Class of the ontology

equals, disjoint, intersects, touches, contains, convers, covered by, and within. Distinct specializations of topological relations also exist, such as the so called alignment relations (horizontally aligned or vertically aligned) and orientation relations (left of, right of, top of and bottom of); these are considered in literature as mereology and parthood relations, as described in detail in [20] and visualized in Figure 4, from different perspectives (object-centered vs observer-centered).

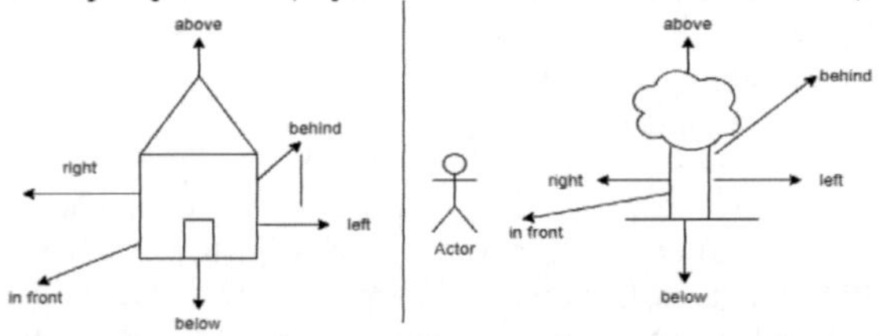

Figure 4. Object centered (left) and observer-centered (right) frames of reference

We focus on specific spatial relations that have to do with the orientation (left/right), existence (in a room) and the distance (far/close/immediate). Thus, in the ontology, an entity that occupies space (e.g., persons, objects) is considered as a SpatialEntity and the occupied space (e.g., a room or a location) belongs to the SemanticSpace representation. These two aspects formulate the respective entity's Spatial Context, which provides information regarding the entity's relationship to the semantic space it is located in. Examples include: in, on, left, right, far, close, etc. The aforementioned concepts are depicted in Figure 7.

B. *Spatial Information Presented in Natural Language*

In Table 1, we present a list of indicative queries that are used in the ontology to provide the user with a natural

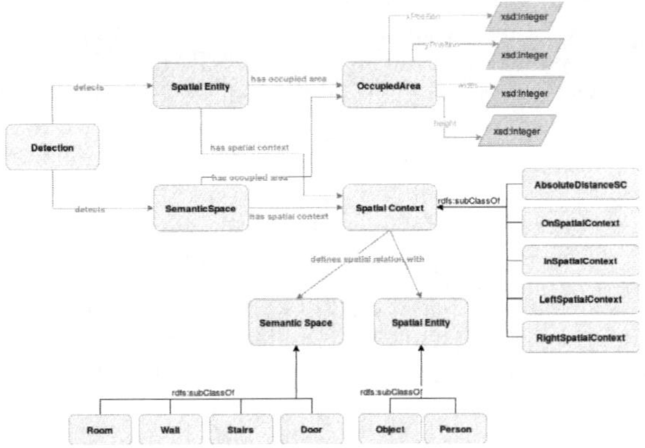

Figure 7. Extended Semantic Spaces and Spatial Contexts of the SUITCEYES ontology

language output regarding spatial information of their surroundings. In these queries, variables are used to cover a broad number of objects and spatial contexts. This kind of inference is achieved by using a set of ontological rules, written in SPARQL/SPIN notation, that run on top of the ontology, whenever a specific query is triggered by the user. Within the context of this scenario, a list of indicative queries have been created that can dynamically change on specific aspects, i.e., to cover different entities of interest, different spatial relations (with respect to the distance of position left/right), etc.

TABLE 1. LIST OF RULES

#	Query	Nat. Lang. Output
1	Where is my <object >?	Your <object > is on your <right/left spatial context> side, <close/far spatial context> to/from you.
2	How many <objects> are on my <right/left spatial context> side?	<# counted> objects, an <object1> and an <object 2> are on your <right/left spatial context> side.
3	Which <objects> are <spatial context> to/from me?	An <object1>, <object2> and object3> are located <spatial context> to/from you.

The variables in the above queries are given a specific value, depending on what the user wants to ask. For example, the first query can be transformed in natural language to: "Where is my laptop" and its output can be: "Your laptop is on your right side, close to you". The implementation of this query is presented as an ontological rule written in SPARQL/SPIN syntax in Figure 5, using synthetic data that we created that include various objects and spatial contexts. Most of our ontological rules use SPARQL CONSTRUCT and DELETE/INSERT commands, in order to create new triples in the ontology and thus enrich the knowledge stored in the schema.

In Figure 6, the implementation of the #3 query is presented, which in natural language translates to "Which objects are close to me?", which using our synthetic data produces the output: "A laptop, a TV and a chair are located close to you". For this query, we have skipped the construct rule, which is the same as the #1 query.

```
CONSTRUCT{
        ?detection sot:producesOutput ?output.
        ?output sot:refersToDetection ?detection.
        ?output a sot:Output.
        ?output sot:hasTextualDescription ?description
}
WHERE{
BIND(sospin:Function_SelectLatestObjectDetection() AS ?detection).
?detection a sot:Detection.
?detection sot:detectsObject ?object.
?object a ?objectType.
FILTER(?objectType = sot:Laptop).
?object rdfs:label ?objectName.
?object sot:hasSpatialContext ?spatialContext.
?spatialContext a ?spatialContextType.
?spatialContextType rdfs:subClassOf sot:spatialContext.
FILTER((?spatialContextType = sot:RightSpatialContext) ||
(?spatialContextType = sot:LeftSpatialContext)).
BIND(sospin:Function_leftRightContext(?spatialContext) AS
?leftright_annotation).
FILTER((?spatialContextType = sot:CloseSpatialContext) ||
(?spatialContextType = sot:FarSpatialContext).
BIND(sospin:Function_CloseFarContext(?spatialContext) AS
?closefar_annotation).
BIND (BNODE() AS ?output).
BIND(CONCAT(?objectName, "is on your", ?leftright_annotation, "
side", ?closefar_annotation, " to you") AS ?description).
}
```

Figure 5. SPARQL/SPIN rule for query #1

```
WHERE{
        BIND(sospin:Function_SelectLatestObjectDetection() AS
?detection).
        ?detection a sot:Detection.
        ?detection sot:detectsObject ?object.
        ?detection sot:detectsObject ?object2.
        ?detection sot:detectsObject ?object3.
        ?object a ?objectType.
        ?object2 a ?objectType.
        ?object3 a ?objectType.
        ?object sot:hasSpatialContext ?spatialContext1.
        ?object2 sot:hasSpatialContext ?spatialContext2.
        ?object3 sot:hasSpatialContext ?spatialContext3.
        ?spatialContext a ?spatialContextType.
        ?spatialContext2 a ?spatialContextType2.
        ?spatialContext3 a ?spatialContextType3.
        ?spatialContextType rdfs:subClassOf sot:SpatialContext.
        ?spatialContextType2 rdfs:subClassOf sot:SpatialContext.
        ?spatialContextType3 rdfs:subClassOf sot:SpatialContext.
        FILTER((?SpatialContextType = sot:closeSpatialContext)
        && (?SpatialContextType2 = sot.closeSpatialContext) &&
        (?SpatialContextType3 = sot:closeSpatialContext)).
        BIND(sospin:Function_CloseFarContext(?spatialContext)
                AS ?closefar_annotation).
        BIND (BNODE() AS ?output).
        BIND(CONCAT("A ", ?objectName, " , a ",
                ?objectName2, " and a ", ?objectName3, ", are
                ". ?closefar_annotation, " to you." AS
                ?description).
}
```

Figure 6. SPARQL/SPIN rule for query #3

C. *Dynamic Update of the ontology with Differential Population Procedure*

In our proposed method, we use an efficient approach to store the incoming data to the ontology. We call this approach "differential population" of the ontology, which means that instead of populating the ontology with every Detection type data coming from every message in the message bus, a method is applied that checks if the incoming detection data is already included in the ontology. If it is, then we change only the timestamp of the existing Detection instance, otherwise we add the new Detection to the ontology. This method is implemented in java code and the pseudocode is presented in Figure 8.

```
Procedure: Differential Population
Data: D (Detection type object obtained from message bus)

for each D_i stored in ontology do
  | if D.detects_object == D_i.detects_object and
  | D.spatialContext == D_i.spatialContext then
  |   | D_i.timestamp = D.timestamp;
  | end
end
```

Figure 8. Pseudocode for the Differential Population procedure

By using this technique, we limit the volume of data that are inserted in the ontology by only applying a simple check each time a message arrives from the message bus. This increases efficiency, considering that in a home environment the same objects could be detected in the same place multiple times (e.g., a TV almost never changes place in a home) and that many sensor systems continuously send data via their sensors.

IV. PROOF OF CONCEPT – THE SUITCEYES USE CASE

The SUITCEYES system tries to enrich the spatial awareness of the deafblind by implementing a solution involving a vest with a processing unit that receives raw information from sensors and actuators (mainly cameras and haptograms), and advanced information through a visual analysis component and a semantic component. An important aspect of the platform is the integration of information coming from the environment (via sensors) and from the system's analysis components (camera feed and visual analysis). The most important sensors of the system are static cameras placed in rooms and cameras integrated a vest that the deafblind user wears (dynamic). In this sense, the ontology is primarily focused on semantically representing aspects relevant to the users' context, in order to provide them with enhanced situational awareness and augment their navigation and communication capabilities. More importantly, the ontology also serves as the connection between environmental cues and content communicated to the user via haptograms. The deafblind user receives the output of the ontology via haptic sensors. i.e., a vest that vibrates in specific patterns on the user's back and a special tablet that has haptic capabilities so the user can form patterns to ask questions regarding their environment. These mechanics and translation patterns from text to haptics are outside of the scope of this paper, which focuses on the information models and rules to form the natural language constructs to be translated. To test the correctness of those constructs formed using the rules, we have used synthetic data received by the SUITCEYES system.

We present the output of the SPARQL/SPIN rule presented in Figure 5 by using the Protégé [21] software. In Figure 9, we present the object and description returned by the query. The description will be used as the output to the user.

object	description
Laptop	"Your Laptop is on your right side, close to you."^^<http://www.w3.org/2001/XMLSchema#

Figure 9. Query output of "Where is mylaptop"

In Figure 10 and Figure 11, we present the objects returned from the query "Which objects are close to me" with their spatial context, and the output that the user will get as textualDescription.

obj	spatialContext	obj2	spatialContext2	obj3	spatialContext3
Laptop	Close	Chair	Close	TV	Close

Figure 10. Objects and their Spatial Context from query #3

queryOutput	textualDescription
QueryOutput	"A Laptop, a Chair and a TV are close to you"^^<http://www.w3.org/2001/XMLSchema#string>

Figure 11. Query output of "Which objects are close to me"

By receiving the expected results regarding our synthetic data, we validated the correctness of our queries.

We also used the same synthetic data to test the differential population procedure. In the example below, we assume that a detection of a laptop in the living room exists in our ontology and that a same new detection arrives from the SUITCEYES system with a more recent timestamp. In Figure 12, we present what the outcome would be without using our differential population method, i.e., the storing of 2 similar detections.

detection	obj_detected	where	timestamp
Detection_1	Laptop	Living_Room	"2020-08-25T13:18:30Z"^^<http://www.w3.org/2001/XMLSchema#dateTimeStamp>
Detection_2	Laptop	Living_Room	"2020-08-25T13:20:00Z"^^<http://www.w3.org/2001/XMLSchema#dateTimeStamp>

Figure 12. Detections saved without the differential population procedure

In Figure 13, the outcome of the same incoming detection is presented, but with using the differential population procedure. Here, only the first detection is stored, with its timestamp field changed to match that of the second, incoming detection.

detection	obj_detected	where	timestamp
Detection_1	Laptop	Living_Room	"2020-08-25T13:20:00Z"^^<http://www.w3.org/2001/XMLSchema#dateTimeStamp>

Figure 13. Detections saved with the differential population procedure

This absence of almost identical detections could save a lot of space in real applications that continuously receive and store data.

V. CONCLUSION AND FUTURE WORK

In this paper, our method for providing spatial awareness to people with deafblindness, using natural language and semantic reasoning with SPIN/SPARQL notation was presented. Also, the differential population technique used for updating the data in the ontology was proposed. Finally, we tested our methods using synthetic data coming from the SUITCEYES platform.

With today's technology, people with deafblindness can be provided with advanced tools that enhance their spatial awareness. For our future work, we aim for the integration of our methods with the SUITCEYES project in two phases. The first phase includes the use of real data provided by sensors, such as cameras and the visual analysis component, and the second phase includes deafblind users' interaction with the system.

ACKNOWLEDGMENT

This work was supported by the European Union's Horizon 2020 Research and Innovation Program through the project SUITCEYES under Grant 780814.

REFERENCES

[1] "SUITCEYES Project." https://suitceyes.eu (accessed Sep. 13, 2020).

[2] G. Fico *et al.*, "Co-creating with consumers and stakeholders to understand the benefit of internet of things in smart living environments for ageing well: The approach adopted in the madrid deployment site of the activage large scale pilot," 2017, doi: 10.1007/978-981-10-5122-7_272.

[3] G. Meditskos, T. G. Stavropoulos, S. Andreadis, and I. Kompatsiaris, "KnowSense: A semantically-enabled pervasive framework to assist clinical autonomy assessment," 2015.

[4] T. G. Stavropoulos, G. Meditskos, S. Andreadis, and I. Kompatsiaris, "Dem@Care: Ambient sensing and intelligent decision support for the care of dementia," 2015.

[5] A. Pliatsios, C. Goumopoulos, and K. Kotis, "A Review on IoT Frameworks Supporting Multi-Level Interoperability{\textendash}The Semantic Social Network of Things Framework," *Int. J. Adv. Internet Technol.*, vol. 13, pp. 46–64, 2020, [Online]. Available: https://www.iariajournals.org/internet_technology/inttech_v13_n12_2020_paged.pdf.

[6] A. I. Maarala, X. Su, and J. Riekki, "Semantic Reasoning for Context-Aware Internet of Things Applications," *IEEE Internet Things J.*, 2017, doi: 10.1109/JIOT.2016.2587060.

[7] J. A. Bateman, J. Hois, R. Ross, and T. Tenbrink, "A linguistic ontology of space for natural language processing," *Artif. Intell.*, 2010, doi: 10.1016/j.artint.2010.05.008.

[8] D. Estival, C. Nowak, and A. Zschorn, "Towards ontology-based natural language processing," 2004, doi: 10.3115/1621066.1621075.

[9] H. Lee and J. Kwon, "Ontology model-based situation and socially-aware health care service in a smart home environment," *Int. J. Smart Home*, 2013, doi: 10.14257/ijsh.2013.7.5.24.

[10] M. Sander, U. Waltinger, M. Roshchin, and T. Runkler, "Ontology-based translation of natural language queries to SPARQL," 2014.

[11] C. Fürber and M. Hepp, "Using SPARQL and SPIN for data quality management on the Semantic Web," 2010, doi: 10.1007/978-3-642-12814-1_4.

[12] C. Villalonga, H. Pomares, I. Rojas, and O. Banos, "MIMU-Wear: Ontology-based sensor selection for real-world wearable activity recognition," *Neurocomputing*, 2017, doi: 10.1016/j.neucom.2016.09.125.

[13] S. Darányi, N. Olson, M. Riga, E. Kontopoulos, and I. Kompatsiaris, "Static and Dynamic Haptograms to Communicate Semantic Content : Towards Enabling Face-to-Face Communication for People with Deafblindness," *SEMAPRO 2019, The Thirteenth International Conference on Advances in Semantic Processing, Porto, September 22-26, 2019.* 2019.

[14] "Multimedia Knowledge and Social Media Analytics Laboratory." https://mklab.iti.gr/ (accessed Sep. 21, 2020).

[15] K. Janowicz, A. Haller, S. J. D. Cox, D. Le Phuoc, and M. Lefrançois, "SOSA: A lightweight ontology for sensors, observations, samples, and actuators," *J. Web Semant.*, 2019, doi: 10.1016/j.websem.2018.06.003.

[16] D. Brickley and L. Miller, "FOAF Vocabulary Specification 0.99," *Namespace Document*. 2014.

[17] M. Lefrançois, "Planned ETSI SAREF extensions based on the W3C&OGC SOSA/SSN-compatible SEAS ontology patterns," 2017.

[18] M. J. Egenhofer, "A formal definition of binary topological relationships," 1989, doi: 10.1007/3-540-51295-0_148.

[19] E. Clementini, J. Sharma, and M. J. Egenhofer, "Modelling topological spatial relations: Strategies for query processing," *Comput. Graph.*, 1994, doi: 10.1016/0097-8493(94)90007-8.

[20] A. C. Varzi, "Spatial Reasoning and Ontology: Parts, Wholes, and Locations," in *Handbook of Spatial Logics*, 2007.

[21] "Protege" https://protege/stanford.edu (accessed Sep. 01, 2020).

Analysis on the Impact of GDPR in Healthcare-related Blockchain Solutions and Guidelines for Achieving Compliance

Christos Kontzinos
Decision Support Systems Lab
National Technical University of Athens
Athens, Greece
e-mail: ckon@epu.ntua.gr

Ourania Markaki
Decision Support Systems Lab
National Technical University of Athens
Athens, Greece
e-mail: omarkaki@epu.ntua.gr

Panagiotis Kapsalis
Decision Support Systems Lab
National Technical University of Athens
Athens, Greece
e-mail: pkapsalis@epu.ntua.gr

Dimitris Askounis
Decision Support Systems Lab
National Technical University of Athens
Athens, Greece
e-mail: askous@epu.ntua.gr

Spiros Mouzakitis
Decision Support Systems Lab
National Technical University of Athens
Athens, Greece
e-mail: smouzakitis@epu.ntua.gr

Haralampos Karanikas
Datamed S.A.
Athens, Greece
e-mail: h.karanikas@gmail.com

Michael Kontoulis
Decision Support Systems Lab
National Technical University of Athens
Athens, Greece
e-mail: mkontoulis@epu.ntua.gr

Alexandros Christodoulakis
Datamed S.A.
Athens, Greece
e-mail: alexandroschristodoulakis8@gmail.com

Panagiotis Dimitrakopoulos
Datamed S.A.
Athens, Greece
e-mail: pdimitrakopoulos@datamed.gr

***Abstract*—Blockchain is an emerging technology that offers decentralised data management capabilities in a distributed ledger. While the initial and main domain of application has been and continues to be that of cryptocurrencies and their use in secure, digital economic transactions, blockchain's inherent properties for security, immutability and transparency make it an excellent solution for various domains that deal with personal or even sensitive data such as the domain of healthcare. At the same time, the radical increase of digitisation in healthcare, as well as automation of processes, robotics, and the rise of medical sensors and devices also led to an unprecedented rise in medical data availability. This fact has brought forth citizen and societal concerns regarding the safety of their medical data and the transparency of data transactions that they engage in, especially given that medical data are sensitive. To address these and other concerns, in the European Union, the European Commission established in 2016 that General Data Protection Regulation (GDPR) to establish user rights concerning their data as well as set out the obligations of an organisation that is responsible for storing, managing and processing said data. Some of the articles of the GDPR, while integral to safeguard a user's rights, are opposite to the technical capabilities of innovative technologies, blockchain being one of them. The scope of this publication is to present blockchain, describe its potential in the healthcare domain, list limitations between blockchain and the GDPR, and finally provide guidelines to reconcile the two, acting as a roadmap for researchers and organisations who are developing blockchain applications in healthcare and other domains.**

Keywords-blockchain; healthcare; GDPR; sensitive data; patient rights.

I. INTRODUCTION

Blockchain is a technology that enables decentralised data management in a distributed ledger [1, 2]. It consists of a number of blocks, which are interconnected and represent a set of transactions [3]. The structure of blockchain involves the distribution of a ledger's blocks in a number of peer-to-peer nodes of a given data infrastructure instead of being stored centrally [4]. Each block, in addition to the data stored in it, contains a cryptographic hash, the hash of the previous block, and the timestamp of its creation. A hash is generated with the help of cryptographic hash functions and is the representation of the block data as it is directly dependant on them. For this reason, it can be used to verify the integrity of data transactions contained in each block. In addition, the hash of each block depends on the hash of the previous block in the chain, which essentially makes the entire blockchain ledger immutable, given that any change to a block will change the hash, not only of that block but of all following blocks as well. In addition, blockchain is not managed by a central authority, and maintains the integrity of the transaction history in the system through consent algorithms that require some form of network majority to validate and accept a system change. Consequently, blockchain solutions empower the users by providing them not only with the capability to manage and share their data as they see fit but also with voting rights that are important to achieve network consensus.

Blockchain first appeared in 2008 and has become widely known for its use as a public account for cryptocurrencies such as bitcoin and Ethereum. Nevertheless, it can also be used in other domains to create permanent, public, and transparent data management systems [5]. In fact, in recent years, the added value of blockchain has been realised by the research community and various organisations that propose and develop innovative solutions in areas other than cryptocurrencies such as health, education, public administration, and supply chains among others. Regardless of the scope of the application, the general operation of a blockchain can be seen in Figure 1.

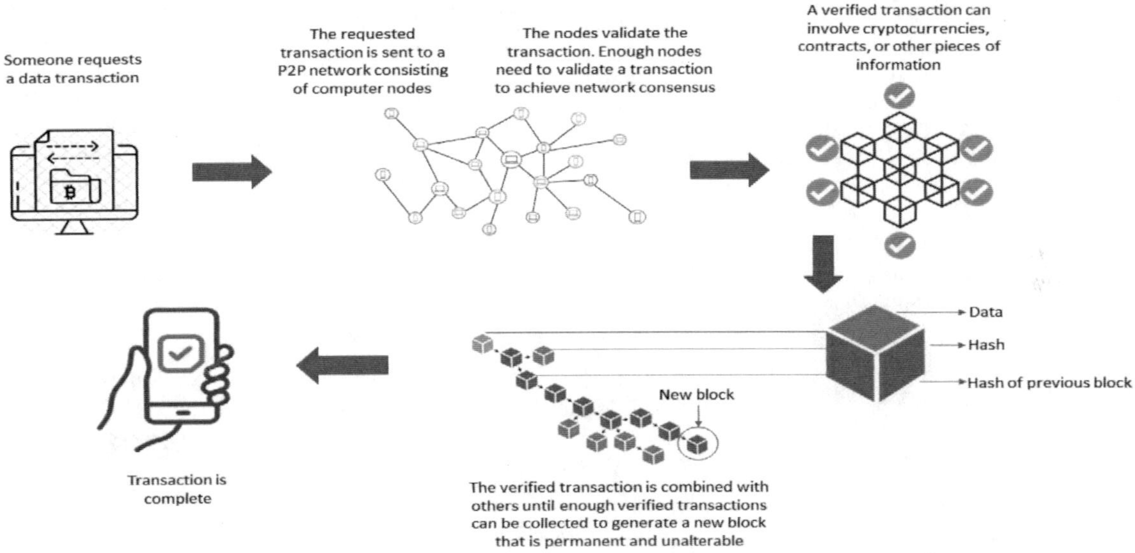

Figure 1. High level operation of a blockchain.

It is widely accepted that blockchain can be an effective and most importantly secure solution in any field involving data transactions. Smart contracts (contracts that can be executed/implemented partially or completely without human interaction) also play a big role not only in providing data protection but also by adding a degree of automation in the management and validation of transactions in the system [6]. Given that in the general case of an open, public blockchain, users have ownership of their data, smart contracts are used to help them define the rights (access, sharing, processing rights etc.) that third party users have on specific pieces of data. Overall, blockchain networks ensure continuous availability, reliability, security, durability, and integrity in a given network or system [4]. The main reason that allows blockchain to have such a wide scope is that at any time, network users can know the system status. To increase security and trust in the network, blockchain includes various mechanisms through which a transaction is secured [4], the main ones being the proof of existence and non-existence (it can be easily verified with certainty if an item exists in the system), the proof of time (when information is stored in the blockchain, the time at which it was added is also stored as part of the transaction's metadata), the proof of order (in case of network congestion, the order in which some requests/transactions were made is stored as well), the proof of authorship (data entries in the blockchain include details of the user who added them) and the proof of ownership (the owner of a specific item is always known).

The main scope of this publication is to present the potential of blockchain for developing secure data management systems in the domain of healthcare, address legal, ethical and societal challenges and obligations and provide guidelines for the development of not only secure but legally compliant blockchain solutions when it comes to managing sensitive medical data.

Section I provides the introduction to the document, explains in short, the operation of a generic blockchain system and describes the scope of the publication. Section II includes a bibliographic research on the potential of blockchain systems in healthcare and describes the current European and national legislations that impact the development of blockchain solutions. Section III lists successful practices and provides guidelines for the development of ethical blockchain systems. Finally, Section IV concludes the document and provides ideas for future work.

II. Blockchain in Health and GDPR Compliance

A. Blockchain in Healthcare

Blockchain is an emerging technology that can be applied in various domains. In healthcare, blockchain's inherent properties for security, immutability, decentralisation, and transparency can help reengineer many everyday processes for both patients and healthcare practitioners, as well as develop and implement more innovative and effective ICT healthcare systems. Blockchain can initially be used as a database, given that in traditional medical/hospital systems, efficient and secure data management comes with high installation and maintenance costs to ensure data safety when it comes to sensitive medical data. This fact can also result in ineffective communication and coordination among systems of different healthcare providers (lack of trust when it comes to data sharing), leading to patients having their medical data split among various ICT platforms in fragmented form. Blockchain significantly reduces these problems by ensuring data integrity and maintainability across all devices and systems connected to the network, since every data transaction leaves a digital trail, thus ensuring visibility of actions, while also reducing the possibility of system malfunction, due to the fact that there is no central point of failure. In other words, in a blockchain network, the creation of a new block of data (which represents a specific number of data transactions) must follow specific rules, usually hardcoded in the system and governed by smart contracts and must also achieve system consensus to be included in the ledger. Blockchain makes users responsible for their data, by also allowing them to define the access rights of third parties to it. Giving control of the data back to the patients and the users of the network and not only to healthcare organisations, ensures their correct use (existence of greater transparency and increased security). As such, blockchain systems are more secure against attacks and eavesdropping, so cases of malicious use are virtually impossible [7].

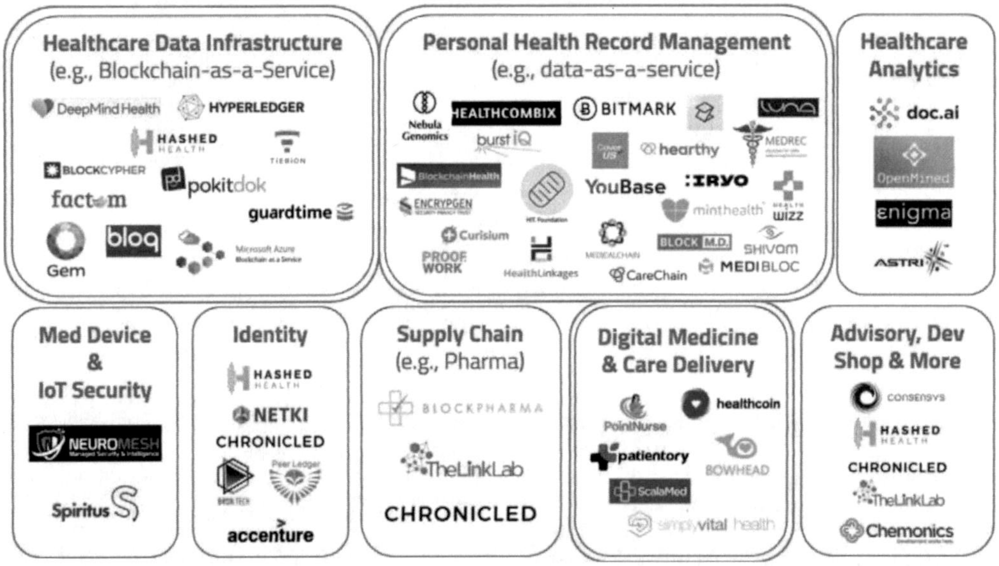

Figure 2. Blockchain Projects and Initiatives in Healthcare [11].

In modern healthcare, the prevailing standard for health data management is the Electronic Health Record or EHR, which represents the amalgamation of health and medical data concerning a patient in digital form. Nevertheless, the low degree of digitisation in the sector, in combination with the increased costs for the adoption of EHR systems hinders their widespread use and implementation in more medical processes. It is considered that the introduction of EHRs in blockchain-based medical platforms could increase overall data interoperability in the domain, and improve communication and coordination among heterogeneous healthcare providers, in addition to reducing costs for managing security and data governance issues [8]. Blockchain offers a robust infrastructure and an additional level of security that can be implemented with no extra cost, compared to traditional systems. This allows the development of value-adding applications and services on top of the blockchain ledger that can in turn facilitate mass analyses of medical data, freeform queries on a patient's data, personalisation of a patient's profile, notifications etc. In traditional systems, such applications usually operate in the isolated ICT environment of each respective healthcare provider to minimise security risks, while analysis of patients' data requires their informed consent. Such issues are minimised when patients have control over their data and decide who can have access to them and for what purpose. Moreover, every action that requires blockchain data leaves a digital trail, thus increasing overall transparency and trust in the system and its stakeholders. Another innovative concept when it comes to blockchain in healthcare, includes the use of cryptocurrencies (peer-to-peer decentralised electronic form of money, which is based on cryptographic principles to ensure network security and authenticate transactions) [9] to facilitate economic transactions in the network. Companies, such as Bowhead Health, have developed such platforms, where users can make their data available to researchers and receive cryptocurrencies as compensation [10]. Finally, blockchain can be used to combat counterfeit drugs and prescriptions. Applying blockchain principles to the pharmaceutical supply chains (usually in conjunction with Internet of Things devices, such as sensors and surveillance systems) increases visibility and reliability throughout the life cycle of a drug. The increased interest in the healthcare domain for blockchain solutions is evident by the sheer number of initiatives and applications, as shown in Figure 2.

B. GDPR and National Legislations

The General Data Protection Regulation (GDPR) 2016/679 [12] is a piece of legislation that was developed to safeguard data protection and privacy for all individuals within the European Union (EU) and the European Economic Area (EEA). It also governs the export of personal data outside of the EU and EEA regions. The GDPR is primarily intended to give individuals control over their personal data and to simplify the rules of compliance for international companies by consolidating regulations within the EU [13]. It was issued on 14 April 2016 and entered into force on 25 May 2018 [14].

According to the GDPR, personal data is defined as information that describes an individual, such as identification, physical characteristics, education, employment, financial status, interests, activities, and habits among others. In addition, the personal data of a person referring to his/her racial or ethnic origin, political views, religious or philosophical beliefs, medical records or health status, social welfare, love life, criminal prosecution, convictions as well as participation in associations/ organisations related to the above are characterised as sensitive. Sensitive data is protected by law with stricter regulations compared to personal data [15] and cannot be used to identify or profile an individual. According to the GDPR, citizens have a number of rights [16] with regard to their personal data, the most important of which are the rights to information, access, correction, deletion, objection, notification, automatic decision making and data portability. To ensure compliance with the above, according to the GDPR, the management of personal data by companies and other organisations should be explicitly defined and governed by specific principles [17].

To begin with, organisations must ensure beyond any reasonable doubt that there are legitimate reasons for collecting and using personal data by obtaining the informed consent of their clients for such purposes. In addition, and regarding data processing, companies must ensure that they accurately store only the data that is necessary for the purpose of the processing for which consent has been obtained (data minimisation principle). The explicit consent of individuals is also required for the transfer and processing of their data by third parties. For this reason, consent forms are used, which state in detail the purposes of data collection, as well as the processing they will undergo. To comply with the above, the GDPR recommends that organisations appoint a data protection officer (DPO), who acts as a single point of contact between customers, the organisation and supervising authorities and provides advice on security measures. and data protection policies. Finally, the GDPR recommends conducting a data protection impact assessment (DPIA) to identify and manage potential threats to personal data, especially in systems that implement innovative/emerging technologies in traditional systems.

As already mentioned, according to the GDPR, medical information about a person is considered as sensitive personal data. As such, it is a given that organisations active in the health sector are responsible for storing and processing medical data by following the basic provisions of the GDPR. However, in most countries of the EU there are some additional fundamental rights regarding patient data, which are defined by national legislations and must be respected by organisations that operate in said countries. For example, in the Greek legislation, according to article 12 of Law 2472/ 97 [18], patients have the right to access their data and receive their official medical files, certified from the healthcare facility where they have been treated. In addition, law 3418/2005 [19] regulates the conditions for granting access rights to a patient's medical record to a relative. This right of access is allowed only in exceptional cases, in which the third party acts as a legal representative, has an official written authorisation, or to protect the vital interests of patients in cases that they are unable to provide their consent (e.g., a patient has lost consciousness). On the other hand, medical professionals, especially when they also act as data controllers, have the obligation to report accidental or illegal damage, loss, alteration, and disclosure (unauthorised) of data to the respective supervisory authorities. To comply with the aforementioned rights and obligations, data controllers must provide patients with the relevant documents that describe the respective regulations and ensure they have their explicit and informed consent (temporary or permanent) for processing their data.

III. Achieving GDPR Compliance in Blockchain Projects and Initiatives

GDPR compliance when it comes to blockchain, concerns the way that this technology is used in various cases and applications and is considered to be a critical issue for the European Commission. Therefore, there are some important issues that need to be addressed, under the GDPR, regarding the protection of personal data in a blockchain. In particular, the structure and storage of data in a blockchain network is done in such a way that it does not allow the deletion or correction of data once it is registered in the chain, according to the immutability property. Therefore, even if the controller of a network can be identified, it becomes impossible for that controller to delete or update the file of a transaction without irreversibly damaging the blockchain ledger, since the hash of each block (which is representative of the data in the block) is also dependent on the hash of the previous block. In other words, blockchain is built entirely on the assurance that transactions will never be forgotten or deleted, aiming to build decentralised trust as well as develop and expand the network of participants.

Nevertheless, the correction and deletion of personal data are two basic rights provided to data owners by the GDPR. To comply with GDPR provisions, organisations developing blockchain applications employ various encryption techniques, in combination with the destruction of the key that provides access to the blockchain data. Another technique that is widely used is the storage of data in central databases and the use of blockchain to store and verify transactions between users of the system, thus minimising the security and privacy risks arising from the need to respect GDPR regulations. Especially regarding the right to be forgotten, non-blockchain storage allows deletion of user data. The fact that the data cannot be deleted from the blockchain does not affect the system in this case, since after data deletion from the centralised storage the only data remaining in the blockchain are the hashes of the transactions, which cannot lead to the identification of the user (especially since the data to which the hashes point have been deleted). The same is true in networks where data storage is done in a decentralised manner, such as in a private cloud or the mobile device of each user.

The aforementioned GDPR obligations do not mean that blockchain is an ineffective solution that is hard to implement and maintain in healthcare ICT systems. In Section II, it was explained that blockchain can actually solve most current challenges and shortcomings. In fact, in a way, blockchain and the GDPR are similar as they both focus on empowering the user and increasing visibility, data privacy and security. The only issue when it comes to developing GDPR-compliant blockchain systems is that of data deletion/correction and that can be bypassed as already explained. However, the fact of the matter remains that blockchain is an emerging technology and its implementation in domains other than cryptocurrencies is relatively new. One the one hand, this means that users and stakeholders of blockchain systems are largely unfamiliar with the technology and oftentimes suspicious of its attributes, which makes sense as the actual algorithmic part that ensures security in a blockchain is not easy to comprehend. On the other hand, and given that non-cryptocurrency blockchains are just now emerging, the technical specifications of implementing such systems have not yet been specified for each distinct domain in a way that ensures legal and ethical compliance and at the same time maximises the effectiveness of the technology. This should be viewed as an opportunity to set out the legal and ethical requirements that must be met in such cases and create roadmaps that will guide blockchain development efforts in the future. While, different areas of implementation have their own requirements and require different roadmaps, the consolidation of legal and ethical obligations by the GDPR means that knowledge generated for a given domain may be transferred and applied in other domains as well.

As such, in healthcare-related blockchain projects the following actions are deemed as necessary to guide the efforts for achieving GDPR compliance.

1. Informed consent: Blockchain initiatives must draft informed consent forms that set out the rights of patients and the legal obligations of the data controller and participating organisations. If user data populate the system, the consent form must explicitly explain the reasons for data gathering and ensure that those reasons are grounded on a legal basis. The same is true for the analysis of user data and their use by third parties.
2. Data Protection Impact Assessment (DPIA): While there are a number of prerequisites that define whether a DPIA must be performed, in healthcare blockchains that deal with sensitive data, a DPIA is considered necessary in almost all cases. The DPIA is an excellent opportunity for blockchain initiatives to pinpoint potential risks and explain how such risks will be resolved. It is imperative to note that a DPIA must be exhaustive when it comes to data privacy risks. For example, even if malicious attacks in a blockchain system are almost impossible, this risk should still be mentioned in a DPIA along with the explanation of how blockchain maintains system security. This should also be seen as an opportunity to educate non-technical stakeholders and legal entities on blockchain's properties and steadily increase trust in the technology.
3. Appointment of a DPO: Every blockchain development effort should have a DPO appointed. It is important that DPOs have legal backgrounds and are knowledgeable on the GDPR, since they validate the measures taken at each step and can also act as a point of contact with national and EU GDPR offices for clarifications and other matters.
4. Privacy-by-design: It is important that the privacy by design principle is followed when developing a medical blockchain, meaning that all security measures must be set out from the start, before the system is implemented. In a medical blockchain, the privacy-by-design principle sets out the following conditions/measures:
    a. Non-blockchain storage: To comply with the users' right of data deletion, all personal and sensitive data must be stored in non-blockchain repositories. In that case, the blockchain adds an additional level of security in the system as it validates all data transactions. In that case, the non-blockchain solutions that may be used by patients for data storage (e.g., cloud, IPFS, mobile devices) must also be set out from the beginning.
    b. Anonymisation/pseudonymisation: Despite the security offered by blockchain, it is considered a good practice to anonymise/pseudonymise personal and sensitive data that can lead to a user's identification.
    c. Patients' rights: It is important that all legal requirements that concern patient rights on their data (access, sharing, deletion etc.) are set out before development starts. This usually entails researching the legal and ethical landscape, stakeholder interviews etc. When such legal requirements are set out, they can be

programmed in smart contracts, which are used to verify whether a data transaction can be performed.

It can be surmised from the above that GDPR compliance requires informed consent, risk assessment, and the design of all security measures before development starts. In addition, blockchain initiatives and consortia should periodically assess the legal and ethical landscape to uncover changes or updates in the respective legislation or even innovative approaches that facilitate compliance.

IV. CONCLUSIONS AND NEXT STEPS

The scope of this publication was to showcase blockchain as an emerging technology that can facilitate the reengineering of data governance practices in healthcare, as well as assess the respective legal and ethical landscape so that guidelines for compliance can be generated. All in all, it can be surmised that the GDPR is not a piece of legislation aiming to create limitations for innovative technologies, but rather to set out the requirements that must be met by technical solutions to ensure privacy and security. In fact, when the GDPR came into practice, many researchers thought that the right to data deletion will mean the end for EU blockchain initiatives. However, that is not the case as it led to blockchain's role as a security infrastructure to come into the forefront of technical solutions instead of its role as a database. At the moment, the end goals of blockchain and the GDPR seem quite similar, both aiming to empower the data owner and bring forth a new age in data management. The guidelines that have been generated to achieve GDPR compliance are not hard to implement and keep track of, while the technical requirements that must be met follow the lines of effective programming practices instead of imposing hard constraints that will affect the end-result. It is imperative that in this period, researchers, legal experts and developers work together to set out roadmaps for compliance in various areas and domains. Such roadmaps will facilitate more ethical technical solutions in the future.

ACKNOWLEDGMENT

This paper was written under the context of the IMPILO project, which is funded by the Greek operational program "Competitiveness, Entrepreneurship, Innovation" - EPANEK (2014–2020), in the national action: "RESEARCH - CREATE - INNOVATE", with project code T1EDK-01382.

REFERENCES

[1] C. Esposito, A. Santis, G. Tortora, H. Chang, and K.K.R. Choo, "Blockchain: A panacea for healthcare cloud-based data security and privacy?" IEEE Cloud Computing 5, pp. 31–37, 2018.

[2] S Nakamoto, "Bitcoin: A peer-to-peer electronic cash system", 2008. Available at: https://bitcoin.org/bitcoin.pdf.

[3] K. Peterson, R. Deeduvanu, P. Kanjamala, and K. Boles, "A blockchain based approach to health information exchange networks." Proceedings of NIST Workshop Blockchain Healthcare, Gaithersburg, pp. 1–10, 2016.

[4] D. Drescher, "Blockchain basics: A non-technical introduction in 25 steps." 1st ed. Apress, Frankfurt, 2017.

[5] K. Kotobi and S.G. Bilen, "Secure blockchains for dynamic spectrum access: A decentralized database in moving cognitive radio networks enhances security and user access.", ieee vehicular technology magazine 13.1, pp. 32-39, 2018.

[6] P. Franco, "Understanding bitcoin: Cryptography, engineering, and economics." 1st ed. Wiley Finance Series, New York, NY, 2014.

[7] C.C. Agbo, Q.H. Mahmoud, and J.M. Eklund, "Blockchain technology in healthcare: A systematic review." Healthcare. Multidisciplinary Digital Publishing Institute, pp. 56, 2019.

[8] A. Petre and N. Hai, "Opportunities and challenges of blockchain technology in the healthcare industry." Medecine sciences 34(10), pp. 852-856, 2018.

[9] T. Polansek, "CME, ICE prepare pricing data that could boost bitcoin". Available at: https://www.reuters.com/article/us-cme-group-bitcoin-idUSKCN0XT1G1, 2016. Accessed in August 2020.

[10] Bowhead Health, Bowhead. Available at: https://bowheadhealth.com/. Accessed in August 2020.

[11] A. Coravos, "Where are the healthcare blockchains?", Available at: https://blog.andreacoravos.com/where-are-the-healthcare-blockchains-8fcf6a3e28f8. Accessed in August 2020.

[12] European Commission, "Regulation (EU) 2016/679 of the European parliament and of the Council of 27 April 2016." European Commission, Brussels. Available at: https://eur-lex.europa.eu/eli/reg/2016/679/oj. Accessed in August 2020.

[13] Council of the European Union, "Presidency of the Council: Compromise text. Several partial general approaches have been instrumental in converging views in Council on the proposal for a General Data Protection Regulation in its entirety. The text on the Regulation which the Presidency submits for approval as a General Approach appears in annex." Brussels, 2015. Available at: https://www.swlaw.com/publications/legal-alerts/2544. Accessed in August 2020.

[14] European Parliament, "Implementation of GDPR since 25 May 2018. Available at: https://eugdpr.org/. Accessed in August 2020.

[15] Administration of Personal Data Protection, "General Data Protection Regulation." Available at: https://www.dpa.gr/portal/page?_pageid=33,213319&_dad=portal&_schema=PORTAL. Accessed in August 2020.

[16] European Commission, "General Data Protection Regulation – chapter 3." GDPR, 2016. Available at: https://gdpr-info.eu/art-3-gdpr/. Accessed in August 2020.

[17] European Commission, "General Data Protection Regulation – chapter 5." GDPR, 2016. Available at: https://gdpr-info.eu/art-5-gdpr/. Accessed in August 2020.

[18] Greek Government Gazette, "Law 2472/1997, Protection of citizens from personal data processing– Right of access (article 12). "SGG 50/A/10.4.1997. Available at: https://www.dpa.gr/portal/ page?_pageid =33,19052&_dad=portal &_schema=PORTAL#12. Accessed in August 2020.

[19] Greek Government Gazette, "Law 3418/2005, Code of medical ethics." SGG 287/A/28.11.2005. Available at: http://www.dpa.gr/APDPXPortlets/htdocs/documentDisplay.jsp?docid=162,112,178,83,91,84,31,147. Accessed in August 2020.

Detection of Health-Related Problems of People with Dementia from Lifestyle Wearables: A Rule-Based Approach

Antonios Pliatsios, Thanos G. Stavropoulos, Dimitris Strantsalis,Vasileios Kassiano, Spiros Nikolopoulos, and Ioannis Kompatsiaris
Information Technologies Institute (ITI)
Centre for Research & Technology Hellas (CERTH)
Thessaloniki, Greece
email: {apliatsios, athstavr, dstrants, vaskass, nikolopo, ikom}iti.gr

Abstract— **In this paper, we describe a rule-based framework for the detection of health-related problems of people with dementia. The framework combines a novel ontology for lifestyle data (steps, sleep duration and heart rate measurements) and health-related problem representation and a novel set of SPARQL Inferencing Notation (SPIN) Rules to infer problems from lifestyle data. Both the ontology and the rule set are designed based on clinical expert knowledge in the field of dementia. More specifically, lifestyle data is acquired from lifestyle wearable devices in the market, making the system affordable and convenient. A model based on Semantic Web technology, Web Ontology Language (OWL), is used to formally represent and integrate sensor measurements, which promotes interoperability with other models and data exchange. SPIN rules offer the benefit of simplicity and flexibility as opposed to other rule representations in the domain. A proof-of-concept scenario is realized, showing data gathered from a real subject and the generation of expected problems by the framework.**

Keywords-*Ontology; Event Detection; Semantic Web; SPIN; Reasoning; Rule-Based Systems; Dementia.*

I. INTRODUCTION

As the world population is rapidly aging, people living with dementia globally amounted to 50 million in 2019 and are expected to triple to 150 million by 2050 [1]. Yet there is no silver bullet for dementia, such as a pharmacological solution. Only holistic and objective information about a patient's health status can drive tailored interventions to alleviate the ailments and slow down the progression of the disease. However, this imposes a huge burden on informal caregivers and healthcare professionals to manually monitor lifestyle, and health-related problems such as movement, sleep and stress.

Lifestyle sensors are a promising and affordable solution to objectively, continuously and affordably monitor patients, but a framework to map and extract clinical health-related problems is needed. The acquisition of knowledge from continuous and heterogeneous data flows is a prerequisite for IoT applications [2]-[4]. Semantic technologies provide integrated tools and methods for representing data and producing new Knowledge from them. Smart environments are increasingly encountered in healthcare technologies at home in actions that create better living conditions for older people by using Internet of Things (IoT) technologies, such as Active and healthy Ageing (AHA) and Ambient Assisted Living (AAL). In this context, human activity recognition plays a main role [5], because it could be considered as a starting point to facilitate assistance and care for the people with dementia. Due to the nature of human behavior, it is necessary to manage the time and adhere to the spatial restrictions. In doing so, semantic technologies enable expressive reasoning over health data, allowing clinical decision support to be realized. Ontologies are used to describe the context elements of interest (e.g., persons, events, activities, location, time), their pertinent logical associations [6], as well as the background knowledge required to infer additional context information.

In this paper, we propose a Semantic framework for Health-related Problem detection that combines ontologies and SPARQL Inferencing Notation (SPIN) Rules [7]. Ontologies are used to provide the common vocabulary for representing activity related contextual information, whereas SPIN rules derive high-level activity interpretations. SPIN is used as standardized declarative language able to address the limitations of the standard OWL Semantic Web technologies mentioned previously. More specifically, the temporal relations among activities are handled by SPARQL functions, whereas the derivation of new composite activities exploits the native capabilities of SPARQL to update the underlying activity model.

The SPIN language was chosen to implement this system because it combines concepts from object-oriented languages, query languages, and rule-based systems to describe the behavior of objects on the web of data and the Internet of Things [8]. In addition, it makes the rules accessible and easy to maintain, extend and share. A suitable Reasoner tool, such as the SPIN API, can extract the extra information generated by the rules and reuse it, for example, in executing a SPARQL query, thus generating new knowledge. These rules apply using SPARQL CONSTRUCT or SPARQL UPDATE requests (INSERT and DELETE). SPIN standards also make it possible to define such rules in higher level domain specific languages, so that rule designers do not have to work directly with SPARQL.

The rest of the paper is organized as follows. Section II presents related work in the domain of ontology-based reasoning architectures in Healthcare field. Section III describes the proposed System architecture that combines OWL ontologies and SPARQL rules in order to derive high-level activity interpretations. Section IV presents the use case scenario that evaluates the proposed architecture. Finally, Section V concludes our work.

II. RELATED WORK

In previous related studies, Semantic Web technologies have been used to represent knowledge from home healthcare systems. Some examples of projects are Knowsense [9], COSAR [10], ACTIVAGE [11], Dem@Care [12], Faber [5], and FallRisk [13]. Table I summarizes their aim and semantic web methods used.

TABLE I. RELATED WORK.

Project	Year	Aim	Methodology
KnowSense	2015	Activity Recognition in Healthcare system	Description Logic Reasoning, (DL) for activity detection and SPARQL queries to extract clinical problems
COSAR	2011	Activity Recognition in in context-aware environments	Ontological reasoning is also combined with statistics
ACTIVAGE	2017	Development of Smart Living solutions for active and healthy aging	Interoperable Ontologies, rule-based reasoning
Dem@Care	2015	Supporting independent life for elderly people with dementia	Interoperable Ontologies, Rules, Reasoning
FABER	2015	Detect abnormal behaviors for medical applications	Simple reasoning on an ontology
FallRisk	2015	Detect falls of elderly in smart homes.	Semantic Reasoning techniques

KnowSense is designed to support monitoring of the activities of elderly people with dementia in controlled and ubiquitous environments. Semantic Web technologies, such as OWL 2, are extensively used in KnowSense to display observations from sensors and specific applications, and to implement solutions to identify activities and problems in everyday life activities (IADLs) with the aim of clinical evaluation in different stages of dementia. The Description (Logic Reasoning, DL) reasoning for activity detection and SPARQL questions are used to extract clinical problems. However, the semantic techniques used by KnowSense cannot be easily extended and reused.

COSAR offers a solution based on the use of ontologies and ontological reasoning combined with statistical inference. Simple patient activities are identified by statistical methods, such as selecting the most likely method compared to others. The ontological reasoning is also combined with statistics to identify complicated activities that are not only detectable by statistical methods.

ACTIVAGE is a large-scale pilot project, aimed at developing Smart Living solutions that have a positive impact on active and healthy ageing. The ACTIVAGE IoT Ecosystem Suite (AIOTES) project, provides a set of techniques, tools and methodologies (rule-based reasoning, interoperable ontologies, etc.) that enhance semantic interoperability at different levels between heterogeneous IoT platforms. The approach uses different reasoning mechanisms that can improve the understanding of heterogeneous patient's data and help to generate new knowledge by providing services to end users.

Dem@care offers a complete system, consisting of heterogeneous sensors, to support the independent life of elderly people with dementia or similar health problems. This approach includes a heterogeneous set of detection methods and technologies, including video, audio, in addition to normal, environmental and other measurements. Semantic technologies (e.g., rule-based reasoning) are used to process and analyze sensor data according to user requirements. This results in feedback and decision support, that is delivered to end users through appropriately designed user interfaces. A variety of clinical scenarios and environments are supported, from short-term experiments in a hospital environment to long-term monitoring and support of daily life at home, for independent living.

FABER is is an ubiquitous system developed to detect abnormal behavior for medical applications. It first calculates events and actions from the available context data by using simple reasoning on ontology. Computed boundaries, actions and events are sent to the knowledge-based inference engine.

The main goal of FallRisk is to detect falls of elderly people in smart homes. It is based on a platform that uses several learning-based fall detection systems. The results of these systems are filtered and entered into an ontology that contains the contextual knowledge. The knowledge, including contextual information about the user, is then used to refine fall detection. The strength of this approach, in addition to the combination of both techniques, is the compatibility with any fall detection technique. However, it deals exclusively with fall detection.

The above systems use semantic rule-based mechanisms and provide solutions for activity and event recognition based on the use of ontologies and ontological reasoning. However, most methods are quite sophisticated and complex to express and to maintain due to rich logic support. For this reason, the SPIN language was chosen by us to create semantic rules. SPIN offers a lot of advantages [6][7]. SPIN rules offer the benefit of simplicity and flexibility as opposed to other rule representations It is based on SPARQL, a well-established query language and protocol, which is well supported by numerous engines and databases. This means that SPIN rules can be directly executed on the databases and no intermediate engines with communication overhead need to be introduced. Moreover, it has an object-oriented model that leads to better maintainable models. Specifically, the SPIN rule engine does not have to check all rules at all times, but instead rules are checked incrementally when new instances of a certain class are inserted (or modified) in the ontology. This leads to better rule execution performance. Furthermore, SPIN is a more promising de-facto industrial standard for the future of combining ontologies and rules, because it builds upon the widespread use of SPARQL.

III. SEMANTIC REASONING APPROACH FOR HEALTH-RELATED PROBLEMS DETECTION

This section presents the proposed Semantic System for Health-related problem detection with the aim of recognizing the activities of the people with dementia through different sensors and producing new knowledge by offering new services to end users of the system such as doctors, health professionals and patients. As shown in the Figure 2, the raw data are collected by users (i.e., the patient with dementia) using various wearable sensors and smart home sensors. Afterwards, raw data are modeled on RDF ontologies and stored in the Knowledge Database (GraphDB) for the purpose of creating the System Knowledge Base of the system. Then, the semantic analysis, which will be presented in the next section (Spin Rule Engine, Ontology and Rule reasoner, etc.), processes and interprets the data, enriching the Knowledge Base of the system.

A. Ontology and Knowledge Base

The proposed approach is built on top of emerging Semantic Web technologies. We started with the definition of system ontology for representing different elements of a healthcare system. The goal of ontology is to semantically visualize all concepts related to activity recognition in healthcare system and acts as a semantic information integration model derived from the system's sensors. A common practice in the development of ontologies is the reuse of existing models, so we're relying on already developed and valid ontologies for developing a part of the supporting ontology. The following are an overview of the existing entities used:

Dem@Care [12]: An ontology to represent experimental protocols of diagnostic support and dementia diagnosis in a controlled environment.

Semantic Sensor Network (SSN) [14]**:** Contains the ontology SOSA (Sensor. Observation, Sampler and Actuator). These ontologies describe semantic sensors, actuators, sampling and their actions. It is a W3C recommendation and OGC application.

SmartHome [15]**:** This ontology is an extension of SSN ontology and focuses on the representation of spatial and time aspects of entities included in spaces with devices belonging to the smart home category.

The system's ontology is expressed in OWL 2 (W3C, 2012), which is a representation language commonly used in the semantic issue community for entity development. Figure 1 shows the hierarchy of entity classes and the hierarchy of its properties. Object attributes are relationships that link classes together, and data attributes link classes to simple values (such as integers, alphanumeric, dates, etc.).

The main classes of ontology are Device, Event, HealthProblem, Person, and Profile. The Device represents the devices of the system. Event is a parent class for different Event-related classes. It has two subclasses Activity and Measurement. Activity contains the information of activities. Measurement includes instances, which represent information of measurements (Calories, Distance, Floor, HeartRate, Movement, Sleep, Steps). The HealthProblem is a parent class for different Health Proplem-related classes. It consists of subclasses HeartProblem, MovementProblem, MultiProblem and SleepProblem. The class Person includes instances, which represents the type of Person of the system (Doctor, Patient). Finally, the class Profile includes information from users' profile (Age, Gender, etc.).

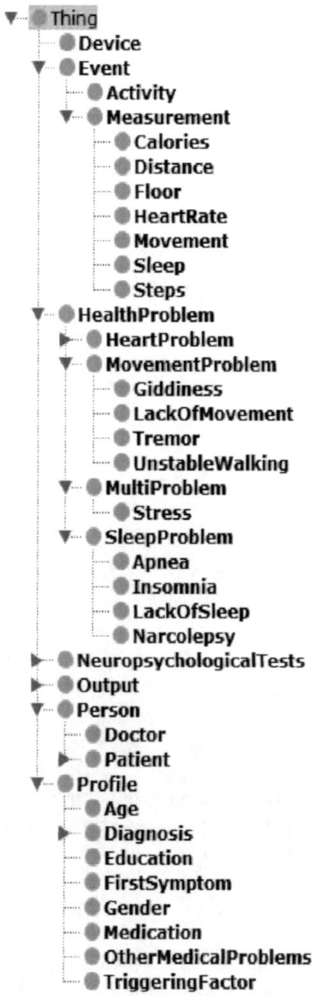

Figure 1. Classes of the proposed ontology.

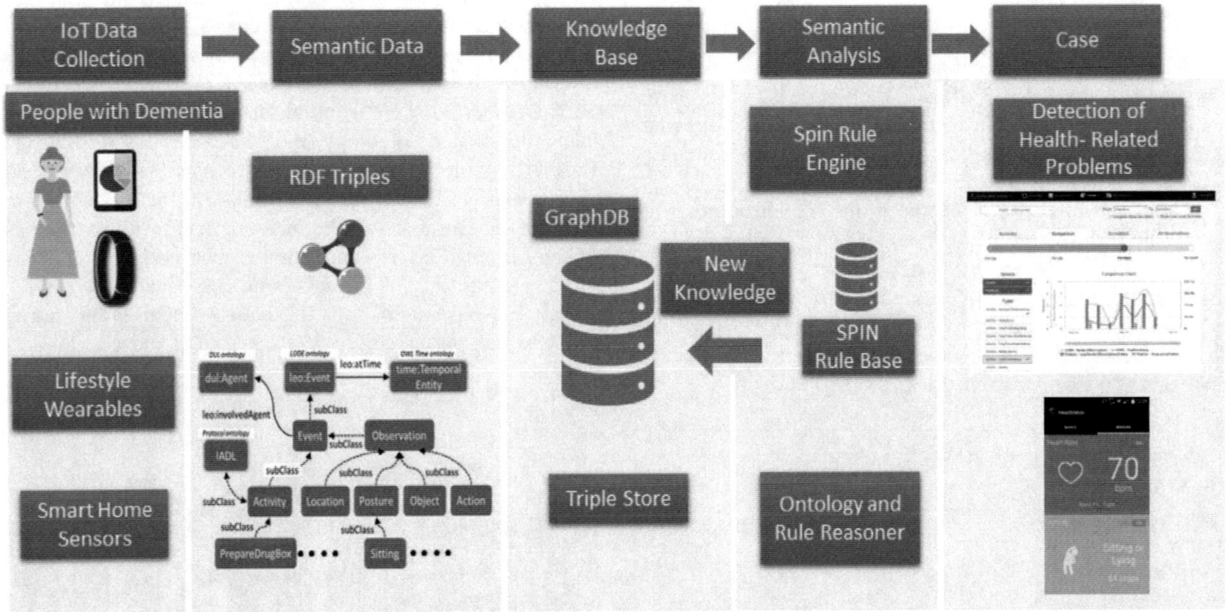

Figure 2. Architecture of the proposed system.

After adding Semantic Web technologies to the raw data and modeling them based on the system ontology, "Semantic Data" are stored in a semantic Graph Database, which constitutes the Knowledge Base of our system. For this purpose, we have chosen GraphDB, an enterprise ready Semantic Graph Database, compliant with W3C Standards. Semantic Graph Databases (also called RDF triplestores) provide the core infrastructure for solutions where modelling agility, data integration, relationship exploration and cross-enterprise data publishing and consumption are important. Querying and reasoning are performed over stored RDF graphs with SPARQL language.

B. Rule Base

Table II presents a sample of the semantic rules created after the collaboration of scientists, doctors, psychologists and patients. The rules contained in this section are a subset of the rule base of the proposed system. In every rule, there are upper and lower limits that control whether a condition is satisfied or not. The numerical values of the limits were decided after consultation of the clinicians and the patient users. Thus, for example, in the first rule, the limit for drawing a conclusion of a user's insomnia problem was set at 1800 minutes. In addition, to conclude that the patient needs to exercise more, his steps are limited to less than 80 in one day.

TABLE II. A PRIORI RULE BASE OF THE DIFFERENT SEMANTIC RULES THAT DESCRIBE THE MODELED ACTIVITIES.

Variables (number)	Rule	Problem
Duration in minutes	Time to fall asleep in a day > 1800	Insomnia
Count of sleep interruptions	Number of interruptions in a day > 10	Restlessness
Duration in minutes	Sleep total duration in a day> 480	Too much sleep
Duration in minutes	Sleep total duration in a day < 300	Lack of sleep
Duration of "Nap" state in minutes	Asleep in Naps > 100 in a day	Increased Napping
Occurrence of "Nap" State, Occurance of "Night Sleep" state	Asleep in Naps end time < 2 hours from Sleep start time	Nap close to bedtime
Time Alseep / Time in bed	Sleep Efficiency < 85	Bad Quality Sleep
Step count, Heart Rate measure, Duration in minutes	Steps < 50 & Heart Rate > 90 (Fat Burn Zone) for duration > 300	Stress or Pain
Heart Rate measure	HR < 60	Low Heart Rate
Step count, Heart Rate measure, Duration in minutes	Steps <1000 & Heart Rate < 80 for duration > 300	Inactivity
Step count, Heart Rate measure, Duration in minutes	Steps < 500 & Heart Rate < 100 for duration > 800	Lack of Movement
Step count	Steps < 80	Lack of Exercise

C. Implementation of the rules with SPIN

We used the TopBraid composer [16], a tool for modeling and developing semantic data applications, to present the SPIN rules. Topbraid allows us to easily develop spin rules in the form of SPARQL queries, which is more readable than regular spin syntax. In practice, the following three code blocks present in SPIN three simple semantic rules that were applied to the system ontology.

The following code block shows the implementation of SPIN rule for Sleep problem "Lack of Sleep". Applying this rule produces the addition of a new property that represents the type of sleep problem "Lack of Sleep" in the objects of the ontology (users of a support system). If the patient's sleep duration is less than 300 minutes, then it is considered that there is a sleep problem (lack of sleep).

SPIN rule for sleep problem "Lack of Sleep".

```
CONSTRUCT {
?p owl:hasSleepProblem "Lack of Sleep "}
WHERE {
  ?p a :Person .
        ?p :duration ?d.
        FILTER (?d <300 )}
```

The following rule in SPARQL and SPIN adds new knowledge to the system Ontology. If the sleep duration of the patient with dementia is greater than 480 then we conclude that there is a sleep problem (too much sleep).

SPIN rule for sleep problem "Too much Sleep".

```
CONSTRUCT {
?p owl:hasSleepProblem "Too much sleep "}
WHERE {
  ?p a :Person .
        ?p :duration ?d.
        FILTER (?d >480 )}
```

The following code block shows the implementation of the simple semantic rule "lack of exercise". If the steps of the patient with dementia are less than 80 and we conclude that there is a lack of exercise.

SPIN rule for problem "Lack of Exercise".

```
CONSTRUCT {
?p owl:hasProblem "lackOfExersice" }
WHERE {
?p a :Person .
        ?p :steps ?st1.
        FILTER (?st11<80)}
```

IV. USE CASE

For the evaluation of the proposed architecture we consider the following use case scenario. A wearable sensor was given to a patient with dementia in order to monitor his activities. The duration of the measurement is 11 days (20-30 November 2019). The initial sensor data was modeled by using the system ontology and stored in the Knowledge Base. Then, the proposed semantic techniques were applied and in particular the semantic rules of the system were checked. Figure 3 shows the measurement of sleep minutes of the patient. Figure 4 presents the measurements of steps per day of the user. This data is processed, and the results, showing the problems that the patient experiences during this time, are produced.

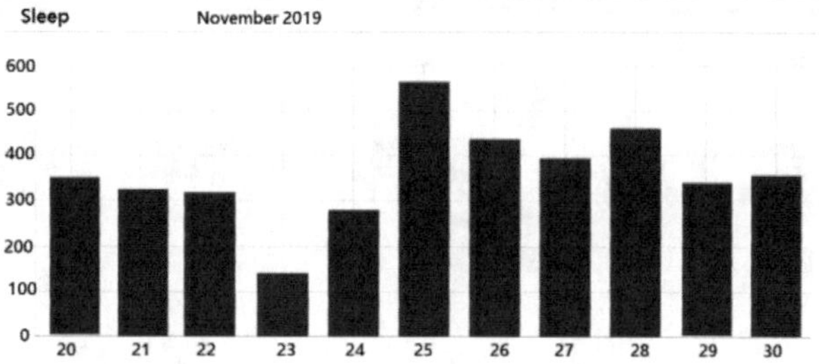

Figure 3. Sleep Minutes of a single patient with dementia whose activity was monitored.

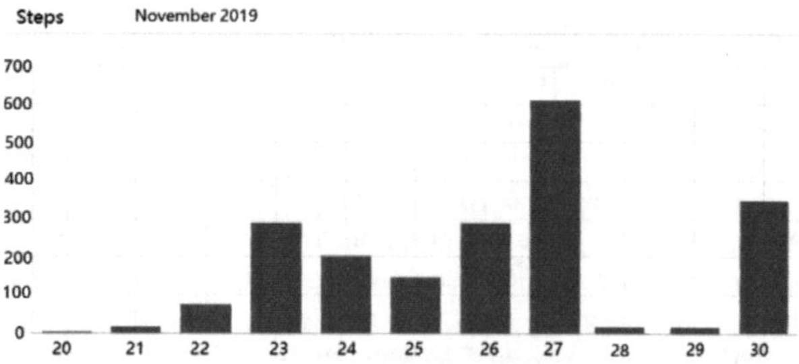

Figure 4. Sleep Minutes of a single patient with dementia whose activity was monitored.

Specifically, as shown in Figure 3, the patient with dementia slept below the limit of 300 minutes (rule 1) on November 23 and 24. The results of these measurements are shown in Figure 5 with the creation and visualization of the "Lack of sleep" problem. In addition, the patient slept above the limit set by rule 2 (480 minutes) on November 25, and this resulted in the creation of the problem "too much sleep". Finally, in Figure 4 it is observed that on November 20, 21, 28, and 29 the patient took a few steps. The result of this measurement is shown in Figure 5, by creating the problem (rule 3) "Lack of exercise". This also shows how the user can easily observe the days with a lot and different problems e.g., the 25th, and the days with a few problems e.g. 20th, 21th, 28th, and 29th, at a glance, without going through the raw data each time.

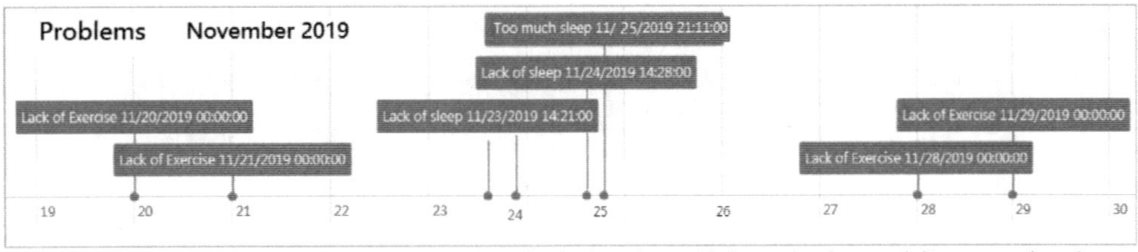

Figure 5. Health-related Problem Detection of a single patient with dementia whose activity was monitored.

V. CONCLUSION AND FUTURE WORK

In this paper, we presented our approach towards the definition of a semantic system for Health-related Problem detection that combines ontologies and SPIN Rules. Architectures related to the proposed framework are listed, and the advantages of using the SPIN language to create semantic rules are presented. The main purpose of the proposed architecture is to generate new knowledge from the original raw data, especially recognition of healthcare problems in the users with dementia. The system is validated through a proof-of-concept use case scenario where a wearable sensor gathers data from a real subject and the framework extracts the expected health-related problems.

As future work, we plan to evaluate the framework in a formal clinical trial with real subjects. Subjects will be recruited in the spectrum of dementia, as well as healthy controls and use the wearables for several months. The framework will be used to extract problems and clinical experts will evaluate its accuracy, usability and usefulness for the disease. In the long run, it will support decision making of the clinicians adjusting their non-pharmaceutical interventions, e.g., a clinician can "prescribe" exercise for lack of activity or relaxation exercises for stress, insomnia and lack of sleep problems.

ACKNOWLEDGMENT

This work was co-financed by the European Union and Greek national funds through the Operational Program Competitiveness, Entrepreneurship and Innovation, under the call RESEARCH-CREATE-INNOVATE, project code: T1EDK-02668 (https://www.ypostirizo-project.gr/).

REFERENCES

[1] Alzheimer's Disease International, "World Alzheimer Report Attitudes to Dementia," *Alzheimer's Dis. Int. London*, 2019.

[2] A. Pliatsios, C. Goumopoulos, and K. Kotis, "Interoperability in IoT: A Vital Key Factor to Create the 'Social Network' of Things," in *UBICOMM*, pp. 63–69, 2019.

[3] A. I. Maarala, X. Su, and J. Riekki, "Semantic Reasoning for Context-Aware Internet of Things Applications," *IEEE Internet Things J.*, vol. 4, no.2, pp. 461-473, 2017.

[4] A. Pliatsios, C. Goumopoulos, and K. Kotis, "A Review on IoT Frameworks Supporting Multi-Level Interoperability The Semantic Social Network of Things Framework," *Int. J. Adv. Internet Technol.*, vol. 13, pp. 46–64, 2020.

[5] M. Ziaeefard and R. Bergevin, "Semantic human activity recognition: A literature review," *Pattern Recognit.*, pp. 2329-2345, 2015.

[6] K. L. Skillen et al., "Ontological user modelling and semantic rule-based reasoning for personalisation of Help-On-Demand services in pervasive environments," *Futur. Gener. Comput. Syst.*, pp. 97-109, vol. 34, 2014.

[7] H. Knublauch, J. A. Hendler, and K. Idehen, "SPIN: SPARQL Inferencing Notation," *W3C Member Submission*, 2011.

[8] C. Fürber and M. Hepp, "Using SPARQL and SPIN for data quality management on the Semantic Web," In International Conference on Business Information Systems Springer, Berlin, Heidelberg, pp. 35-46, 2010.

[9] G. Meditskos, T. G. Stavropoulos, S. Andreadis, and I. Kompatsiaris, "KnowSense: A semantically-enabled pervasive framework to assist clinical autonomy assessment," In SWAT4LS, pp. 122-131, 2015.

[10] D. Riboni and C. Bettini, "COSAR: Hybrid Reasoning for Context-aware Activity Recognition," *Pers. Ubiquitous Comput.*, vol. 15, pp. 271–289, 2011.

[11] G. Fico *et al.*, "Co-creating with consumers and stakeholders to understand the benefit of internet of things in smart living environments for ageing well: The approach adopted in the madrid deployment site of the activage large scale pilot," pp. 1089-1092, 2017.

[12] T. Stavropoulos, G. Meditskos, S. Andreadis, and I. Kompatsiaris, "Dem@Care: Ambient sensing and intelligent decision support for the care of dementia," In SWAT4LS, pp. 229-230, 2015.

[13] F. De Backere *et al.*, "Towards a social and context-aware multi-sensor fall detection and risk assessment platform," *Comput. Biol. Med.*, vol. 64, pp. 307-320, 2015.

[14] M. Compton *et al.*, "The SSN ontology of the W3C semantic sensor network incubator group," *J. Web Semant.*, 17, pp. 25-32, 2012.

[15] M. Alirezaie *et al.*, "An ontology-based context-aware system for smart homes: E-care@home," *Sensors (Switzerland)*, vol 17, no. 7, pp. 1586, 2017.

[16] "Enterprise Data Governance — TopQuadrant, Inc," 2019. https://www.topquadrant.com/ accessed Aug. 13, 2020.

The Use of Electronic Signature in Processes and Applications of the Croatian Agency for Medicinal Products and Medical Devices

Arian Rajh, Dubravka Sudic, Katarina Gvozdanovic
Agency for Medicinal Products and Medical Devices
Zagreb, Croatia
e-mail: arajh@halmed.hr, dsudic@halmed.hr, kgvozdanovic@halmed.hr

Abstract—**In this paper, we describe the Information Technology (IT) solution for electronic signature in the Agency for Medicinal Products and Medical Devices of Croatia (HALMED). Electronic signature with various certificates is currently enabled in the Windows setting and tested across HALMED's Digital Archival Information System. HALMED's plans include the digital transformation of the main processes.**

Keywords—digital transformation; Digital Signature Services; electronic signature; eSignature building block; paperless agency; Qualified Electronic Signature; Digital Archival Information System.

I. INTRODUCTION

The European Telecommunications Standards Institute (ETSI) explains electronic (digital) signature as "essentially the equivalent of a hand-written signature, with data in electronic form being attached to other electronic subject data (Invoice, Payment slip, Contract, etc.) as a means of authentication" [1]. EU Regulation on Electronic Identification and Trust Services (EU Regulation 910/2014 of 23 July 2014, eIDAS) states that an electronic signature is "data in electronic form which is attached to or logically associated with other data in electronic form and which is used by the signatory to sign" [2].

eIDAS differentiates electronic signatures, Advanced Electronic Signatures (AES), and Qualified Electronic Signatures (QES). An Advanced Electronic Signature is used under control of the signatory, it identifies the signatory, and it is associated with the signatory without any uncertainty. It can be used to show if the signed content has been changed. A Qualified Electronic Signature creation device creates a QES, which can legally replace a handwritten signature. All EU countries should recognize QES based on a qualified certificate and issued by a qualified trust service provider. The provider should be listed on the European Trusted List [3]. A qualified certificate for electronic signatures holds data that clearly represents the qualified trust service provider and the natural or legal person. An electronic record signed with QES should not require supplementary evidence in courts. This represents a giant step forward for using electronically signed records and working paperless in the EU. EU has also launched an eSignature building block through its Connecting Europe Facility (CEF) instrument to facilitate the use of electronic signature [4]. eSignature building block helps IT-solution providers to develop their solutions in conformance with eIDAS regulation. ETSI provides a free online tool for electronic signature conformance verification [5].

The Agency for Medicinal Products and Medical Devices (HALMED) is the Croatian authority in charge of the regulation and services related to medicinal products, medical devices, homeopathic medicinal products, and veterinary medicinal products. HALMED is responsible for permitting marketing authorizations for the registration of medicinal products and homeopathic medicinal products, for parallel imports of medicinal products, for granting authorizations for manufacturers of medicinal products, for granting authorizations for wholesale distribution of medicinal products, for brokering of medicinal products, and for the retail sale of medical devices. HALMED performs laboratory analyses of medical devices and quality control procedures for medicinal products and homeopathic medicinal products as the official laboratory for the Republic of Croatia. It is responsible for the inspection of the production of active substances, excipients, and finished medicinal products. It approves the entry and import of medicinal products. It monitors adverse reactions in clinical trials and on the marketed products, and conducts pharmacovigilance of medicinal products and vigilance of medical devices. HALMED also advises marketing authorization holders, the Minister of Health, and the public, and cooperates with other EU and international medical agencies within its scope of work as well [6].

In Section II of this paper, we describe processes with electronic signatures in HALMED. In Section III, we discuss the solution and standards that it uses, from both practical and scientific viewpoint. Further work and conclusion are provided in Sections IV and V.

II. PROCESSES IN HALMED WITH ELECTRONIC SIGNATURES

For the time being, HALMED has implemented a general signing procedure with electronic signatures. That means that any record can be signed in its Windows environment as long as it is saved in PDF form (as shown in Fig. 1).

The solution for signing records enables HALMED's employees to verify signatures and signed records. This is the standard function of eSignature software solutions [7]. The next upgrade of this solution should enable the electronic signing of any record in MS Word or PDF directly from HALMED's Digital Archival Information System's (DAIS) environment. This functionality is in the early testing phase. DAIS should automatically convert MS Word records to PDFs before signing. The solution is currently used with internal certificates, issued by HALMED's organization to its employees, but it can work with any certificate for qualified electronic signatures. The solution was tested with several Croatian trusted certificates. However, to work paperless, HALMED should digitally transform its processes.

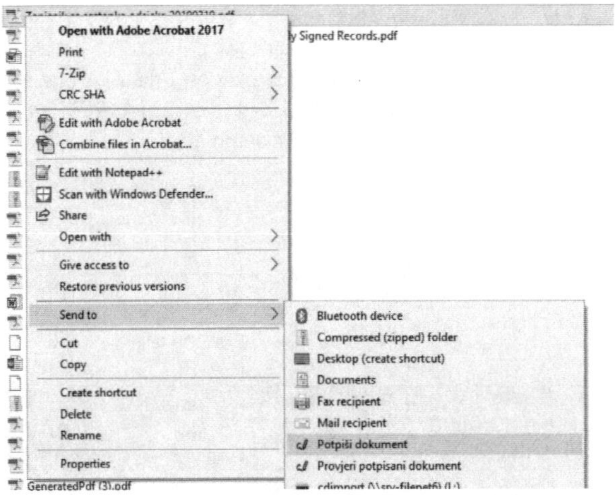

Figure 1. The process of signing records in the Windows environment.

Next, we describe and explain the IT solution for electronic signatures installed in HALMED more closely.

III. Solution For Signing with Electronic Signature Implemented in HALMED

The IT solution for electronic signatures in HALMED was developed and provided by Ericsson Nikola Tesla company (ENT, www.ericsson.hr). ENT has been working closely with HALMED on various occasions. This IT company won the EU-financed project of developing HALMED's DAIS in 2013. This IT solution for signing electronic records represents a logical upgrade of DAIS. The signing solution was based on CEF eSignature building block to ensure its compatibility with eIDAS regulation and to decrease the level of legal risk in HALMED's work with electronically signed records.

CEF eSignature building block was provided to assist "public administrators and businesses to accelerate the creation and verification of electronic signatures" [8]. The eSignature building block was established upon the following standards [9]-[13]:

- ETSI EN 319 132 XML Advanced Electronic Signatures (XAdES)
- ETSI EN 319 122 CMS Advanced Electronic Signatures (CAdES)
- ETSI EN 319 142 PDF Advanced Electronic Signature Profiles (PAdES)
- ETSI EN 319 162 Associated Signature Containers (ASiC)
- ETSI TS 119 612 v2.1.1 Electronic Signatures and Infrastructures (ESI), Trusted Lists

The building block includes Digital Signature Services (DSS), open-source software for the creation and validation of eIDAS-conformant electronic signatures. IT-solution providers can use DSS according to the GNU Lesser General Public Licence 2.1. The current DSS version is 5.5 [14].

The IT solution for signing electronic records works with certificates compatible with RFC 5280 and X509 standard, RFC 7292 and PKCS#12/Personal Exchange File (PFX) standards [15]- [16]. The procedural and technological basis for working with electronic signature is Public Key Infrastructure (PKI). PKI connects public keys with natural or legal subjects via registration procedure. X509 is a standard that determines public key certificates and revocation lists. PKCS#12 is used for the binding of various cryptography objects, like private keys and certificates, to a file.

The IT solution for electronic signatures enables using internal or other certificates, e.g., certificates issued by a trusted service, as well as choosing a business role or a reason for signing an electronic record [17]. This is the first advantage of the solution; it can work with various certificates. Certificates should be compatible with X509, RFC 5280, PKCS#12 and PFX standards. The solution in its web-part uses RFC 8446 Transport Layer Security (TLS) cryptographic protocol for safe communication over the internet [18]. The solution also contains its local installation. As the web part of the solution communicates with the service installed locally, the TLS certificate should be issued for the localhost domain. The solution adds the table with data about signing and the signer at the end of the records. A property file contains definitions of signatories' roles, and an administrator in HALMED can change this file. The table contains the following data: signatory, role, date, time, and time zone.

DAIS is a wide-range IT system and HALMED's digital archives, developed by ENT for HALMED under the EU-financed project in 2013 and 2014. It was established on the IBM FileNet platform, and it contains several modules. The basic module is Content Navigator; it enables working with any electronic records and starting various record-driven processes (see Fig. 2 that shows the folder view and organization in HALMED's DAIS, and the folder in the example contains action plans and analyses records).

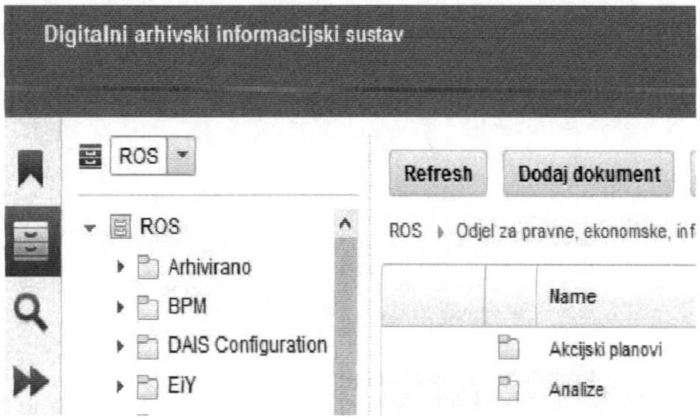

Figure 2. Content Navigator.

The second module is Enterprize Records. Enterprize Records is a records management and archival management module, and it is connected with HALMED's archival (metadata) management system Pismohrana (the Archive). IT Company Omega-Software developed Pismohrana in 2012 (www.omega-software.hr). It is a process-driven application based on professional standards of the International Council of Archives (www.ica.org/en). The purpose of the DAIS Enterprize Records module is to capture archival metadata from Pismohrana application and protect the records, or to execute an activity upon the records, triggered by Pismohrana application. Enterprize Records module works with EMC2 (from the names of the IT company founders Egan, Marino, Conolly, and Curly) Isilon archival storage and moves archived records from active HP 3PAR storage to slower, but more protected Isilon storage. DAIS also works with an integrated PDF/A file converter.

The third module is a module for ingest of ISO 14721-compatible [19] submitted packages with records and their metadata into the DAIS repository (see Fig. 3 that shows the analysis, migration and export of packages functions). By using this module, records are prepared for description in Pismohrana application and transported into the DAIS repository, and their metadata are transferred and linked to newly-established digital objects in the repository. Metadata are used for the retrieval and management of archived records.

Figure 3. The migration module.

The fourth module is related to business processes and management of workflows. The next module is a module related to the creation and management of HALMED's Quality Management System. This module enables the role-defined creation of Standard Operative Procedure (SOP) records and related templates and reports. It notifies SOP authors and their managers when an SOP needs to be reviewed and updated and enables automated distribution and archiving of new versions upon their completion. HALMED's employees are signing new SOPs by using the previously developed functionality of signing electronic records. This functionality works with the same internal certificates. The difference is related to the visual appearance of the table with signatures and the fact that it is placed at the beginning of the record (see Fig. 4 and 5 - Fig. shows the header of SOP records' template).

AGENCIJA ZA LIJEKOVE I MEDICINSKE PROIZVODE		
NASLOV:		

Figure 4. Table for signatures in an SOP record.

Figure 5. The role-dependent signing of SOP records in the Quality Management module, by using integrated functionality for signing electronic records.

The next module is used for the administration of HALMED's internal projects, like the development and customizations of business applications, digitization, and other projects. The signing function is not integrated with this module, but HALMED has developed plans to enable electronic signatures in this module in the next few years. In this time, HALMED's project managers and the members of the groups for the supervision of HALMED's projects use the signing functionality in the Windows environment, as shown in Fig. 1. Finally, there is a module for the administration of the DAIS system and its users.

IV. Further Work

The near phase of development of HALMED's options with electronic signatures comes down to the implementation of electronic signature across the DAIS environment, i.e., across all modules except the admin module. The ability to use the solution across at least two different environments is the second advantage. Various business applications (for approval of medicines, for medical devices, for inspections) connect with DAIS, so the solution should automatically facilitate electronic signatures in these specific business environments.

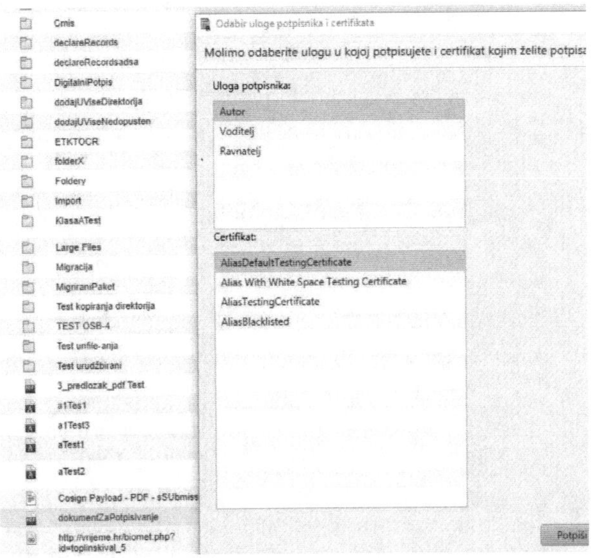

Figure 6. Signing of an electronic record in the DAIS pre-production environment (figure taken from the functional specification).

Ericsson Nikola Tesla developed this functionality of signing in DAIS for HALMED, and currently, it is in the testing phase. Fig. 6, taken from Ericsson Tesla's functional specification, shows the selection of record, signer's role, and a certificate from the list of available certificates. Fig. 7 shows the saving of electronically signed records as the main version in the DAIS repository. However, solely the implementation of electronic signature across the entire DAIS system will not enable HALMED to work paperless.

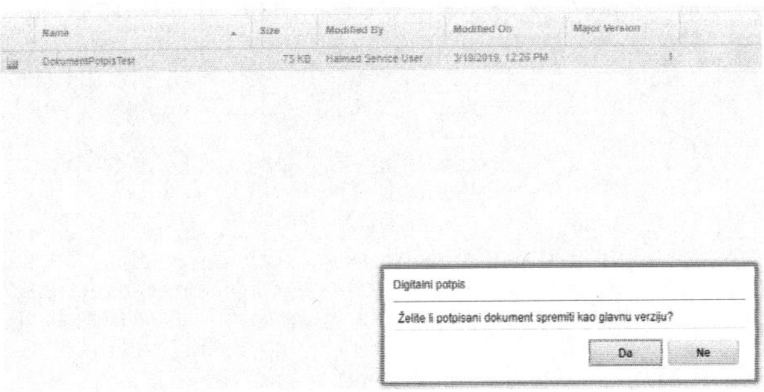

Figure 7. Saving the electronically signed record as the new main version, a figure from the functional specification.

The plans for further development include business process reengineering for main HALMED's business processes, starting with granting marketing authorization for medicinal products. For this reason, HALMED has gathered a group dedicated to turning the marketing authorization process more comfortable for its clients and its employees. The final goal is to make this process completely paperless. The components of the digital transformation of HALMED's marketing authorization process are the following:

- Reengineering of the business process. The re-engineered and streamlined marketing authorization process should be established upon new activities with electronic records and resources, and these activities cannot "mimic" the existing paper-based process.
- HALMED representatives with process-significant roles should be equipped with legally valid certificates.
- Marketing authorization holders (as HALMED's clients) need to be prepared for the new processes. For this reason, HALMED works on establishing a registration procedure and portal for its clients. Further guidelines should be produced for the clients in the next period.
- Supportive resources should also exist in electronic form. This has already been ensured on the EU-level as the dossiers for medicinal products are being created and transferred to medicines agencies in the electronic Common Technical Document (eCTD) format [20]. HALMED has actively participated in this EU initiative since 2008, it has enabled work with electronic eCTD resources since 2010 on the national level, and since 2013 on the EU level.
- DAIS should enable archiving the electronically signed records. Additional customization of DAIS was planned for the 2020/2021 period. The debate is run on possible utilization of blockchain and distributed ledger technologies for ensuring the information on the authenticity of electronically signed records in a particular period [21] [22].

Another example of the digital transformation of the business process, which occurs in HALMED right now, is the reengineering of the support process related to employee vacations. For this reason, HALMED has initiated the customization of its ERP and HR (Human Resources) system (developed by the local company Irata, www.irata.hr). This process will also use electronically signed records.

V. Conclusion

HALMED has taken the preliminary step when enabling the functionality of electronic signing. The solution is based on the CEF eSignature building block, it is compatible with EU practice, and it works with various certificates. The innovative application of relevant standards and the usability on Windows and FileNet primary platforms, as well as on HALMED's business platforms, represent both the scientific and practical contribution to the electronic signature subject matter. After the digital transformation of the marketing authorization process, what HALMED is currently solving, other core and support processes will follow. Establishing the generic process of signing electronic records is not sufficient for the successful functioning of a paperless organization – the complete digital transformation of selected business processes should be performed. In this sense, technology can assist the digital transformation of an organization, but the actual work on the processes cannot be dodged.

ACKNOWLEDGMENT

The authors have presented the case implemented at the Agency for Medicinal Products and Medical Devices of Croatia (HALMED); hence, HALMED has funded the project and, therefore, contributed to this research.

REFERENCES

[1] "Digital Signature," n.d., URL: https://www.etsi.org/technologies/digital-signature [retrieved: February, 2020].

[2] EU Regulation 910/2014 of 23 July 2014 (eIDAS), URL: https://eur-lex.europa.eu/legal-content/EN/TXT/HTML/?uri=CELEX:32014R0910&from=EE#d1e2373-73-1 [retrieved: February, 2020].

[3] "Trusted List Browser," 2019, URL: https://webgate.ec.europa.eu/tl-browser/#/ [retrieved: February, 2020].

[4] "Introduction to the Connecting Europe Facility. eSignature building block," 2016, URL: https://www.esens.eu/sites/default/files/building_block_dsi_introdocument_esignature_v0.0.20_3.pdf [retrieved: February, 2020].

[5] "ETSI Signature Conformance," n.d., URL: https://signatures-conformance-checker.etsi.org/pub/index.shtml [retrieved: February, 2020].

[6] "HALMED Activities," 2019, URL: http://www.halmed.hr/en/O-HALMED-u/Osnovni-podaci-i-dokumenti/Djelatnosti/ [retrieved: February, 2020].

[7] C. Kościelny, "An Electronic Signature and Hash Functions," Springer, Berlin, Heidelberg, 2013, chapter 5, pp. 127-145, in Kościelny, C., Kurkowski, M., and Srebrny, M., Modern Cryptography Primer, ISBN: 978-3-642-41386-5.

[8] "CEF Digital About. Definitions," n.d., URL: https://ec.europa.eu/cefdigital/wiki/display/CEFDIGITAL/CEF+Definitions#CEFDefinitions-eSignatureDefinitions [retrieved: February, 2020].

[9] ETSI EN 319 132 XML Advanced Electronic Signatures (XAdES). URL: https://www.etsi.org/deliver/etsi_en/319100_319199/31913201/ and https://www.etsi.org/deliver/etsi_en/319100_319199/31913202/ [retrieved: February, 2020].

[10] ETSI EN 319 122 CMS Advanced Electronic Signatures (CAdES). URL: https://www.etsi.org/deliver/etsi_en/319100_319199/31913201/ and https://www.etsi.org/deliver/etsi_en/319100_319199/31912201/ [retrieved: February, 2020].

[11] ETSI EN 319 142 PDF Advanced Electronic Signature Profiles (PAdES). URL: https://www.etsi.org/deliver/etsi_en/319100_319199/31914201/ and https://www.etsi.org/deliver/etsi_en/319100_319199/31914202/ [retrieved: February, 2020].

[12] ETSI EN 319 162 Associated Signature Containers (ASiC). URL: https://www.etsi.org/deliver/etsi_en/319100_319199/31916201/ and https://www.etsi.org/deliver/etsi_en/319100_319199/31916202/ [retrieved: February, 2020].

[13] ETSI TS 119 612 v2.1.1 Electronic Signatures and Infrastructures (ESI), Trusted Lists. URL: https://www.etsi.org/deliver/etsi_TS/119600_119699/119612/02.01.01_60/ts_119612v020101p.pdf [retrieved: February, 2020].

[14] "Digital Signature Services (DSS). The current release," 2019, URL: https://ec.europa.eu/cefdigital/wiki/display/CEFDIGITAL/DSS [retrieved: February, 2020].

[15] "Internet X.509 Public Key Infrastructure Certificate and Certificate Revocation List (CRL) Profile," n.d., URL: https://tools.ietf.org/html/rfc5280#section-4.2.1.9 [retrieved: February, 2020].

[16] K. Moriarty, M. Nystrom, S. Parkinson, A. Rusch, and M. Scott, "PKCS#12: Personal Information Exchange Syntax v.1.1.," 2014, URL: https://tools.ietf.org/html/rfc7292 [retrieved: February, 2020].

[17] Ericsson Nikola Tesla, "FS01 Nadogradnja funkcionalnosti potpisivanja dokumenta. Funkcionalna specifikacija (FS01 Customization of the function of signing records. Functional specification)," 2019, unpublished (technical document of the software upgrade project).

[18] E. Rescorla, "The Transport Layer Security (TLS) Protocol Version 1.3.," 2018, URL: https://tools.ietf.org/html/rfc8446 [retrieved: February, 2020].

[19] ISO 14721:2012. Space data and information transfer systems – Open archival information system (OAIS) – Reference Model (reviewed and confirmed in 2018). URL: https://www.iso.org/standard/57284.html [retrieved: February, 2020].

[20] "eCTD," 2019, *URL*: http://esubmission.ema.europa.eu/eCTD%20NMV/eCTD.html [retrieved: February, 2020].

[21] "Work program of the ISO/TC 307 Blockchain and distributed ledger technologies technical committee," n.d., URL: https://www.iso.org/committee/6266604/x/catalogue/p/0/u/1/w/0/d/0 [retrieved: February, 2020].

[22] V. L. Lemieux, "A typology of blockchain recordkeeping solutions and some reflections on their implications for the future of archival preservation," in Proceedings of the 2017 IEEE International Conference on Big Data (Big Data), Boston, MA, Dec. 2017, pp. 2271-2278, doi: 10.1109/BigData.2017.8258180

The Core of Design Thinking and its Impact on Digital Transformation in Healthcare

Niels F. Garmann-Johnsen[a,b], Migle Helmersen[a]
[a]NORCE Research, [b]University of Agder
Kristiansand, Norway
e-mail: niels.f.garmann-johnsen@uia.no, mihe@norceresearch.no

Tom R. Eikebrokk[b]
[b]University of Agder
Kristiansand, Norway
e-mail: tom.eikebrokk@uia.no

Abstract—**This study contributes to creating awareness about the potential of Design Thinking in healthcare. The review of the literature showed that this innovation-philosophy and paradigm has had a high impact on several sectors, especially the Information Technology sector, something that in its turn, impacts healthcare. Further research is needed to find solutions to remaining issues, like how to orchestrate dialog and co-creation with system-users, patients, different kinds of healthcare employees, and, e.g., policymakers. Although popular in practice, there has been a gap in academic literature, especially in Information systems, regarding the impact of Design Thinking in this context. Doing a literature review based on a central source of Design Thinking paradigm, Kees Dorst, the authors explore examples of such impact and generalize a picture of the state of art in this field. The impact of further research is then briefly discussed.**

Keywords-healthcare; Design Thinking; Abduction.

I. INTRODUCTION

All Recent design science research has advanced our understanding of the value of Design Thinking methodologies (Design Thinking) and of using the design process for public policy innovations [1].

Dorst's article «The core of 'Design Thinking' and its application» [2] lays out the theoretical foundation for Design Thinking. This fundament builds on Abduction as the third inference method of social science research (the first two being Induction and Deduction, respectively), and abductive reasoning is often described as "the core of Design Thinking" [2]. The term 'Abduction' stems from Charles Sanders Peirce (1839–1914), the founder of American Pragmatism philosophy [3], and describes an approach to science which commences with one or more observations and then seeks the simplest and most likely explanation.

There are two forms of Abduction, relevant to Design Thinking, explanatory and innovative [4]. In explanatory Abduction, the environment is scanned for truly surprising facts. While the rule may be known in other contexts, it is generally not familiar to the current one. Innovative Abduction produces an explanation (the design concept) for the desired value, the function, and an explanation (the form) for the design concept. "That is, we infer a new rule to explain the observation" (Op.cit.).

Design Thinking has several characteristics as a methodology and gives a new formula for creating value. Design Thinking can lead to radical innovation through 'reframing the challenge' [5]. The 'Thing', what to invent or build, and 'Working principle', how to invent or do something, lead to the aspired values. By iteratively exploring the problem space and solution space, the best combinations of the best 'what' and 'how' to achieve the desired values are discovered. This design principle is comparable to the LEAN ontologies for avoiding wasting resources in innovation and production. For example, Ries in his book "LEAN start-up (…)" emphasizes "doing the right things" before and over "doing things right" [6].

Design Thinking can be traced back to 1950-60, but was defined as a term by Rowe in 1987 [7]. Design Thinking builds on the elicited practice of industry designers [1] and aims at finding surprises; unexpected user- or consumer needs and desires, as basis for product- and service innovations [8]. Organizations should focus not only on "high end" users (the most demanding customers, those who pays the highest premium today), but be concerned with the needs of ordinary users as overshot consumers are the segment were new competitive market entrants' hits. This is an analogy to Christensen's studies of disruptive innovation patterns, that also underpin this point [9] along with "Blue Ocean Strategies" [10]. Design Thinking can also be an analogy to Soft Systems Methodology [11], as Design Thinking is well suited for complex, ´wicked´ problems or problematic areas [12].

Design Thinking as a scientific approach to design research has also been subject of critique in academia. There is a lack of rigour to research in Design Thinking [13]. Arguably, more rigid standards and protocols are needed; indeed, a protocol may be needed for designing innovation in general [14].

Research problems

The whole of the healthcare sector needs rapid transformation to sustain a high, human-centred quality while increasing the production of services, due to an expected aging of the population in many industrialized countries. Digitalization of services and Digital Transformation of the whole sector are looked upon as both a goal and a tool to achieve this efficiency. 'Digitalization' is a term that means implementing new technologies together with changed business- and service models. 'Digital Transformation' encompasses the content of the term 'Digitalization',, but in addition, it also means the parallel comprehensive change to the organization, its clients, and the society.

Design Thinking is emerging as a methodology suited for facilitating these innovation processes, for human-centred design in the highly complex sector of healthcare. In this article, we examine the following questions:

RQ1. How is Design Thinking defined in healthcare, and what kind of Design Thinking methods are applied?

RQ2. What are the Design Thinking designer practices for dealing with complexity in healthcare?

RQ3. What are potential pitfalls in applying Design Thinking in healthcare, and what kinds of caution should be applied?

RQ4. What are the potential positive outcomes of applying Design Thinking methods, for the care providing organizations and for care-receivers?

The rest of this article is organized as follows: In Section 2, the methods applied to answer these research questions, namely, a literature search and analysis, are explained. Following this, in Section 3, the results of the search, and subsequent answers to the research questions are shown. In Section 4, the state of art is briefly summed up followed by a discussion of what are the remaining or new questions regarding this area. Finally, the article ends with propositions on avenues for further research.

II. METHOD

A literature review was chosen as the method for eliciting answers to the research questions. Dorst's often cited 2011-article [2] was chosen as a starting point for a forward search conducted in January 2020 with the Google Scholar search engine. This resulted in 1089 books and articles.

A secondary search within these identified 254 books and articles using the term ′healthcare′. Screening these, we found 44 articles and books of relevance to the research questions. The articles omitted, although containing the search word, were found not concerned with healthcare, Design Thinking, or the combination of these two subjects. The 44 remaining articles were analyzed using the Nvivo application (Nvivo 12) for text marking and coding. Text-passages that could share light on the research-questions were coded under nodes created for each research question, with appropriate sub-nodes, marking the found answers and coded with keywords for node-names, related to the first set of nodes [the research questions). As a result, the nodes became placeholders for concepts or themes after the principles of Webster and Watson for literature reviews [15] and resulted in themes organized after concepts and insights. This procedure stimulated the validity of the coding process, since several cited articles, independently of one another, underpinned the same concept. Conflicting concepts could also be identified, contrasted and lifted for discussion and further research. The next section provides a summary of the results of this analysis.

III. RESULTS

The results from reviewing the literature start with how Design Thinking is defined in healthcare, and with an overview of Design Thinking methods used. Then, Design Thinking designer practices for dealing with complexity in healthcare were identified along with potential pitfalls in applying Design Thinking in healthcare, and what literature says on how to avoid these. Finally, the results identify potential positive outcomes of applying Design Thinking methods, for the care-providing organizations as well as for care-receivers.

A. Definitions of Design Thinking in healthcare, and Design Thinking methods applied

Tim Brown, president and CEO of IDEO, a consultancy company pioneering ideation and process innovation in healthcare, defines Design Thinking as ′human-centred approach to innovation that draws from the designer's toolkit to integrate the needs of people, the possibilities of technology, and the requirements for business success′. As such it is useful for dealing with open complex problems as in Information Technology (IT) and business development [16].

Guerra and Tripp [17] compare different design methods for large scale information infrastructures and find that the more traditional methods [Stakeholder management, Community participation, Charette design, Lean design and construction, and Value sensitive design) lack emphasis on especially the later stages in Hasso Platner's roadmap of the Design Thinking process. Hasso Platner's Design Thinking process is here defined as the steps: Empathize (with users), Define (the user-problem), Ideate (seek potential solutions), Prototype and Test [17]. The inclination of Design Thinking to put forward prototyped solutions for the users to test, before investing and implementing, seems unique for Design Thinking methodology. This way, Design Thinking assures that the design process is based on accurate assumptions and understandings of the problem at hand [18]. With such credentials, what can explain resistance to the application of Design Thinking? If a special outcome is desired, Design Thinking may pose a problem, ′because creative events or the emergence of creative ideas cannot be predicted and rating ideas according to creativity is not straight forward′ writes Dorst and Cross in 2001, according to Garde [19]. Design Thinking is an intuitive decision-making practice. In environments characterized by fast-paced technological change contradictory, interdependent or changing requirements and information that may be inadequate or incomplete, like is often the case in healthcare, intuitive decision-making practices dominate over those which are evidence-based [20].

Design Thinking can be used as a framework for co-creation with both patients and employees in healthcare institutions. Design Thinking is a human- and needs-centric approach to innovation that is well aligned with the needs of the very labour-intensive healthcare sector [21]. Co-creation can be used by hospitals, e.g., to redesign whole departments [22]. Design Thinking has been implemented in many different organizational settings. In general, if relatively rapid change is needed, as in healthcare, a change- or innovation culture is also needed. Creating an innovation culture is a dynamic process in which areas of tension and fundamental innovation dilemmas should meet, rather than follow a recipe to implement role models towards success criteria [23]. Prud'homme van Reine identifies nine innovation dilemmas that organizational cultures face: including holistic vs. segmented views of challenges, competition vs. partnership, Consistency versus Pragmatism, etc. Identifying and balancing these tensions may be necessary to maintain an innovative culture, but Design Thinking can promote such organizational development by emphasizing long term holistic values over short-term individual goals.

From the perspective of social innovation, many authors have seen the potential of Design Thinking to improve the quality of healthcare and public transportation. At the same time, many have been advocating for Design Thinking to be taught in universities to help students to become innovative professionals. Moreover, it has also been applied to industrial contexts, such as Small and Medium-sized Enterprises (SMEs) and large organizations [24].

Until recently, Design Thinking was not well received by academia, as the term does not give a clear indication of what field of research it should belong to [25]. Design Thinking alongside ´Lean Startup´ [6] belongs to a brand of learning that can be perceived as experiential learning [26]. To accommodate such learning a multiple of Design Thinking models have emerged over the last two decades, like e.g., "The 5C model" with 62 method cards [27]. Another method ´Actor mapping flags´ is used in the project "InnArbeid", a project for providing mentally challenged youth with vocational employment [28]. A four layer-model of insights into human needs has been devised [5]. Behind such methods lies a common epistemological origin in terms of innovative abductive reasoning [29].

An overview of different modes of Design Thinking is provided by Kleinsman et al. [30], called "Description of the four images of Design Thinking". It shows the role of Design Thinking in Purpose-driven innovation, Vision-driven innovation, Experience-driven innovation and Value-driven innovation. Cards and other artefacts of visualization and gamification of the idea-development process are employed in workshops with user representatives.

Some alternatives to Design Thinking as an approach to innovation, in healthcare and other sectors, are used in developing countries and emerging economies. For example, the TRIZ-model - teoriya resheniya izobretatelskikh zadatch, literally: ´theory of the resolution of invention-related tasks´ was created by the Soviet inventor and science-fiction author Genrich Altshuller (1926-1998) and colleagues, starting in 1946. Conferences are regularly held on this theme [31]. Still, in these proceedings, there are also examples of the application of Design Thinking in a healthcare environment. Here, TRIZ is combined with Design Thinking, in the design of an exoskeleton specialized in the assistance of hemiplegic patients during their re-education (Op. cit.).

B. How Design Thinking designer practices are dealing with complexity in healthcare

Under headers such as 'Design Thinking' and 'strategic design' practitioners and researchers advocate design as a way for dealing with complex problems within diverse fields such as business, the environment, and health care [27]. For improvement work in healthcare, there is a growing interest in applying Design Thinking [32]. Experienced designers systematically change their understanding of the problem space through framing the design problem at hand [33]. The core of the free-flowing design-practice entails expanding on what problem needs to be solved, as well as expanding on what type of solution might address the evolved problem. These two processes co-evolve, meaning that they iteratively inform each other [34]. Free flowing means that ´expert design practice shows that even the desired outcome can mutate with the adoption of a new frame, enabling designers much more freedom to step away from the initial paradox´ (Dorst, according to Op.cit.). This is based on the logic of Abduction [35]. Using ethnographical methods like interviewing and shadowing sessions, where clinical practices are observed, designers highlight and prioritize the value they find in the ambiguity present in the organizational culture [36]. Designers then start from the aspired value. Based on this knowledge they infer a suitable working principle (the rule), and finally, they propose an object (precondition) which can produce this working principle, to deliver the aspired value [29]. Such reasoning thus demystifies the genius of entrepreneurship [37]. This also lowers the risk of new investments and inventions. Many technologies have failed when introduced to the market, as a result of lacking or not having the proper Abduction in their reasoning [38]. So, when using Design Thinking, before creating solutions, efforts are put into framing the problem to be solved [39]. Sometimes small changes are not enough, and radical changes are required to achieve the aspired values. New frames that support radical innovation might be introduced by gaining new perspectives from outsiders or developed by insiders through thematic exploration (Op. cit.). The activities of framing (to set up a first problem description) and reframing (to put the initial perspective under scrutiny and change it) describe what skilled innovation teams excel at [40].

Humans with long term conditions might feel that healthcare treats them as a condition rather than as a whole person. Taking a holistic view on such experiences, Design Thinking creates a new context, a better formulation of the problem in a different area and that helps to shift peoples' thinking into areas from where the problem was always unsolvable into where it is suddenly solvable [41]. Framing and reframing should be understood as a ´(novel) standpoint from which a problematic situation can be tackled´ [18].

One of the main principles that public innovation practices can borrow from design is the activity of (re-) framing problems [42]. The frame serves as a working hypothesis for how the solution should work in order to achieve an aspired value ('why'). In this way, the designer creates both a new way of understanding the problem as well as a new way of acting within this problem in order to construct a new meaning [43]. A poor definition of the problem and its causes may falsely direct resources towards trying to solve the wrong problem [44]. Design Thinking has thus emerged as an innovative context framework to obtain a holistic picture of the state-of-the-art and to determine advantages for change [45]. The advantage of taking a holistic perspective first is that it tends to broaden the perspective taken and thus avoids short-sighted design biases, e.g., redesigning a tool rather than the activity-flow itself [19]. Figure 1 below illustrates how abductive reasoning reduces complexity and risk through abductive reasoning, framing and reframing the problem.

Thies [33] reports the case of the Primary Care Unit (PCU) in the County Council of Värmland, Sweden, where an appointment to see a doctor was hard to get. Normal waiting times for non-acute appointments were around 4-6 weeks. Acute meetings were taken care of the same day. However, the number of timeslots per day for acute meetings was limited, which highly influenced the workflow at the PCU, as well as the patients seeking help. A better IT-system for booking could be part of the solution, but it would be deceptive to range this as the whole solution. The underlying problems were bigger, with causes coming from the different actors' different perspectives and conflicting patterns of action. The value of the service designer lies in creating a holistic understanding of the problem as a basis for designing appropriate measures [34]. Complexity is enhanced by legal requirements stating that all patients should undergo an assessment before booking or being sent home.

Patient focus groups are often involved in governance of hospitals, e.g. planning new facilities., but decisions to be made by a clinic administration needs to consider clinical considerations as well as legal requirements that may be at odds with patient focus groups' perceptions of desirability. Transparency around the decision-process, making it more public may help to legitimize decisions. Design Thinking may help in this process [22].

Design Thinking can help in co-designing a holistic approach to achieving wellbeing as a value, in facilities for living and dying (palliative care) with dementia [46]. Design Thinking and user-centred design-process programs have also helped in creating successful new eHealth applications, e.g., The Connected Care start-up. Running since 2009, the start-up has been offering a self-healthcare solution for managing sleep disorders. ′In 2017, the 25+ employee-sized company was operating in 10 countries before it was acquired by a global electronics company a year later′ [47]. Hardy et al [48] reports on how ′inclusive, user-centred design research′ can improve therapies for Psychosis, through the development of the application, ′SlowMo′.

C. *Some potential pitfalls in applying Design Thinking in healthcare, and what kinds of caution should be applied*

There are requirements for a successful implementation of a new technology in healthcare, even if the application seemingly is designed after Design Thinking principles. The Leavitt's diamond theory for change management applies and states that implementation of new technologies must be accompanied by change in structure, tasks and people.

Orlowski et al. [49] reports on a case study using Design Thinking in redesign of an initial Design Thinking service innovation—the Nurse Knowledge Exchange (NKE). This strategy aimed at improving nursing communication and handover (between shifts) in the organization's hospitals.

The process, as in most applications of DT, was rapid and expert-led (i.e. controlled from start to finish by the design team), and it called on end users, who included staff from all organizational levels, but no patients, for contributions at various stages—particularly during interviewing/observing and field testing. The end-result was NKEplus. The authors described heavy resistance to implementation of the NKEplus strategy outside of the pilot site. The organization was used to, that Design Thinking-based innovations and change normally were coupled with training support and formal changes to work roles and position descriptions. The rest of the case study details re-implementation of NKEplus, a process that resulted in higher uptake and buy-in for NKEplus organization-wide.

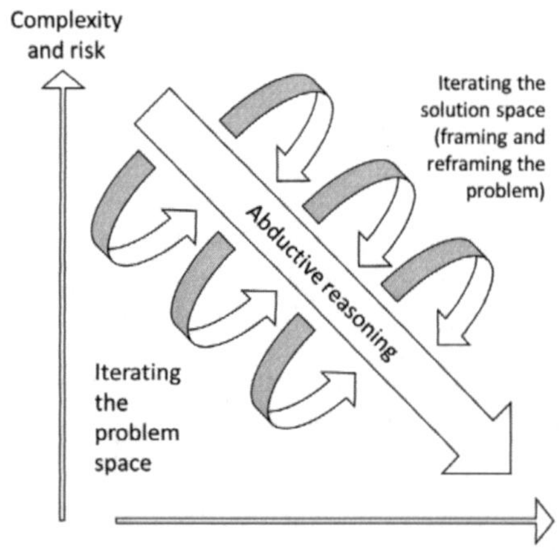

Figure 1. Abductive reasoning in innovation processes as in healthcare

Identifying the right venue and creating a friendly atmosphere may enhance the probability for a positive outcome. In the case of the ′Wellfayre′ (a program for creating welfare innovations) thoughts about branding and the name ′WellFayre′ (...) gave the idea of taking paper cups, cakes and juice, in keeping with a 'country village fayre' aesthetic. This was given a warm welcome (see figure 20). Each participant was comfortable, and conversation flowed freely. When deciding on the venue, the participants' health conditions were taken into consideration, i.e. their need for lifts instead of stairs [41].

Research on the role of Design Thinking in healthcare needs to consider the role of the designers. By observing what designers do, Cross summarized design abilities as "resolving ill-defined problems, adopting solution-focused cognitive strategies, employing abductive or appositional thinking, and using nonverbal modelling media" [39]. Furthermore, studies about design knowledge and expertise also defined them in terms of how designers think and work through their tools, approaches, and ′artefacts′ [39]. To be effective, a distance to the subject at hand may be needed.

A designer relies on hers or his personal integrity. An example: In the context of management research, studies have explored the effects of empathy on individuals and its influence on a group or organization. Empathy may be followed by an emotional contagion that can influence group behavior. Customer service representative may feel a degree of stress given the constant low--grade effect of listening to customers' problems or negative feedback [25]. This issue could arguably underpin the case for using external resources for design research, as they may both have a degree of legitimacy, and a distance to the problems, needed to do an objective assessment that does not over-influence the solution-space.

D. The potential positive outcome of applying Design Thinking methods, for the care providing organizations and for care-receivers

Design Thinking may be a forerunner for other, more rigorous ontologies including Business Process Management and Modelling, Enterprise Architecture development [50] or ITIL [51]. Design Thinking has the potential to take you far into construction, like in making Ambient Intelligent Systems; ′The design process included concept generation and evaluation. In both we followed techniques that supported the design practice. In the generation phase we followed the Design Thinking (…) framework′ [52].

Putting humans first may be the hallmark of Design Thinking. The advantage of design-led innovation is its creation of opportunities based on emotion-rich innovations in the product or services value, as perceived by the user. Empathy is 'the ability to see and experience through another person's eyes, to recognize why people do what they do' [53]. Lupton [54] gives us a narrative from a typical process; ′In a different project, I worked with designers to generate design artefacts for using in a participatory design workshop on digital health. The participants were drawn from healthcare consumer and practitioner groups, industry, and government agencies. They first took part in a group activity involving mapping the landscape of digital health technologies to determine which technologies were being used and the social relationships involved. The groups presented the maps to the other groups, explaining their choices and highlighting the positive and negatives aspects of the current digital health technology landscape. The other two activities asked participants to imagine future opportunities for new digital health devices or software, including making storyboards that inserted these technologies into narratives′ [54].

Using the ability to focus on emotional and social rewards, Design Thinking methods may improve public health. ′In one particular project focused on diabetes management, an IDEO team discovered that traditional clinical goals of diabetes management (such as losing weight and controlling blood sugar levels) to prevent further progression of the disease were inadequate in motivating many patients to make healthy changes to their habits and routines. By contrast, setting social and emotional goals (such as being able to walk a 5-kilometre tour or to dance with one's daughter at her wedding) were highly motivating. As a result of this insight, the team made setting personal life goals (social and emotional) a core service element of the product they designed, changing the patient's mindset from prevention to promotion in the process′ [55]. This way, Design Thinking may create whole new experiences from the user's journey. (Design Thinking) tries to reframe the relationship between the user and the context in which the product/service are experienced. Philips Electronics, for example, developed Ambient Experience for Healthcare, a breakthrough application for reducing the anxiety and stress for patients - and especially for kids - when they undergo medical scans with Computed Tomography (CT) or Magnetic Resonance Imaging (MRI). By rearranging the lay-out of medical devices, introducing cartoons, video and relaxing images into the room, and using sound and interactive walls, the company fostered a new vision of the user experience′ [56].

IV. CONCLUSION

After In this section the authors sum up the answers to the research questions (RQs) as detailed in the previous section and discuss what other topics of interest remains unanswered. Regarding RQ1 (see Section 1), this study has answered the research questions as follows: Design Thinking is defined in healthcare as a human- and patient-centric approach to ideation and innovation, and several Design Thinking methods have been applied. This article has provided a few examples. The Design Thinking practices for dealing with complexity, "wicked problems", like often found in healthcare innovation dilemmas, makes the methodology appropriate in this context. Design Thinking methods for dealing with complexity involves abductive reasoning, framing and reframing of the problem, as shown above as a response to RQ2, and illustrated in Figure 1. There are potential pitfalls in applying Design Thinking in healthcare, and such an approach does not annihilate the needs for normal change management procedures, like getting all employees on board as involved and engaged [57], and employing proper governance methods. This would be the short answer to RQ3., but literature reports many positive outcomes of applying Design Thinking methods, for the care-providing organizations as well as for patients. This article has touched on a few examples, as a response to RQ4. Unanswered questions remain around what degree of standardization of the methodology is warranted. On one hand, a strict protocol may ease adoption of Design Thinking methods. On the other hand, a too rigid protocol may make Design Thinking lose its flexibility and agility and make it less intuitive as an instrument for a dialog between experts and novices within a certain domain.

But Design Thinking is no 'silver bullet' and ideas may have to come from a lot of sources. Let us use the Primary Care Unit [33] as a speculative, imagined example of the importance of iterating between problem and solution. Here you could also mention Kim and Mauborgne's 4-action framework [10] – where the designer should ask what (what steps in the process) should be strengthened, reduced and simplified, eliminated and innovated, respectively - here for example innovating a digital advance diagnosis? The statutory analogue preliminary assessment of patients creates queues and obviously has little perceived value to patients, but requires a lot of time and resources. It steals, among other things, resources from the treatment (let's assume then that the caregivers who make the assessment are also the ones who will assist with surgical intervention, and which is then also a scarce resource). In addition, the physician performs his or her own assessment, so that the pre-assessment is quickly duplicated. A team of process innovators and designers with access to digital expertise might suggest a step ahead of the service, an online self-diagnostic tool, perhaps linked to artificial intelligence, that could collect data and speed up the admission session (serving as decision support for both) those who do the pre-assessment and for the doctor. Of course, such an idea had to be tested and piloted, in a dialogue with all stakeholder groups, as the details are quality-critical,, but similar solutions do exist, e.g., the Norwegian IT-service company Diagraphit [58], offer pre-appointment diagnosis tools.

Further research could consist of case studies within concrete problematic areas in healthcare and public health. Such studies could inform both researchers and practice and devise how Ideally, health-institutions could shape co-creation arenas (real and virtual) where all stakeholders can meet, where ideas developed according to Design Thinking principles, which can then be pursued handing over plans and ideas to more rigid tools and methods.

Limitations of this study

A literature study like this is not an exhaustive overview over all relevant aspects and the authors may have overlooked sources that might provide new insight of relevance to the research subject.

REFERENCES

[1] J. de Koning, E. Puerari, I. Mulder, and D. Loorbach, "Landscape of participatory city makers: A distinct understanding through different lenses," Formakademisk. 2019;12(2).

[2] K. Dorst, "The core of 'Design Thinking'and its application," Design studies. 2011;32(6):521-32.

[3] S. Psillos, "An explorer upon untrodden ground: Peirce on Abduction," Handbook of the History of Logic: Elsevier; 2011. p. 117-51.

[4] A. Dong, M. Garbuio, and D. Lovallo, "Generative sensing in design evaluation," Design Studies. 2016;45:68-91.

[5] M. Van Der Bijl Brouwer and K. Dorst, editors, "How deep is deep? A four-layer model of insights into human needs for design innovation," 9th International Conference on Design and Emotion 2014: The Colors of Care; 2014.

[6] E. Ries, "Lean Startup: Schnell, risikolos und erfolgreich Unternehmen gründen," Redline Wirtschaft; 2014.

[7] I. Luka, "Design Thinking in pedagogy," The Journal of Education, Culture, and Society. 2014(2):63-74.

[8] J. Vink, B. Edvardsson, K. Wetter-Edman, and B. Tronvoll, "Reshaping mental models–enabling innovation through service design," Journal of Service Management. 2019.

[9] C. M. Christensen, M. E. Raynor and R. McDonald, "What is disruptive innovation," Harvard Business Review. 2015;93(12):44-53.

[10] W. C. Kim and R. A. Mauborgne, Blue ocean strategy, expanded edition: How to create uncontested market space and make the competition irrelevant, Harvard business review Press; 2014.

[11] P. B. Checkland, "Soft systems methodology," Human systems management. 1989;8(4):273-89.

[12] S. Weedon, "The Core of Kees Dorst's Design Thinking: A Literature Review," Journal of Business and Technical Communication. 2019;33(4):425-30.

[13] P. Micheli, S. J. Wilner, S. H. Bhatti, M. Mura, and M. B. Beverland, "Doing Design Thinking: Conceptual review, synthesis, and research agenda," Journal of Product Innovation Management. 2019;36(2):124-48.

[14] R. Anwar, S. Z. Abidin, and O. H. Hassan, editors, "In-vitro design protocol: Artificial situation strategy uses to comprehend designers' thought," MATEC Web of Conferences; 2016: EDP Sciences.

[15] J. Webster and R. T. Watson, "Analyzing the past to prepare for the future: Writing a literature review," MIS quarterly. 2002:xiii-xxiii.

[16] A. Ganova, "A case study for the evolution 6^2 model," 2017.

[17] M. A. Guerra and S. Tripp, editors, "Theoretically comparing Design Thinking to design methods for large-scale infrastructure systems," DS 89: Proceedings of The Fifth International Conference on Design Creativity (ICDC 2018), University of Bath, Bath, UK; 2018.

[18] B. F. Nielsen, "Framing humanitarian action through Design Thinking," 2015.

[19] J. A. Garde, "Everyone has a part to play: games and participatory design in healthcare," 2013.

[20] R. McLaughlan, "Learning from evidence-based medicine: exclusions and opportunities within health care environments research," Design for Health. 2017;1(2):210-28.

[21] N. F. Garmann-Johnsen, M. Helmersen, and T. R. Eikebrokk, "Digital Transformation in Healthcare: Enabling Employee Co-Creation through Web 2.0," 2018.

[22] M. Kucmanic and A. R. Sheon, "What Critical Ethical Values Guide Strategic Planning Processes in Health Care Organizations?," AMA journal of ethics. 2017;19(11):1073-80.

[23] P. Prud'homme van Reine, "The culture of Design Thinking for innovation," Journal of Innovation Management. 2017;5(2):56-80.

[24] D. Paula and K. Cormican, editors, "Understanding Design Thinking in Design Studies (2006-2015): A Systematic Mapping Study," DS 84: Proceedings of the DESIGN 2016 14th International Design Conference; 2016.

[25] A. Llamas, "Human-centered innovation processes. The Case of Design Thinking in-," 2015.

[26] M. B. Ramsgaard and M. E. Christensen, "Missing links between lean startup, Design Thinking, and experiential learning approaches in entrepreneurship education," 2016.

[27] Y. Rahim, I. Refsdal, and R. S. Kenett, "The 5C model: A new approach to asset integrity management," International Journal of Pressure Vessels and Piping. 2010;87(2-3):88-93.

[28] K. Rygh and S. Clatworthy, "The Use of Tangible Tools as a Means to Support Co-design During Service Design Innovation Projects in Healthcare. Service Design and Service Thinking in Healthcare and Hospital Management," Springer; 2019. p. 93-115.

[29] G. Lamé, B. Yannou, and F. Cluzel, editors, "Analyzing RID methodology through the lens of innovative Abduction," 2018.

[30] M. Kleinsmann, R. Valkenburg, and J. Sluijs, "Capturing the value of Design Thinking in different innovation practices," International Journal of Design. 2017;11(2):25-40.

[31] R. Benmoussa, R. De Guio, S. Dubois, and S. Koziołek, "New Opportunities for Innovation Breakthroughs for Developing Countries and Emerging Economies: 19th International TRIZ Future Conference," TFC 2019, Marrakesh, Morocco, October 9–11, 2019, Proceedings: Springer Nature; 2019.

[32] M. Marshall, D. De Silva, L. Cruickshank, J. Shand, L. Wei, and J. Anderson, "What we know about designing an effective improvement intervention (but too often fail to put into practice)," BMJ Qual Saf. 2017;26(7):578-82.

[33] A. Thies, "On the Value of Design Thinking for Innovation in Complex Contexts: A Case from Healthcare," IxD&A. 2015;27:159-71.

[34] A. Thies, "Understanding Complex Problems in Healthcare: By Applying a Free-Flowing Design Practice," Stockholm University, Department of Computer and Systems sciences; 2016.

[35] J. Vallverdú, A. Puyol, and A. Estany, "Philosophical and Methodological Debates in Public Health," Springer; 2019.

[36] Dorland, A. R. E., editor, "Doing Design Thinking: An Ethnography of the Digital Graphic Design Studio," 2018: Arts.

[37] M. Garbuio, A. Dong, N. Lin, T. Tschang, and D. Lovallo, "Demystifying the genius of entrepreneurship: How design cognition can help create the next generation of entrepreneurs," Academy of Management Learning & Education. 2018;17(1):41-61.

[38] L. Møller and R. Bennyson, "Project title: Design Adapted Technology for Healthcare [Project sketch]," In press 2015.

[39] M. A. Pfannstiel and C. Rasche, "Service Design and Service Thinking in Healthcare and Hospital Management," Springer; 2019.

[40] Å. Ericson, J. Wenngren, J. Lugnet, H. Kaartinen, S. Pieskä, J. Vähäsöyrinki, et al., "Innovationstrial Internet," Luleå tekniska universitet; 2018.

[41] M. Gill, "Exploring wellness," 2014.

[42] M. Van der Bijl-Brouwer and R. Watson, editors, "Designing for the deepest needs of both public service consumers and providers; Innovation in mental health crisis response. DS 80-1, Proceedings of the 20th International Conference on Engineering Design (ICED 15), Vol. 1: Design for Life, 2015.

[43] K. L. Knudsen and L. M. Haase, editors, "The construction of meaning in design-driven projects," DS 89: Proceedings of The Fifth International Conference on Design Creativity (ICDC 2018), University of Bath, Bath, UK; 2018.

[44] J. E. Reed and A. J. Card, "The problem with plan-do-study-act cycles," BMJ Qual Saf. 2016;25(3):147-52.

[45] C. Battistoni, C. Giraldo Nohra, and S. Barbero, "A Systemic Design Method to Approach Future Complex Scenarios and Research Towards Sustainability: A Holistic Diagnosis Tool," Sustainability. 2019;11(16):4458.

[46] K. Seemann and D. Barron, editors, "Proceedings of the International Conference on Design4Health," Design4Health 2017; Melbourne, Australia.

[47] N. Sturkenboom, E. Baha, R. Price, M. Kleinsmann, and D. Snelders, editors, "A CHAT approach to understand framing in digital service innovation," Proceedings of the Design Society: International Conference on Engineering Design; 2019: Cambridge University Press.

[48] A. Hardy, A. Wojdecka, J. West, E. Matthews, C. Golby, T. Ward, et al., "How inclusive, user-centered design research can improve psychological therapies for psychosis: development of SlowMo," JMIR mental health. 2018;5(4):e11222.

[49] S. Orlowski, B. Matthews, N. Bidargaddi, G. Jones, S. Lawn, A. Venning, et al., "Mental health technologies: designing with consumers," JMIR human factors. 2016;3(1):e4.

[50] T. J. Blevins, J. Spencer, and F. Waskiewicz, "TOGAF ADM and MDA," The Open Group and OMG. 2004.

[51] S. Sahibudin, M. Sharifi, and M. Ayat, editors, "Combining ITIL, COBIT and ISO/IEC 27002 in order to design a comprehensive IT framework in organizations," 2008 Second Asia International Conference on Modelling & Simulation (AMS); 2008: IEEE.

[52] M. Pavlovic, S. Kotsopoulos, Y. Lim, S. Penman, S. Colombo, and F. Casalegno, editors, "Determining a Framework for the Generation and Evaluation of Ambient Intelligent Agent System Designs," Proceedings of the Future Technologies Conference; 2019: Springer.

[53] J. Schweitzer, L. Groeger, and L. Sobel, "The Design Thinking mindset: An assessment of what we know and what we see in practice," Journal of design, business & society. 2016;2(1):71-94.

[54] D. Lupton, "Towards design sociology," Sociology Compass. 2018;12(1):e12546.

[55] L. Thompson and D. Schonthal, "The social psychology of Design Thinking," California Management Review. 2019:0008125619897636.

[56] C. Cautela and M. Simoni, "The Role of Design in Design Intensive Start-ups," International Journal of Design Sciences & Technology. 2019;23(2):145-58.

[57] N. F. Garmann-Johnsen, M. Helmersen, and T. R. Eikebrokk, "Worklife Ergonomics in eHealth Co-Creation Governance - “You can’t manage what you don’t measure”," eTELEMED; 2018; Rome; 2018.

[58] DiaGraphIT, "GoTreatIT® - documentation of daily medical treatment, benchmarking and quality registers in one work operation," https://diagraphit.no/en/2020 [cited 2020 February]; Homepage (Eng. version) for the GoTreatIT solution. Available from: https://diagraphit.no/en/.